FISH AND FISHERY PRODUCTS

Composition, Nutritive Properties and Stability

FISH AND FISHERY PRODUCTS

Composition, Nutritive Properties and Stability

Edited by

A. RUITER

Department of the Science of Food
of Animal Origin
Faculty of Veterinary Medicine
Utrecht University
The Netherlands

CAB INTERNATIONAL

CAB INTERNATIONAL
Wallingford
Oxon OX10 8DE
UK

Tel: +44 (0)1491 832111
Telex: 847964 (COMAGG G)
E-mail: cabi@cabi.org
Fax: +44 (0)1491 833508

A catalogue entry for this book is available from the British Library.

ISBN 0 85198 927 6

Typeset by Colset Pte Ltd, Singapore
Printed and bound in the UK by Biddles Ltd, Guildford

Contents

Contributors vii

Preface ix

1. Fish Production 1
JAN-WILLEM DE WILDE AND ANDRIES KAMSTRA

2. Edible Species 31
S.J. DE GROOT

3. Composition and Nutritive Value of Fish Proteins and Other Nitrogen Compounds 77
NORMAN F. HAARD

4. Composition and Nutritive Value of Fish and Shellfish Lipids 117
R.G. ACKMAN

5. Vitamins in Fish and Shellfish 157
SANTOSH P. LALL AND MONINA P. PARAZO

6. Macro and Trace Elements in Fish and Shellfish 187
SANTOSH P. LALL

7. Quality and Deterioration of Raw Fish 215
H.K. DAVIS

8. Hygiene and Safety of Seafood 243
D.M. Gibson

9. Contaminants in Fish 261
Adriaan Ruiter

10. Freezing and Chilling of Fish 287
Cees Keizer

11. Fish Products 315
Zdzisław E. Sikorski, Asbjørn Gildberg and Adriaan Ruiter

12. Fish Meal and Fish Oil – Not Only By-products 347
Walther Schmidtsdorff

Index 377

Contributors

Ackman, R.G. *Canadian Institute of Fisheries Technology, Technical University of Nova Scotia, PO Box 1000, Halifax, Nova Scotia B3J 2X4, Canada.*

Davis, H.K. *CSL Food Science Laboratory, Torry, Aberdeen, UK.*

Gibson, D.M. *43 Brighton Place, Aberdeen AB1 6RT, UK.*

Gildberg, A. *Norwegian Institute of Fisheries and Aquaculture Inc., PO Box 2511, N-9002 Tromsø, Norway.*

de Groot, S.J. *Brederoodseweg 49, NL-2082 BS Santpoort-Zuid, The Netherlands.*

Haard, N.F. *Institute of Marine Resources, College of Agricultural and Environmental Sciences, University of California, Davis, California 95616, USA.*

Kamstra, A. *Netherlands Institute for Fisheries Research (RIVO-DLO), PO Box 68, NL-1970 AB IJmuiden, The Netherlands.*

Keizer, C. *Netherlands Institute for Fisheries Research (RIVO-DLO), PO Box 68, NL-1970 AB IJmuiden, The Netherlands.*

Lall, S.P. *Biological Sciences Branch, Department of Fisheries and Oceans, Government of Canada, PO Box 550, Halifax, Nova Scotia B3J 2ST, Canada.*

Parazo, M.P. *Canadian Institute of Fisheries Technology, Technical University of Nova Scotia, PO Box 1000, Halifax, Nova Scotia B3J 2X4, Canada.*

Ruiter, A. *Department of the Science of Food of Animal Origin, Faculty of Veterinary Medicine, Utrecht University, PO Box 80.175, NL-3508 TD Utrecht, The Netherlands.*

SCHMIDTSDORFF, W. *Technological Laboratory, Ministry of Fisheries, Technical University of Denmark, Building 221, DK-2800 Lyngby, Denmark.*

SIKORSKI, Z.E. *Technical University Politechnika Gdańska, Department of Food Preservation, Ul. Gabriela Narutowicza 11-12, PL 80-952 Gdańsk-Wrzeszcz, Poland.*

DE WILDE, J.-W. *Agricultural Economics Research Institute (LEI-DLO), PO Box 29703, NL-2502 LS Den Haag (The Hague), The Netherlands.*

Preface

This scope of this book is to provide the reader with information on fish and fishery products in terms of their chemical composition, safety and quality, and value as a nutrient source, and their preservation. In the past, such information has been available from several handbooks. From 1922 onwards the German handbook *Der Fisch* (Rudolf and Clara Baader, Lübeck) was issued, the last volume (No. V) appearing in 1960; and from 1961 to 1965 the four volumes of *Fish as Food*, edited by Georg Borgstrom, became available. Other books also appeared on the scene at that time, most of these being cited in this book.

Since then, much more knowledge on fish and fishery products has been obtained and a number of new products from fish have been developed. Freezing of fish became of paramount importance and new processing techniques and new methods for quality assessment have appeared on the scene. The abundance of n-3 polyunsaturated fatty acids in seafood lipids was established and its significance for human health was thoroughly studied. Better methods of analysis revealed more information concerning the chemical composition of seafood, but also the presence of environmental contaminants.

As a result, several other new books have been published. Many of these deal particularly with technological developments, while others focus on topics such as the manufacture of special fishery products, the role of fish in human nutrition, etc. This book is intended primarily as a compilation of general information regarding fish and fishery products, in which older as well as more recent data find their place, but with special attention paid to the composition, nutritive value, safety and quality aspects. It is meant as a handbook for food researchers in this particular field; as an introduction for people who will become involved in research on fish and fishery products; for students of food science and technology; and for food inspectors. Furthermore, the book may be helpful in courses dealing with fish and fishery products and related topics.

The book starts with a chapter on fish production written by de Wilde and Kamstra. World catches, catching methods and problems with respect to aquaculture are reviewed. The book continues with de Groot's contribution in which fish, molluscan and crustacean species commonly eaten by humans are described and data about catches are summarized.

In the next four chapters (3 to 6) overviews are given with respect to proteins and other nitrogen-containing components (Haard), lipids (Ackman), vitamins (Lall and Parazo) and inorganic matter (Lall) as fish constituents. Much knowledge about the chemical composition of fish, molluscs and crustacea is available now and these chapters are intended as global overviews.

Freshness, deterioration and quality assessment of fish are discussed in Chapter 7 by Davis, with the emphasis on spoilage patterns and on measurements of deterioration.

The next two chapters are devoted to safety of fish from microbiological and chemical viewpoints, respectively. Gibson, in his chapter, discusses pathogenic microorganisms naturally present in the equatic environment as well as those which may contaminate fish during processing, while some toxins are also discussed. In the chapter on contaminants most of the emphasis is placed upon those components that have been brought into the environment, such as heavy metals and organochlorine compounds, and which contaminate fish as well.

The vulnerability of fish urged people to develop preservation methods from early times. This led to a number of characteristic products. For fish, this may be true more than for other classes of food of animal origin and, as a result, these products find their place in this book. Chapter 11, written by Sikorski, Gildberg and Ruiter, deals with both traditional and novel fishery products. Dried, salted, smoked and canned fish are discussed briefly as well as fish minces, fish hydrolates and enzymes from fish. We have refrained, however, from devoting more separate chapters to these products, as there are several good books available in which they are discussed in detail.

Cooling and freezing are the most important methods for extending the shelf-life of fish and therefore warrant a separate chapter. In Chapter 10 a treatise is presented on cooling, freezing and on the equipment used for these techniques, in which Keizer gives the view of the engineer.

The engineer's viewpoint is also given in the final chapter (12), in which the preparation of fish meal and fish oil is discussed by Schmidtsdorff. As a considerable part of world catches is transformed into these products, they merit discussion in this book. Schmidtsdorff explains that production of fish meal is not just the result of cooking, pressing and drying, and that both fish meal and fish oil should be considered as more than just by-products.

In summary, the book aims to provide global information on fish and fishery products, and the large number of references cited offers the opportunity to extend the reader's knowledge further. I hope that this book will be useful to scientists and others involved or interested in this field.

I am indebted to Mr Paul Ph. Deley of the Faculty of Veterinary Medicine, Utrecht University, for preparing most of the drawings for the book. I also want to thank my colleagues Hannes Bon, Piet Moerman and Henk Pieters for their valuable comments.

Adriaan Ruiter

Fish Production 1

JAN-WILLEM DE WILDE[1] AND ANDRIES KAMSTRA[2]
[1]*Agricultural Economics Research Institute (LEI-DLO), The Hague, The Netherlands;* [2]*Netherlands Institute for Fisheries Research (RIVO-DLO), IJmuiden, The Netherlands.*

From his very early stages of development man has caught fish to provide for his need for food. For millennia he contented himself with serving his direct needs and those of his family. Such subsistence fisheries still exist in the poorer parts of this world; but, as methods of curing were found to overcome its perishable nature, fish became a tradable commodity. As a consequence, fish production has gradually developed into an industry all over the world, reaching high levels of sophistication in some instances. Fishing, essentially a form of hunting, still provides the majority of fish. But catches from nature are approaching their limits, and aquaculture is rapidly increasing its contribution.

World Catches

World fish production has grown tremendously in the second half of the 20th century. From a little over 20 million tonnes (Mt) at the beginning of the 1950s, production came close to 100 Mt at the end of the 1980s (FAO, 1993b). The term production is used deliberately as the very comprehensive statistics on world fisheries from FAO (United Nations Food and Agriculture Organization) do not (yet) distinguish between catches from nature and aquaculture production.

In the 20 years between 1950 and 1970 world catches trebled to 65 Mt, an average annual rate of growth of over 5.5%. Initially this was connected to the post-war recovery of the economies of the developed world. Two technological developments contributed to achieving such high rates of growth. First, synthetic fibres became available as material for nets, making them stronger, lighter and rot resistant, allowing larger

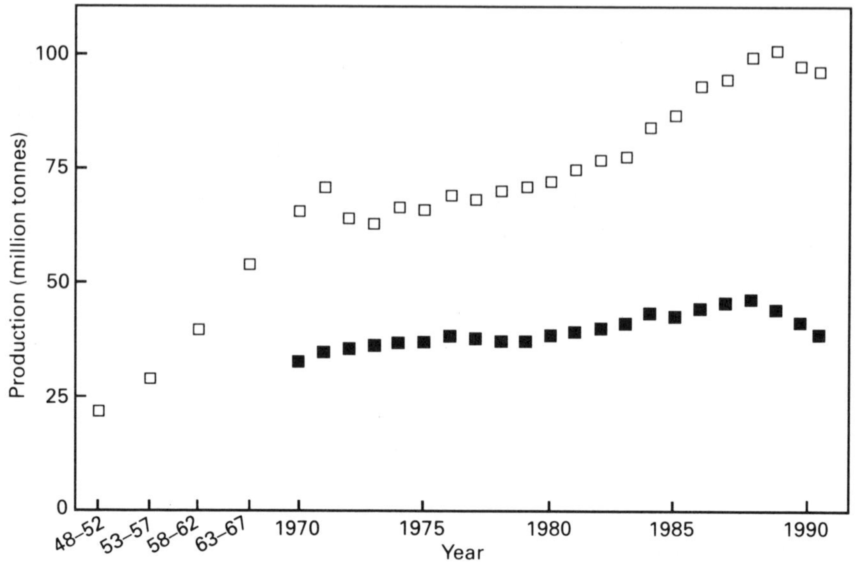

Fig. 1.1. Development of world fish production since World War II. □, Total world catches; ■, developed countries.

nets and consequently larger catches to be handled. Secondly, freezing at sea was introduced, allowing vessels to spend more time at sea and explore more distant fishing grounds (Robinson, 1984).

During the 1960s the high rate of growth was sustained mainly by developing countries, as the traditional fisheries of the developed world were nearing the limits of their potential. In the early 1970s, the failure of the Peruvian anchovy fishery caused a major drop in fish production by developing countries, from which they recovered only slowly, but during the 1980s the rate of growth was up to 5% again. In 1985 the fish production of the developing countries surpassed that of the developed world. In the latter, from 1970 to 1988 growth averaged around 1.5%, but from then onwards has declined, dropping back to the 1980 level by 1991 (Fig. 1.1). Table 1.1 shows 11 countries producing over 2 Mt of fish annually, providing for two thirds of total world production.

Most countries catch their fish in nearby waters, and a few, e.g. China (45–50%) and India (35–40%), obtain a major part from inland resources. However, countries such as Japan, the former USSR, North Korea, Spain and (as far as tuna is concerned) the United States have large ocean-roaming fleets, almost operating worldwide.

Such fleets have increasingly met restrictions of their activities since the mid-1970s when the Third UN Conference on the Law of the Sea established Exclusive Economic Zones (Storer, 1977; FAO, 1983). Coastal

Table 1.1. Major fish producing countries; 1989–1991 average catches and contribution to world production.

Country	Catch (1000 t)	Contribution (%)
China	12,150	12.4
Japan	10,305	10.5
(Former) USSR	10,277	10.5
Peru	6,891	7.0
Chile	5,884	6.0
USA	5,698	5.8
India	3,824	3.9
Indonesia	3,059	3.1
South Korea	2,850	2.9
Thailand	2,728	2.8
Philippines	2,206	2.2
Other countries	32,316	32.9
World total	98,189	100.0

Source: FAO (1993b).

states are allowed to exert sovereign rights on the exploitation of natural resources in a zone extending 200 miles (approx. 370 km) offshore. This does not necessarily lead to an exclusion of all other countries from fishing in those waters, as there is an obligation to let others tap the resources if the coastal state itself is unable to do so. In most cases the coastal state will require something in exchange for the right to fish in its waters. This may vary from a simple payment of fees to comprehensive development schemes with joint ventures, extension programmes and phasing-out schedules.

About 15 Mt of fish is produced in inland waters. Here in particular the distinction between fishing and culture is quite often not very clear. Asia accounts for more than two thirds of inland fish production and most of this comes from culture: China alone, with around 5 Mt, accounts for about half of that. Africa, with 1.8 Mt, makes a substantial contribution, followed by the former USSR with a production of 1 Mt. The inland fish production of other continents is relatively insignificant.

Fish production from marine areas is still almost exclusively the result of fishing. This provides for around 83 Mt of fish. The origin of these catches is given in Table 1.2.

Productivity is not evenly spread among marine waters (Gulland, 1971). Northern waters yield more fish than southern, and the central and tropical waters generally have the lowest yield by surface area. The reason for this is the uneven food distribution in the seas. Basically the food

Table 1.2. Catches by marine areas; 1989–1991 averages.

Area	Catch (1000 t)	Area ($km^2 \times 10^6$)
Atlantic Ocean		
North[a]	12,680	22.1
Central[b]	7,223	31.6
South	3,771	36.2
Total	23,673	89.9
Indian Ocean		
Western	3,400	30.2
Eastern	2,824	29.8
Total	6,224	60.0
Pacific Ocean		
North	28,675	28.0
Central	9,064	82.1
South	15,607	58.4
Total	53,346	168.5
Southern Ocean		
Total	402	35.3
Total marine areas[a]	83,465	353.7

[a]Excluding the Arctic Sea (no catches recorded).
[b]Including the Mediterranean and Black Sea.
Source: FAO (1993b).

situation in a certain area is determined by the presence of nutrients to sustain the primary production of phyto- and zooplankton. This is the start of the food chain of all marine life. In this respect large oceanic areas are like deserts, with very low production. On the other hand there are three types of highly productive waters:

- shallow waters over continental shelves;
- upwelling areas and divergencies off the western continental coasts and along the equator, where cold, nutrient-rich waters rise to the surface;
- temperate and sub-Arctic waters of the North Pacific, the North Atlantic and the Southern Ocean.

In these areas fish are concentrated and consequently caught. Examples are the North Sea fisheries, annually yielding some 5 t of fish per square kilometre; the Peruvian anchoveta fishery once yielding 10 Mt and now

Table 1.3. World catches by major groups; 1989–1991 averages.

Species group	Average catch (1000 t)
Freshwater fishes	12,035
Diadromous fishes	3,130
Marine fishes:	
demersal fishes	18,685
small pelagic fishes	36,100
large pelagic fishes	5,000
miscellaneous	10,192
Total	69,976
Crustaceans	4,674
Molluscs	8,062
Miscellaneous aquatic animals	313

Source: FAO (1993b).

still realizing half of that; and the huge Alaska pollack fishery with an annual catch of 6 Mt.

The division of world fish production by major groups of species is given in Table 1.3. Diadromous fishes are species spending part of their life cycle in fresh water and part in a marine environment. The principal species group of these, the salmon(id)s, are mostly caught in the North Pacific and in the northeast Atlantic, to a total amount of over a million tonnes.

The bottom-dwelling demersal marine fishes are primarily species of the northern waters. Flatfish and cod-like fishes to a total of 13 Mt are mainly caught in the North Atlantic and the North Pacific and little elsewhere. The fisheries for these species are those of the traditional fishing industries of East Asia, North America and Europe. The catches of red-fishes, basses and congers of nearly 6 Mt are more evenly spread over the world with northern waters contributing less than 60%.

Small pelagic fishes are shoaling species, living midwater between bottom and surface. With total catches at around 35 Mt this is the largest group of species. The main fishing areas for these species are found in the eastern Atlantic, from north to south, and in the northwest and southeast Pacific. Japan, Peru, Chile, the former USSR and the USA are major producers of small pelagics, but numerous other countries have a greater or smaller stake in these fisheries.

The fisheries for the larger, ocean-roaming pelagic fishes – tunas, bonitos, billfishes, sharks and rays – are primarily in the tropical and subtropical zones around the world. Large producers are Japan and the USA, and a number of developing countries such as Mexico and Indonesia are increasing their catches.

Catches of up to 10 Mt of marine fishes are reported annually without further specification, mainly by developing countries. Major contributions are from China, North Korea and Thailand, combining to nearly half of the total.

Of the often highly valued crustaceans, shrimps and prawns (2.6 Mt) are the most important species, and fisheries for them are found almost all over the world. Complementary to this, the culture of tropical shrimps is on the increase. Crab production comes second at just over 1 Mt. The once very promising krill fishery in the Southern Ocean is stuck at a catch level of under 400,000 t.

Fishing for cephalopods (squids, cuttlefishes and octopuses) is mostly an offshore activity. The main fisheries are to be found in the eastern Central and southwest Atlantic, and in the northwest, western Central and southeast Pacific. Apart from local fisheries, long-distance fleets of the larger fishing nations pursue these species. The catches have increased considerably during the 1980s, from 1.2 to 2.5 Mt.

The 'proper' molluscs have a static nature and are caught mostly in inshore waters. In the production of molluscs, culture has played an important role for a long time, particularly with oysters (1.0 Mt) and mussels (1.3 Mt), and interest in it is still expanding. But clams, cockles and arkshells (1.5 Mt), and scallops and pectens (0.8 Mt), are still mainly caught from nature.

A review of a dozen species with the highest catches is given in Table 1.4. The majority belong to the small pelagic group. Of these the herring-like fishes are most prominent, with five species and combined catches of over 16 Mt. Only parts of the small pelagics are used for human consumption. Most of the landings serve as raw material for reduction into fish meal and oil. This industry absorbs close to 30% of total world catches (FAO, 1993c).

Alaska pollack more or less fits into the same category, as a considerable part of it goes for meal, and most of the balance is processed into surimi, a denatured fish protein originating from Japan.

Species for human consumption are represented by cod, carps and skipjack tuna only, with a combined catch of 5.5 Mt. The 12 species together contribute 32% to world catches.

The FAO Marine Resources Service estimates the potential world catch of fishing from nature at 100 Mt (FAO, 1989). Total world production has already reached this volume, but including 13% from aquaculture. Most traditional fisheries are fully or even over-exploited. Apart from single species with relatively small potential, substantial scope for increased catches is offered by just a few species groups in a few areas. These are

Table 1.4. Catches of 12 principal species; 1989–1991 averages.

Species		Catch (1000 t)
Alaska pollack	*Theragra chalcogramma*	5650
Japanese pilchard	*Sardinops melanostictus*	4515
Anchoveta	*Engraulis ringens*	4399
South American pilchard	*Sardinops sagax*	4325
Chilean jack mackerel	*Trachurus murphyi*	3779
Atlantic cod	*Gadus morhua*	1540
Atlantic herring	*Clupea harengus*	1514
European pilchard	*Sardina pilchardus*	1495
Silver carp	*Hypophtalmichtys molitrix*	1398
Chub mackerel	*Scomber japonicus*	1398
Skipjack tuna	*Katsuwonus pelamis*	1360
Common carp	*Cyprinus carpio*	1174

Source: FAO (1993b).

primarily cephalopods and tunas in the Pacific and the Indian Ocean and then the numerous deep sea species living on the verges of the continental shelves, which until recently have escaped attention.

Increases in fishing intensity yield higher catches only up to a certain level. At this level the breeding and growing potential of a stock is being used optimally, with catches at the maximum sustainable yield. Further increases in fishing pressure generally result in declining catches in the long run, as the heavy fishing is affecting the growth potential. This is called overfishing in its milder form of growth overfishing. In the more serious form it is also eating into the breeding potential of stocks and this can easily lead to their depletion.

Shoaling pelagic species, in particular, are vulnerable to heavy overfishing and history has shown some examples of near extinction (Atlanto-Scandian and North Sea herring, North Sea mackerel and Peruvian anchoveta). Demersal species in the moderate and colder zones generally are considered to be less vulnerable, as fishing for them will become uneconomic long before they have been depleted. In the warmer zones a more cautious approach is called for, as the numerous species are relatively small in numbers and in reproductive power, and heavy fishing can easily endanger them.

Fluctuations in fish stocks and catches are not caused solely by heavy fishing. Changes in natural conditions can influence the behaviour and the state of fish stocks considerably: often the negative effects of one factor enhance those of another, leading to a situation where the primary cause cannot be distinguished. A well-known example is the collapse of the Peruvian anchovy fishery in the beginning of the 1970s mentioned earlier.

A strong 'El Niño' – the recurrent phenomenon of warm water masses disturbing the Pacific current system and thereby influencing the weather worldwide – struck at a time when heavy fishing had made the anchoveta stock vulnerable. A similar thing happened to the North Sea herring: heavy overfishing brought the stocks down to a very low level, but, in spite of a stop to fishing, recovery was slow, as recruitment failed due to a hydrographic anomaly (Corten, 1984).

A third factor increasingly affecting fish stocks and catches is environmental pollution. This ranges from heavy metals and non-biodegradable chemicals through oil and fertilizing effluents to solid debris. Part of these 'only' make contaminated fish unfit for consumption. Others are weakening or killing fish, or reducing their fertility. Eutrophication, resulting from an excess of fertilizing agents, causes recurrent algal blooms which may be poisonous, killing fish directly, or can cause de-oxygenation, a regular phenomenon in the Baltic Sea now. Although the impact of this factor on catches is not yet very substantial, it may become so if drastic measures fail to be taken in this field.

Catching Techniques

In his age-long experience in catching fish, man has developed a wide variety of fishing methods. Every method is devised to catch the fish which the fisherman is aiming at in the most efficient way, under the locally prevailing conditions, including the level of technology available to him. Fish behaviour is intensely studied by fishermen and this knowledge is an integral part of his skills and a dominant factor in the design and adaptations of fishing gear.

The fish catching methods of the world have been extensively described by Andres von Brandt (1984). He also classified the countless methods into a system of 16 principal categories based on the actual manner of capture.

From the end of the 1950s and throughout the 1960s the FAO was very active in disseminating information and knowledge of the use, design and construction of fishing gear (Kristjonsson, 1959; Anon., 1963, 1972) and boats (Traung, 1955, 1960, 1967). Later FAO concentrated more on supporting extension work in developing countries (FAO Training Series, e.g. Noël and Ben-Yami, 1980, and Ben-Yami, 1988). FAO adopted von Brandt's classification of fishing methods as a basis for its International Standard Statistical Classification of Fishing Gear (Nédélec, 1982). In the following we will draw upon this, ordering the methods more or less by level of sophistication and energy requirement.

Simple, traditional fishing methods

Fishing without gear is still being practised to catch slow-moving species, e.g. gathering shellfish by hand and diving for sponges or for pearls: clamps, rakes and tongs serve to lengthen the fisherman's arm and to strengthen, widen or improve his grip.

Fish can be stunned or stupefied before gathering, by hitting them with a club or a stone. A modern way of stupefying fish is by inducing an electrical field in the water.

Scoop nets and bag nets kept stationary in a stream or moved through the water, catch the fish by filtration. They are widely used in river and estuarine fisheries and in some cases the gear has grown into large and sometimes intricate devices, called gape nets and stow nets. These are the stationary forerunners of trawls.

Spears, arrows and harpoons are examples of old hunting gear, used commercially only in a few fisheries. Whaling is a well-known example of the use of harpoons fired by guns. Another way of increasing the reach of arm and hand is cover pots and lantern nets. They are conical, dome or bucket-shaped devices, put by hand over single fish in shallow, mostly turbid water. The trapped fish is taken out through an opening in the top. The method is still used in fishing paddy fields and (community) ponds.

The cast net, well known from pictures of exotic tourist resorts, works on the same principle. Basically a conical net with a weighted edge, cast skilfully to give a wide spread, it sinks to the bottom rapidly, covering the fish that happen to be underneath. The net is generally hauled by a centre line, sometimes connected to pockets along the edge for improved retainment of the catch. Cast nets are widely used in artisanal (small-scale) fisheries.

The action of lift nets is exactly the opposite of that of falling gear: the net is brought into position in the water or at the bottom and after some waiting time is lifted, catching the fish that happen to be above it. The shape varies from square, practically flat, to all kinds of more or less shallow bags. They are hung from frames, poles or gallows, fixed ashore or in the water, or from rafts or boats, and are operated by lines. Sizes range from about one to several hundred square metres in some East Asian commercial fisheries.

Traps

A wide variety of traps is used in all kinds of fisheries all over the world. The gear is stationary and the fish have to enter it by themselves. Once they are in, they are generally prevented from going out again by retaining devices. Traps are a kind of gear where a knowledge of fish behaviour is of particular relevance.

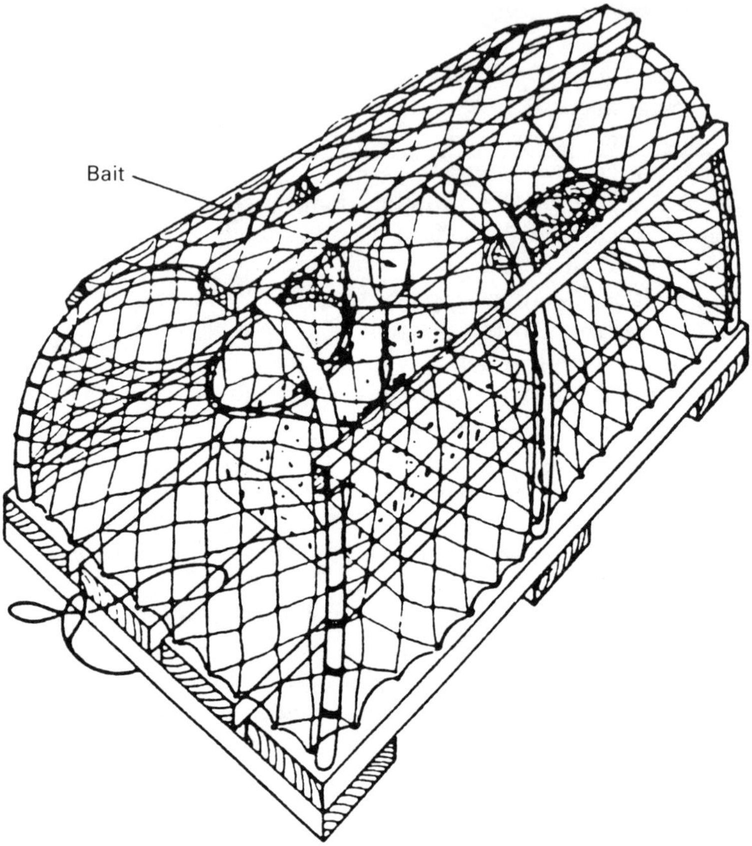

Fig. 1.2. Lobster pot.

One of the most simple traps is to offer the fish a hiding place, like the ceramic jars for octopus fishing, used in the Mediterranean as well as in eastern Asia.

'Pots', cage-like devices widely used to catch crustaceans, work differently, as they are mostly baited to lure the fish in (Fig. 1.2). They are usually set in a row, connected by a line, with marker buoys at the ends. In modern high value fisheries, accurate electronic navigation systems help to locate and find the pots. Mechanical means, like hydraulic pot haulers, help to increase the number of pots that can be handled, as do fast boats, which lose little time steaming to and from the fishing grounds.

Fyke nets are tapered cylinders of netting (kept in shape by a number of hoops), provided with funnels. They are kept in position by anchors or stakes. Fish, once they have entered, can only get deeper into the trap and will end up in the last conical part, which can be opened. Fykes are used in European inland fisheries, e.g. to catch eel.

Pound nets and weirs are large-scale stationary trapping systems, consisting of fences guiding the fish into one or more successive labyrinth-like chambers. They are generally used in shallow water to catch all kinds of fish, from sardines to tuna.

Hooks and lines

The catching principle of hooks and lines is clear: fish trying to eat the bait get hooked and, generally retained by a barb, are hauled ashore or aboard by a line attached to the hook. The bait may be live or dead or artificial, depending on the species aimed at, or the hook may not be baited at all, and sometimes there is not even a hook but just the bait (bobbing). Generally the type of bait and the location and time are related to the target species and the size of the hook to its size.

The most simple method in this category is handlining, with or without a pole or rod. It is the common gear for sports fisheries. Fishing without a rod, with one or more baited hooks attached to the hand-held line is a very common commercial method in developing countries.

A special development of handlining is pole and line fishing for tuna. In this common fishing method, specially designed boats go out with a rather large crew. On spotting a shoal of tuna, the fish are brought into a feeding frenzy, by feeding them dosed quantities of live bait and enhancing the effect by spraying water on the surface, which also conceals the boat from the fish. The feeding tuna are prepared to snap at anything moving and are caught on bare hooks without barbs, swept aboard and unhooked in one flowing movement.

The handlining methods have also all been mechanized and more or less automated, with electrically or hydraulically driven reels, jigging machines and pole-and-lining machines.

Towing baited hooks on lines is called trolling. In commercial fishing for tuna, three or more lines are trailed from long outrigger booms on both sides of the boat and some more from the stern. The lines can have multiple hooks and paravanes may be used to keep them in position.

In longlining, numerous hooks on short lines (snoods) are hitched or clipped to a long line, which is set at or close to the bottom, fixed by anchors or weights, or drifting at or close to the surface, with marker buoys at intervals. This traditional method had a revival when largely automated systems were brought onto the market for small as well as larger boats. These systems take over most of the tedious labour of baiting and clearing and allow the setting of large numbers of hooks (up to 50,000 per day) with small crews. Bottom lines are, for instance, used to catch cod and drifting lines for tuna.

Gill, tangle and trammel nets

Netting is generally a passive way of fishing, where fish have to move into the nets by themselves to get caught. Therefore the system requires a thorough knowledge of fish behaviour, in particular of migration patterns.

Gill nets are single-layered nets, having excellent size selectivity by retaining fish at the gills. Species selectivity depends on the make of the net and on the time and place where it is used. Set at or close to the bottom, they are kept in position by anchors with marker buoys. The nets are spread vertically by floats on the upper seam and weights on the lower seam. Often several nets are placed in a file or 'fleet'. Such set nets are used in a wide variety of fisheries all over the world. An interesting example is the North Sea wreck fishery for cod, utilizing their daily feeding rhythm. Setting the nets close to and over wrecks and hauling them without damage requires great skill, and accurate echo-sounding equipment is practically indispensable.

Gill nets can also be positioned drifting on or close to the surface, e.g. the traditional method for catching herring and other small pelagics, but this has been taken over now by other bulk fishing methods. Drift nets of modern synthetic fibres and construction have found wider application in catching salmon, tuna and other large pelagic species.

Large-scale drift netting operations in the Pacific attracted the attention of nature conservationists because of accidental by-catches of marine mammals, sharks and birds. Even if these were small on a per haul basis, the sheer size of the operations – hundreds of vessels, each setting tens of kilometres of drift nets, shaping so-called 'walls of death', each night – made for unacceptable losses. These protests led to a Resolution of the United Nations, appealing for a worldwide ban on large-scale drift netting from mid-1992 onwards. To date, most countries have complied with the Resolution, so that drift netting is now a relatively small-scale operation.

Set and drift nets may also catch fish by entangling them. Such tangle nets are usually hung more loosely than proper gill nets.

Trammel nets are constructed specifically for entangling. They are triple-layered, with large-meshed outer layers and a very loosely hung smaller meshed inner layer. Fish swimming into the net get entangled easily, pushing the inner layer through a mesh of the outer, thus forming a pocket. Trammel nets are used in various, mostly demersal fisheries, e.g. for turbot and for sturgeon.

Hydraulic haulers, net transporters with chutes and net storage drums have brought a great deal of mechanization into netting operations. However, the clearing of the nets still can be a tedious and labour-intensive operation.

Seine nets

Instead of waiting passively for the fish to enter a net, the wings of the net can be moved forward, in order to herd the fish together in the area covered and eventually catch them in a final scooping movement. This method is called seining and it is widely used in freshwater and coastal fisheries, and also in some North Sea fisheries. The wings of the net may be largely replaced by lines or ropes and the net may be provided with a bag or 'cod end' to retain the fish caught.

The beach seine, operated from ashore, is a traditional method of bulk fishing, still very much used in developing countries. Generally the gear is set in a semicircle from shore to shore with the help of a boat. Net and ropes can have a combined length of several kilometres. The foot rope of the net scrapes the bottom, while the head line is lifted to the surface by floats. Hauling is mostly done by manpower, requiring tens of people depending on the size of the seine. During hauling the wings are gradually brought together, herding the catch into the bag or bunt.

Seining can also be done with one or two boats. The most advanced development is that of single-boat seining by the Danes and the Scots (Thomson, 1981). Here the lines, which may be several kilometres long, are set roughly in the shape of a kite, with a buoy at the (bottom) starting point and the relatively small net at the top. In hauling, the lines are herding the fish together in the enclosed area and in the final stage they are swept up in the net (Fig. 1.3). In Danish or anchor seining, the buoy is anchored and the vessel stays anchored on hauling. In Scottish seining, or fly dragging, the buoy is left drifting and the gear is hauled with the vessel slowly moving forward. The method requires rather smooth grounds and aims specifically at groundfish. Mechanization with rope reels and power blocks has extended the life of the low energy method, but it is losing ground to trawling.

Surrounding nets

Shoaling pelagic fish need not be herded, as they are already close together. Therefore it is sufficient to put a net around them, dry them out and scoop the fish up – of course it is not as simple as that, but this has become one of the major fishing methods of the world. Searching for shoals of target species is done in various ways: modern electronic echo-sounding techniques have greatly improved the possibilities of finding shoals, and the most recent development is remote sensing from space (Le Gall, 1989). Setting and closing the net quickly enough to prevent the fish from escaping depends on the speed of the boat(s), the net setting and the design and construction of the gear. The curtain-like nets have floats at the top and a ballasted 'lead line' at the bottom to make them sink fast. Two main

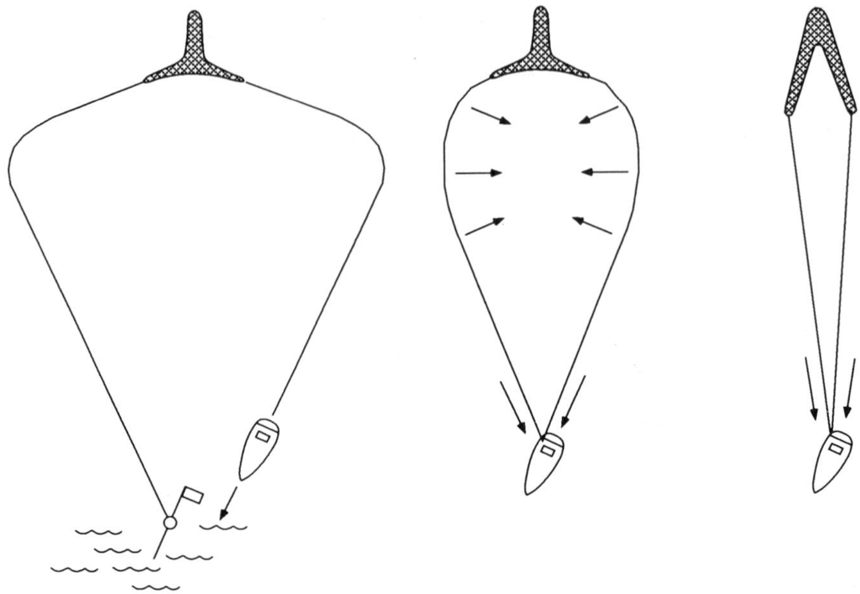

Fig. 1.3. Boat seining.

types can be discerned: without and with purse lines to close the bottom of the net.

The lampara net is the most representative of the surrounding nets without a purse line. It has the shape of a dustpan with large wings, the foot rope being much shorter than the head rope. After setting the net, generally by one boat circling as widely around the shoal as the length of net permits, the wings are pulled together. In this way the net is drawn under the fish and escape prevented. When the fish are concentrated sufficiently by hauling the net aboard, they are brailed out with scoop nets. Light is often used to attract fish, which has given the lampara net its name. This type of net is mostly used in small-scale and artisanal fisheries.

Purse seines, where the bottom of the net is closed with a pursing line slung through rings hung from the foot rope, are used in several large-scale fisheries (Fig. 1.4). Here the foot rope generally is as long as, or even longer than, the head rope. The net may be operated by one boat or two boats. The design, dimensions, materials and mesh sizes may vary widely (Ben-Yami, 1992). The huge size of some purse seines – up to two kilometres long and 300 metres deep – is only made possible by the modern light and strong synthetic fibres. These can be handled by large, powerful and highly mechanized vessels only. Electronics not only helps to find shoals of fish, but also to monitor and control the surrounding operation. Purse seines also come in smaller sizes, down to a level where

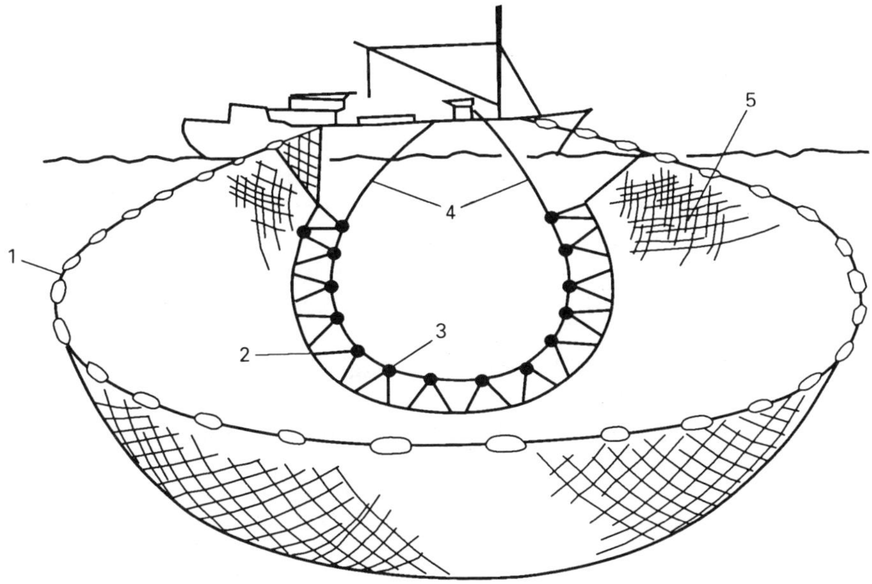

Fig. 1.4. Purse seining. 1, Float line; 2, lead line; 3, purse ring; 4, purse line; 5, bunt.

they can be set from rowing or outboard powered boats and hauled by hand. Two-boat seining for tuna is practised in Japan.

Purse seining for tuna by large tuna clippers from the USA and some other countries is one of the most advanced fishing methods of the world. These vessels hunt for tuna worldwide and freeze their catches to extend trip duration. The purse seine is set with the help of a small motorized boat, the 'skiff', keeping one end in position, while the seiner is streaming it out around the shoal. After hauling the purse line, the net is hauled on one end by a powerblock hung from a boom and stacked on deck, ready for the next set.

The method has raised criticism from conservationists because of unacceptable by-catches of porpoises. Methods have been developed to reduce these accidental catches and further efforts are being made to bring them down to zero.

The Norwegian method is another major purse seining system, used for catching small pelagics. The purse seine is stacked in a bin in the stern and is set with a buoy at one end, without help of a skiff. The net is hauled by a triplex (three-drum) hauler mounted near the bridge and transported aft to the bin by an auxiliary powerblock. Catches of several hundred tonnes in one haul are common and they generally are pumped aboard.

If left uncontrolled by regulatory measures, the huge fishing power of these systems may easily lead to overfishing. Due to their shoaling behaviour, pelagic species may rather easily be fished to virtual extinction, as eventually even the last shoal might be found and caught.

Trawls and dredges

The fishing method that has most profited from modern technological developments is that of dragging the gear through the water. Without efficient mechanical propulsion systems working on cheap fuel, this would hardly be feasible, but, as energy costs since World War II have in general gone down rather than up, these methods have gained a dominant position in world fisheries.

Dredges, used for mollusc fishing throughout the world, are probably the least sensitive to energy prices. They generally consist of a rigid low frame of up to a few metres wide, with a raking or scraping device at the lower edge and a bag-like construction (sometimes completely rigid) attached to it, to retain the catch. Dragged over or partly through the bottom, they rake up the shellfish, and are hauled intermittently to bring the catch aboard.

In trawling, the catching principle is to filter a volume of water and retain the fish above a certain size, dependent on the mesh size of the cod end. Basically this is not a very selective method, either for size or for species.

Trawls are funnel-shaped nets, generally with wings of some length, towed on the bottom or midwater by two warps. For a maximum volume of water filtered, the opening of the net should be as large as possible. Hydrodynamic lift generated by the belly of the net partly sees to this. The vertical opening can be increased by floats at the head rope and ballast at the foot rope. The horizontal spread can be provided by a couple of methods (Garner, 1988).

Beam trawls, used to catch flatfish or shrimp, are kept open horizontally by a beam provided with a device to keep the beam free from the bottom and secure the vertical opening, which is of minor importance. The beam may vary in length from 3 up to 18 metres. Beam trawling may be done with a single gear or with two, towed from the ends of outrigger booms. High powered double beam trawling is a very productive method of flatfish fishing.

Pair trawling with two vessels on equal course, towing the net between them, is another way of opening up a trawl horizontally. This is a very efficient fishing method, as little towing power is lost to spread the gear, so the nets may be bigger than a single vessel of equal total power would be able to tow. However, the method requires perfect cooperation between the skippers, and handling the large gear working closely together makes the method rather sensitive to bad weather and seas. Successful pair trawling fisheries have developed in the North Sea for cod and whiting and

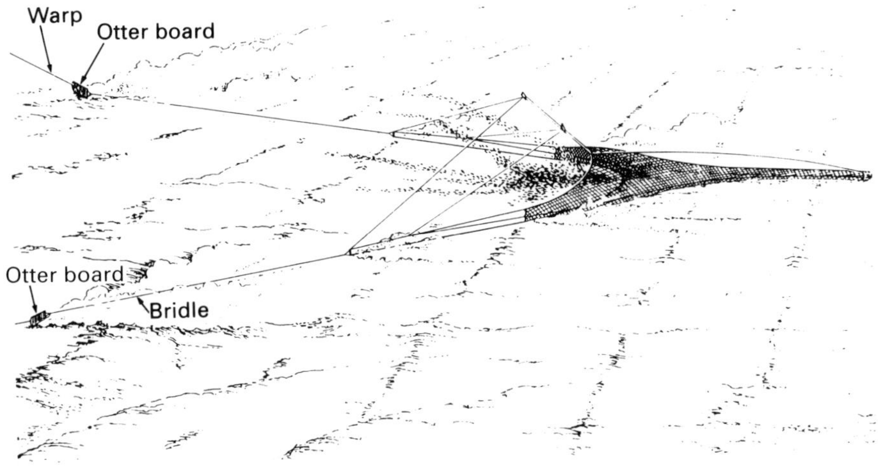

Fig. 1.5. Bottom trawling.

also for herring. Electronic equipment, for fish finding and navigation as well as communication, contributes substantially to this success.

The common way to spread the net is to mount otter boards between the warps and the wings of the net. Working essentially as hydrofoils, the boards generate a sideways pulling force opening up the net. A wide variety of otter boards has been developed, ranging from simple flat boards to rather sophisticated hydrofoils.

In bottom or demersal trawling the boards move over the ground, which provides for part of the shearing force (Fig. 1.5). The boards and the mud stirred up have a herding effect on the fish, so, to enlarge the area covered without needing a bigger net, the wings are provided with bridles connecting them to the boards. For extra opening height the headline may be provided with hydrofoil-like devices, for instance small boards or, more modern, cloth kites. The resistance of the nets may be reduced by using large meshes in the fore-net and gradually reducing mesh size to the desired one in the cod end.

Generally in demersal fisheries a good bottom contact of the ground rope is paramount, in order to prevent fish from escaping underneath. A wide range of ground tackles are used to serve this goal, depending on the target species and the character of the fishing grounds. They may vary from a simple wire, hitched directly to the netting, to the heavy and complicated bobbin gears used on rough grounds in North Atlantic fisheries.

In shrimp fisheries, where the fished area is of more importance than the volume of water filtered, twin otter trawls are used. Connected to a sledge at the inner wings and a pair of otter boards at the outer, two small trawls are towed from a common crowfoot by a single warp. Usually double gears are towed from the top of outrigger booms.

The size of gear that can be handled has been changed dramatically by the invention of stern trawling, in combination with the availability of modern synthetic materials (Hjul, 1972). Formerly trawls were set and hauled from the side of the vessel, requiring rather complicated manoeuvring and handling. Setting and hauling the gear over the stern appears to be a more natural procedure and it has driven out classic side trawling to a great extent.

Two main systems of hauling the gear have developed. The older one is to pull the gear practically completely up a stern ramp on each haul. This is still normal practice in large-scale demersal trawling. But the net can also be hauled partially before bringing in the bag or cod end. For this a ramp is not needed and the bag can either be lifted on the aft deck, or midships in the traditional position and way. The size of the gear can be independent of the size of the vessel. This system is now used in all kinds of trawl fisheries. On all but the smallest vessels, drums for net storage have greatly improved handling of the gear.

In semipelagic fishing the boards are on the ground and the net is hovering at a little distance above it. Actual midwater trawling requires otter boards that provide sufficient lift and are stable in midwater (Garner, 1978). In most high performance pelagic trawl fisheries a so-called Süberkrüb board is used; this is rectangular, with curved section, and has excellent characteristics. Large pelagic trawls have a size and shape comparable to that of the Eiffel Tower, with very large meshes or just lines in the fore-net (Fig. 1.6). Catches of over 100 tonnes are frequently made in a few minutes' actual trawling time, taking only part of a shoal.

With the help of echo sounders and sonar to spot and follow the shoals, directed fishing is possible, by varying the direction and the depth of the trawl. The operation can even be automated by integral fishing systems, controlling direction and speed of the vessel and the length of the warps on the basis of information from sonar and sounders on the movements of a chosen shoal.

Aquaculture

Fish catching is basically a hunting activity, and it accounts for the bulk of total aquatic production. Aquaculture is the farming of aquatic animals and implies some form of intervention in the rearing process to enhance production (e.g. regular stocking, feeding, protection from predators, etc.). Farming implies individual or corporate ownership of the stock being cultivated (FAO, 1993a). In statistics the distinction between fishing and some forms of farming is vague: for example, in the stocking of common

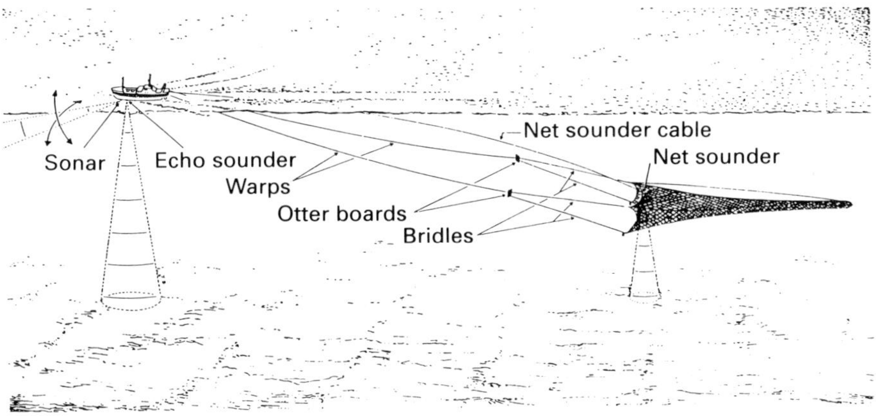

Fig. 1.6. Midwater trawling.

fishing grounds or ocean-ranching. Aquaculture involves an incredible diversity of species ranging from seaweed, molluscs, crustacea and finfish to exotic species such as frogs and crocodiles. The total number of species cultured currently lies approximately around 200.

This section on aquaculture is mainly focused on the culture of fish and only briefly touches on that of other groups. In general it can be said that animal production revolves around 'breeding, feeding and weeding'. These subjects will be treated therefore in more detail with regard to fish farming.

History and status of aquaculture

Although aquaculture and especially rearing of fish is perceived by the general public as a 20th century invention, its roots go back to ancient times. Some form of pond fish culture probably existed in China in 1000 BC (Liao, 1988). In Europe, the Romans were the first to employ techniques which could be categorized under farming. In the work of the classical historians Pliny and Seneca, fish ponds and holding facilities are mentioned as being especially in fashion among the nobility. The farming of salmonids, now comprising a large part of the finfish production in Europe and North America, was greatly stimulated by the invention of artificial fertilization of trout eggs by Jacobi in 1765, later rediscovered by Remy and Gehin in 1842. This was the start of the artificial culture of salmonids. Fish culture has only recently been introduced in North and South America and Africa. Kirk (1987) provides an excellent overview of the often turbulent history of marine fish culture in Europe and North America.

Global aquaculture production in 1991 is estimated at 12.7 Mt, which is 13% of the total fisheries production (FAO, 1993a). The total value

Table 1.5. Aquaculture production by major groups (× 1000 t).

	1984		1991		No. of species
Fish	4335	(43.0%)	8741	(52.7%)	112
Crustaceans	228	(2.3%)	806	(4.9%)	25
Molluscs	2045	(20.3%)	3095	(18.7%)	42
Seaweeds	3446	(34.2%)	3904	(23.5%)	9
Others	21	(0.2%)	32	(0.2%)	6

Source: FAO (1993a).

of this production is approximately 28 billion (10^9) US$. Aquaculture has grown rapidly during the last decade. Between 1984 and 1989 production expanded 70% while in the same period capture fisheries only increased 14%. Table 1.5 gives an overview of aquaculture production by species group.

Large increases in fish production were caused mainly by the rapidly expanding production of Atlantic and Pacific salmon and grass carp. Although global production of crustaceans, mainly high-value marine shrimps, is limited, this group has shown spectacular growth. In Asia, production leapt from 26,000 t in 1975 to 480,000 t in 1988, creating a 'shrimp fever'. Global production of molluscs can be divided roughly into oysters (one third), mussels (one third), scallops, cockles and clams. Production of scallops has shown considerable growth during recent years. Kelp (*Laminaria* spp.) accounts for 50% of the production of seaweed.

The farmed finfish mainly consist of freshwater species (83%), while 13% can be considered diadromous. Marine species represent only 4% of farmed fish production. Most of the species cultured can be considered non-carnivorous and they mainly belong to several cyprinid species (Fig. 1.7).

Roughly half of the production of carnivorous species is that of salmon and trout. There is also a significant production of catfish, yellowtail and eel. The increase during the last decade in production volume of Channel catfish (*Ictalurus punctatus*) and Atlantic salmon (*Salmo salar*) has been impressive. Annual production of catfish has risen from a few tonnes in the early 1960s to 175,000 t in 1991.

The culture of salmon in Norway is undoubtedly one of the biggest success stories in aquaculture. It is hard to imagine that production leapt from a mere 150 t in 1971 (Edwards, 1978) to 150,000 t in 1991.

Although fish farming is often seen as a way of producing cheap protein for 'the masses', this is only true for the production of non-carnivorous fish. Production of carnivorous fish can be viewed as a kind of protein upgrading in which cheap animal protein (fish meal) is converted to a high-value luxury product.

Aquaculture has a long-standing tradition in Asia and it is there that

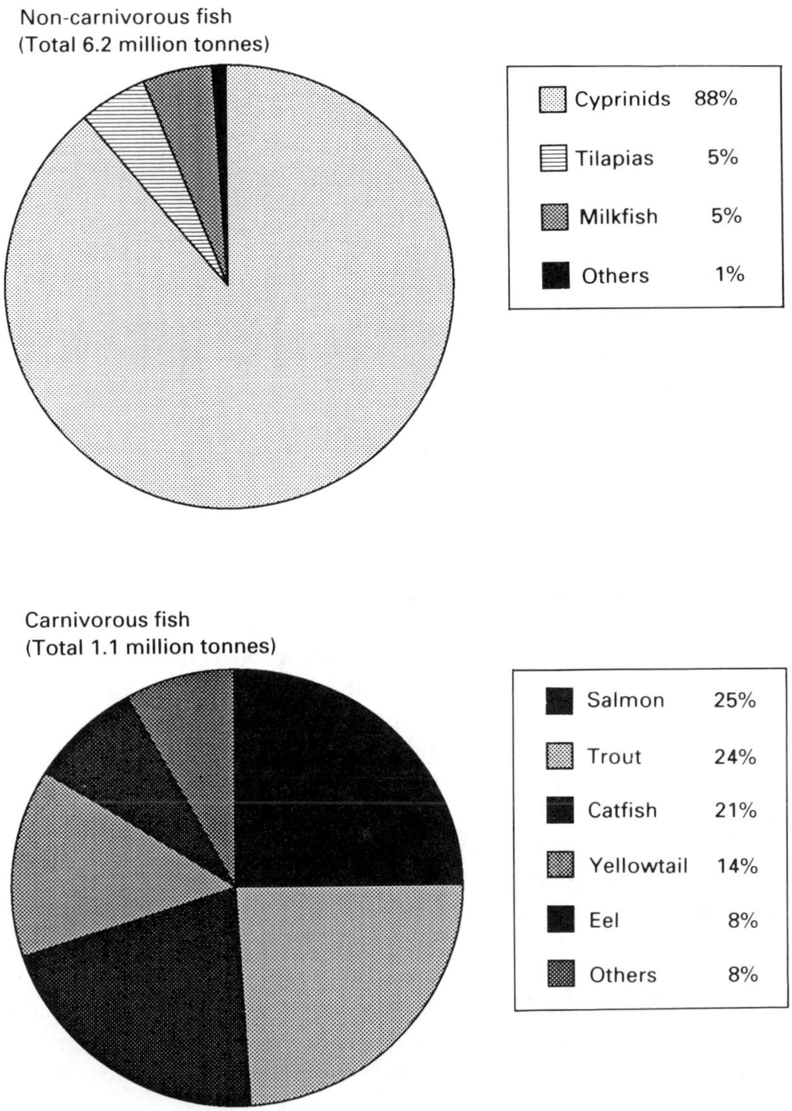

Fig. 1.7. Breakdown of finfish aquaculture into major species groups. Source: FAO (1992).

the bulk of the production is realized (Table 1.6). The production by value of China and Japan combined already comprises half of the global production.

New (1991) gives a detailed overview of aquaculture statistics and recent developments, which should be consulted by the reader interested in more information.

Table 1.6. The 'top ten' aquaculture producers by value in 1991.

	10^6 US$	Percentage
1. China	9,505	33.5
2. Japan	4,567	16.1
3. Indonesia	1,564	5.5
4. India	1,539	5.4
5. Taiwan[a]	1,104	3.9
6. Thailand	859	3.0
7. Philippines	828	2.9
8. Former USSR	797	2.8
9. South Korea	719	2.5
10. USA	541	1.9
World total	28,374	

[a]Estimate.
Source: FAO (1993a).

Types of production and rearing systems

Given the enormous diversity of species cultured, one can imagine the variety of rearing systems and the problems in finding a single classification system for these. One of the most useful classifications of culture systems is that based on the degree of control exerted over the culture process. Systems can thus be classified on a spectrum from extensive systems, in which for example only stocking of seed material is employed, to intensive systems, in which the environment, the feeding and reproduction are completely controlled in order to enhance productivity.

Species low in the food chain, such as seaweed and molluscs, are usually farmed in an extensive way. Only the young stages are grown in protected places while the final grow-out phase is executed in a natural environment. Some shellfish such as mussels and oysters are grown on ropes, thereby utilizing the whole water column. Production in these cases is totally dependent on primary production of algae, thus limiting the carrying capacity of a site.

Extensive production of carp in ponds is probably one of the oldest forms of fish culture. In temperate regions a productivity in the range of 100 to 200 $kg\,ha^{-1}\,yr^{-1}$ can be reached with monocultures of carp without any additional inputs. Stocking a variety of species (polyculture), which can utilize different food items naturally produced in a pond, can greatly increase production (Table 1.7).

A next step towards intensification is the enhancement of primary production by fertilization with organic manure or inorganic fertilizers. In temperate regions a productivity in the range of 500 to 1000 $kg\,ha^{-1}\,yr^{-1}$ can be reached this way with monocultures of carp.

Table 1.7. Fish yield ($t\,ha^{-1}$) in relation to stocking and feeding in Israel.

	Monoculture (carp)	Polyculture (carp, tilapia, silver carp or grey mullet)
Fertilized, without supplemental feeding	1.2	3.0
Fertilized, with supplemental feed	3.5	9.4

After Sarig (1988).

An interesting adaptation on this level is the so-called 'integrated fish farming', in which plant material or waste from domestic animals such as pigs or ducks is utilized in the fish pond. In this system, especially popular in China, the manure and uneaten feed from the domestic animals is utilized directly for fish production. For example, using the excreta and uneaten feed from one laying hen, a bodyweight of 2 kg of fish can be produced in a year (Yang *et al.*, 1992).

Further intensification requires the addition of external feed sources since natural production in a water body is limited. Such feed can consist of cereals in the case of non-carnivorous fish or artificial compound feeds which are fed as pellets. At this level of intensification, special measures are required to maintain a sufficient water quality, since fish consume oxygen and produce toxic metabolites and therefore pollute their own environment. The oxygen level is the first limiting factor and therefore aeration has to be applied under circumstances of high density. Using ponds and aeration, productivity reaches levels in the range of 6000–20,000 $kg\,ha^{-1}\,yr^{-1}$. This type of culture is commonly applied for production of carp, tilapia, catfish, eel and shrimp.

Even higher densities and productivity are possible using large flows of water while fish are kept in ponds or tanks. These flow-through systems are often used for culture of trout and have a productivity in the range from 25 to 100 $kg\,m^{-3}\,yr^{-1}$. Under ambient conditions in northern Europe the production of 1 t of 200 g trout typically requires a flow of 4.5–7.5 $l\,s^{-1}$.

Apart from ponds or tanks, fish can also be kept in pens or cages which are suspended in a natural water body. Oxygen is provided by water movement (tide) through the cage. This type of aquaculture requires sites protected from storms and currents, such as in fjords. Salmon is cultured almost exclusively in this type of system.

A relatively new development is that of 'recirculation systems', in which the effluent from the tanks is treated and subsequently reused in the rearing process. The most advanced types of these systems require little water, thereby enabling heating of water to optimal temperatures. High initial investments and operational problems are bottlenecks with this type of system, which is used in western Europe, e.g. for culture of eel and African catfish. Productivity in these systems can reach the incredible level of 1000 kg m^{-3} yr^{-1} with African catfish.

General aspects of fish culture

Reproduction

A reliable and constant source of stocking material is one of the first requirements in fish farming. Therefore, much research effort is devoted towards artificial propagation and the subsequent rearing of larval fish. Although reproduction of most species cultured today is to a large extent controlled, the culture of some species is still largely dependent on the catch of fingerlings from nature. Some noteworthy examples of the latter case are eels (*Anguilla anguilla*, *A. japonica*), milkfish (*Chanos chanos*) and yellowtail (*Seriola quinqueradiata*).

Maturation and spawning in fish are basically controlled by external stimuli resulting in a stepwise release of hormones through different endocrine glands which eventually control gonadal activity.

Maturation can be induced, for example, by temperature (carp) or photoperiod (salmonids). By manipulating the photoperiod (Bromage and Cumaranatunga, 1988) it has been possible to advance or retard spawning of species such as rainbow trout (*Oncorhynchus mykiss*) and turbot (*Scophthalmus maximus*), thereby enabling a year-round production of fingerlings.

Although maturation can be induced in many species naturally, the final release of eggs often has to be induced artificially. One or two injections with carp pituitary extract (CPE) or human chorionic gonadotropin (HCG) will usually result in ovulation 6 to 12 hours later.

The onset of maturation generally causes a considerable deterioration of fish quality and loss of growth. For some species such as trout special techniques have been developed to produce sterile fish for ongrowing operations.

The size and number of eggs produced per female differ considerably among species. Salmonids produce relatively large eggs with a diameter ranging from 4 to 6 mm but consequently have a low fecundity, i.e. of the order of 800 to 3000 eggs per kg of fish. On the other hand, there are many marine species which have egg diameters in the range of 0.8 to 1.2 mm but can produce up to four million eggs per kg of fish.

Problems in the production of fingerlings are closely related to egg size. Larvae from salmonids, for example, are large at hatching, can be fed directly with artificial feed and have a high survival. However, the tiny larvae from many marine species need very small food particles, often in the form of specially cultured live plankton. Moreover, the composition of this plankton is of vital importance and it is therefore often enriched with certain species of algae. In practice survival from egg to fingerling is usually in the order of a few per cent with marine species.

Control of reproduction has created possibilities for genetic improvement of certain characteristics such as growth rate, dress-out percentage or disease resistance by selective breeding. Although the heritability of most interesting traits in fish is relatively low (Tave, 1986) as compared to those of domestic animals, this disadvantage is partly offset by the large number of offspring, which enables a high selection intensity. Apart from the culture of salmonids the application of genetics to fish culture can still be considered in its infancy.

Feed and feeding in finfish culture

As mentioned before, in many forms of fish culture additional feeding is applied to supplement the natural production of a given water body. The farming of carnivorous fish relies even more on external feed sources. At first, usually cheap industrially fished species or trash fish were used, but these have been replaced gradually by compound feeds. Compound feeds have considerable advantages over other feed types with regard to transport, storage, handling and hygiene. It is obvious that the production of compound feeds requires a detailed knowledge of the nutritional requirements of the fish. In aquaculture research, therefore, much effort is devoted towards nutrition and feeding of fish. Moreover, feed costs represent a large cost item in the budget of a farm.

The protein level required in the diet for maximum growth of fish is 35–50%. At present the major part of the protein in compound feed for intensive fish culture is supplied by high quality fish meal. Attempts to replace part of the fish meal with alternative protein sources such as soyabean meal have met variable success (Tacon and Jackson, 1985).

Qualitative amino acid requirements have been studied in several species of commercial interest. These studies have shown that fish require ten 'essential' amino acids, i.e. arginine, histidine, isoleucine, leucine, lysine, methionine, phenylalanine, threonine, tryptophan and valine. Protein, being the most expensive item in the feed, should be utilized as efficiently as possible for somatic growth. The level of non-protein energy (fats and carbohydrates) in relation to the protein level determines to what extent protein is used for catabolic purposes.

Fish are capable of synthesizing most fatty acids from simple precursors such as acetate. However fish are incapable of *de novo* synthesis of 18:2n-6 18:3n-3, 20:5n-3 and 22:6n-3 (Kanazawa, 1985). In contrast with terrestrial animals which require n-6 fatty acids, most fish species require n-3 fatty acids. In general, fish have a lower body temperature than mammals, which makes incorporation of n-3 fatty acids, having lower melting points than n-6 fatty acids, more suitable. The capability to elongate or to desaturate fatty acids differs considerably among species.

Although fish do not have a structural requirement for carbohydrates, this ingredient is usually included in compound feeds. One of the reasons for the inclusion of carbohydrates is its excellent binding properties and its function in expanded feeds. The capacity of fish to utilize carbohydrates varies widely. Although carnivorous fish such as salmonids do not encounter carbohydrates in their natural diet, it has been shown that these species are able to utilize carbohydrates to some extent.

Like other animals, fish have certain requirements for vitamins and minerals in their diet. Although the basic requirements in this respect are presently known for most important commercial species, there are still a lot of open questions in this field. Part of the mineral requirement can be covered directly from the external environment.

The reader interested in more detailed information should consult, for example, Cowey *et al.* (1985), Hepher (1988) and Steffens (1989).

Another interesting aspect of feed formulation is the addition of certain pigments in feeds for salmonids. Under natural conditions, salmonids develop orange or reddish muscle tissue due to deposits of mainly astaxanthin derived from certain food items. During the last few months before slaughtering farmed salmon or trout usually receive feeds with 50 to 100 mg of astaxanthin or the closely related canthaxanthin per kilogram (Putnam, 1992).

Having developed a compound feed with an appropriate composition, it is of course important to deliver the feed in a way suited to the feeding behaviour of the species involved. In feed manufacturing and feeding aspects such as stability and behaviour (floating, sinking) of the pellet, pellet size and feeding method have to be taken into consideration.

Diseases

Fish diseases are a very important aspect of modern fish farming given the enormous impact on profitability. Fish diseases are caused by a wide range of organisms including protozoan and metazoan parasites, fungi, bacteria and viruses. Nutrition or water quality related problems must also be taken into account in this respect. Diseases can be controlled to some extent by prevention, for example by imposing strict measures on trade and movement of live fish. Some bacterial diseases such as vibriosis or furunculosis in salmon can be controlled by vaccination. However, vaccination does not give complete protection while development of vaccines is time-consuming and expensive.

Bacterial diseases are commonly treated with antibiotics or chemotherapeutics. The number of antibacterial drugs used internationally is estimated at 50 (Rφed, 1992).

There are many problems associated with control of fish diseases, e.g. residues in the final product (see Chapter 9), development of resistance and residues in biota or sediment around fish farms. Residues in the final product can be avoided by observing a proper withdrawal period.

There is a large body of literature dealing with fish diseases and for further reading one could consult, for example, Sindermann (1970), Roberts (1978) or Austin and Austin (1987).

The future of aquaculture

Fish production from capture fisheries will probably reach its limits during the coming decade while the demand for fish products will rise due to the increase in world population. Therefore, demand for fish products will have to be met increasingly with aquaculture products. Future developments in this respect are not easy to forecast but global aquaculture production will probably rise to 25 Mt by the year 2000 (New, 1991). Most of this increase in production will come from farming of freshwater species. The farming of luxury species will increase through further intensification and a broadening of the number of species cultured. However, the latter form of aquaculture could be seriously constrained by environmental issues and increasing competition for limited aquatic resources.

Acknowledgements

The authors thank the FAO for permission to reproduce Figs 1.2 and 1.3 from FAO Fisheries Technical Paper No. 222 (Nédélec, 1982), and The Netherlands Fish Product Board for permission to reproduce Figs 1.4, 1.5 and 1.6 from de Boer (1984).

References

Anon. (1963) *Modern Fishing Gear of the World*, Vol. 2. FAO/Fishing News (Books), London, 603 pp.

Anon. (1972) *FAO Catalogue of Fishing Gear Designs*. FAO/Fishing News (Books), West Byfleet, UK, 155 pp.

Austin, B. and Austin, D.A. (1987) *Bacterial Fish Pathogens: Disease in Farmed and Wild Fish*. Ellis Horwood, Chichester, UK, 364 pp.

Ben-Yami, M. (1988) Attracting fish with light. *FAO Training Series*, No. 14. FAO, Rome, 72 pp.

Ben-Yami, M. (1992) *Purse Seining Manual*. Fishing News Books, Oxford, UK, 480 pp.

de Boer, E.J. (1984) *Visserijmethoden* (Fishing methods). Het Visserijschap, Rijswijk, The Netherlands, 178 pp.

Brandt, A. von (1984) *Fish Catching Methods of the World*, 3rd edn. Fishing News Books, Farnham, UK, 418 pp.

Bromage, N. and Cumaranatunga, R. (1988) Egg production in the rainbow trout. In: Muir, J. and Roberts, R.J. (eds), *Recent Advances in Aquaculture*, Vol. 3. Croom Helm, London, pp. 63–138.

Corten, A. (1984) The recruitment failure of herring in the Central and Northern North Sea in the years 1974–1978, and the mid-1970s hydrographic anomaly. *ICES Mini Symposium C.M.1984/Gen:12*. ICES, Copenhagen, 18 pp.

Cowey, C.B., Mackie, A.M. and Bell, J.G. (1985) *Nutrition and Feeding in Fish*. Academic Press, London, 489 pp.

Edwards, D.J. (1978) *Salmon and Trout Farming in Norway*. Fishing News (Books), Farnham, UK, 228 pp.

FAO (1983) Report of the Expert Consultation on the conditions of access to the fish resources of the exclusive economic zones. *FAO Fisheries Report*, No. 293. FAO, Rome, 209 pp.

FAO (1989) Review of the state of world fishery resources. *FAO Fisheries Circular*, No. 710, Rev. 6. FAO, Rome, 55 pp.

FAO (1991) *FAO Fisheries Circular*, (Provisional) No. 815, Rev. 3. FAO, Rome, 136 pp.

FAO (1992) *FAO Fisheries Circular*, No. 815, Rev. 4. FAO, Rome, 206 pp.

FAO (1993a) *FAO Fisheries Circular*, No. 815, Rev. 5. FAO, Rome, 213 pp.

FAO (1993b) *Yearbook of Fishery Statistics*, Vol. 72: Catches and Landings, 1991 (and previous issues). FAO, Rome, 653 pp.

FAO (1993c) *Yearbook of Fishery Statistics*, Vol. 73: Commodities, 1991. FAO, Rome, 396 pp.

Garner, J. (1978) *Pelagic and Semi-Pelagic Trawling Gear*. Fishing News Books, Farnham, UK, 60 pp.

Garner, J. (1988) *Modern Deep Sea Trawling Gear*. Fishing News Books, Farnham, UK, 96 pp.

Gulland, J.A. (1971) *The Fish Resources of the Ocean*. Fishing News (Books), West Byfleet, UK, 255 pp.

Hepher, B. (1988) *Nutrition of Pond Fishes*. Cambridge University Press, Cambridge, UK, 388 pp.

Hjul, P. (ed.) (1972) *The Stern Trawler*. Fishing News (Books) Ltd, West Byfleet, UK, 221 pp.

Kanazawa, A. (1985) Essential fatty acid and lipid requirement of fish. In: Cowey, C.B., Mackie, A.M. and Bell, J.G. (eds) *Nutrition and Feeding in Fish*. Academic Press, London, UK, pp. 281–298.

Kirk, R. (1987) *A History of Marine Fish Culture in Europe and North America*. Fishing News (Books), Farnham, UK, 192 pp.

Kristjonsson, H. (ed.) (1959) *Modern Fishing Gear of the World*. FAO/Fishing News (Books), London, 607 pp.

Le Gall, J.-Y. (ed.) (1989) Télédétection satellitaire et pêcheries thonières océaniques. *FAO Document technique sur les pêches*, No. 302. FAO, Rome, 148 pp.

Liao, I. (1988) East meets West: An Eastern perspective of aquaculture. *Journal of the World Aquaculture Society* 19(2), 62–73.

Nédélec, C. (1982) Definition and classification of fishing gear categories. *FAO Fisheries Technical Paper*, No. 222. FAO, Rome, 51 pp.

New, M.B. (1991) Turn of the millennium aquaculture, navigating troubled waters or riding the crest of the wave? *World Aquaculture* 22(3), 28–49.

Noël, H.S. and Ben-Yami, M. (1980) Pair trawling with small boats. *FAO Training Series*, No. 1. FAO, Rome, 77 pp.

Putnam, M. (1992) A review of the nature, function, variability and supply of pigments in salmonid fish. In: Pauw, N. De and Joyce, J. (eds) *Aquaculture and the Environment*, EAS Special Publication No. 16. EAS, Ghent, Belgium, pp. 245–263.

Roberts, R.J. (1978) *Fish Pathology*. Baillière Tindall, London, 318 pp.

Robinson, M.A. (1984) Trends and prospects in world fisheries. *FAO Fisheries Circular*, No. 772. FAO, Rome, 25 pp.

Røed, H. (1992) Demands for environmental testing of antibacterials to be used in Norwegian fish farming and environmental solutions. In: Pauw, N. De and Joyce, J. (eds) *Aquaculture and the Environment*, EAS Special Publication No. 16. EAS, Ghent, Belgium, pp. 265–275.

Sarig, S. (1988) The development of polyculture in Israel: a model of intensification. In: Shepherd, J. and Bromage, N.R. (eds) *Intensive Fish Farming*. BSP Professional Books, Oxford, UK, pp. 302–332.

Sindermann, C.J. (1970) *Principal Diseases of Marine Fish and Shellfish*. Academic Press, New York, 369 pp.

Steffens, W. (1989) *Principles of Fish Nutrition*. Ellis Horwood, Chichester, UK, 384 pp.

Storer, J.A. (1977) Law of the Sea negotiations and extended jurisdiction. In: Anderson, L.G. (ed.) *Economic Impacts of Extended Fisheries Jurisdiction*. Ann Arbor Science Publishers, Ann Arbor, Michigan, USA, pp. 105–110.

Tacon, A.G.J. and Jackson, A.J. (1985) Utilization of conventional and unconventional protein sources in practical fish feeds. In: Cowey, C.B. Mackie, A.M. and Bell, J.G. (eds) *Nutrition and Feeding in Fish*. Academic Press, London, UK, pp. 119–145.

Tave, D. (1986) *Genetics for Fish Hatchery Managers*. AVI Publishing, Westport, USA, 299 pp.

Thomson, D.B. (1981) *Seine Fishing*. Fishing News Books, Farnham, UK, 208 pp.

Traung, J.-O. (ed.) (1955) *Fishing Boats of the World*. FAO/Fishing News (Books), London, UK, 579 pp.

Traung, J.-O. (ed.) (1960) *Fishing Boats of the World*, Vol. 2. FAO/Fishing News (Books), London, UK, 781 pp.

Traung, J.-O. (ed.) (1967) *Fishing Boats of the World*, Vol. 3. FAO/Fishing News (Books), London, UK, 648 pp.

Yang, H., Fang, Y. and Chen, Z. (1992) Integrated fish farming in China and the allocation of resources. *World Aquaculture* 23(1), 61–68.

Edible Species **2**

S.J. de Groot
Brederoodseweg 49, Santpoort-Zuid, The Netherlands.

Introduction

Estimates of the number of fish species range between 15,000 and 40,000 (Cohen, 1970). Most species living in the coastal zone and shallow seas are fished. A first selection of the fish takes place at sea, depending on their size and market value. For this reason large quantities of fish are discarded before actual landing of the fish takes place. A complete overview of all fishes caught and landed can be found in the FAO yearbook volumes *Fishery Statistics, Catches and Landings* (e.g. 1991). In this chapter only a very limited selection is given of species landed for human consumption, including shellfish and crustaceans. They will range from cold and temperate water to tropical species, inhabiting both fresh and marine water.

But, before describing individual species, some attention has to be paid to how fish species are distributed in the waters of the world. Based on minimum estimates, Cohen (1970) concludes that there are about 50 jawless fishes (Agnatha), about 555 cartilaginous fishes (Chondrichthyes) such as sharks and rays, and about 21,000 teleost or bony fishes (Osteichthyes). Further, he concludes that about 40% of the fish species inhabit fresh water and about 60% the marine environment. Some species move from the sea to fresh water to reproduce (anadromous fish such as salmonids). Others move to the sea from the fresh water (catadromous fish such as eel).

Most marine species inhabit the shore and continental zone (0–200 m depth); in warm water about 8000 species and in cold water 1130 species. These include perciform fishes and, in colder waters, gadoids and flatfishes as well. The continental slope and deep sea benthic (200 m depth and more) harbours 1280 species (6.4%), the epipelagic (above 200 m) 255 species (1.3%) and the deep pelagic (below 200 m depth) 1010 species (5.0%). Important fishes in the epipelagic zone are scombroid fishes and in the deep pelagic the clupeoid fishes.

Fish Species Landed for Human Consumption

Species landed show an enormous variation in size, shape and appearance. The most important consumption species are discussed below. In most cases English (E), French (F) and German (G) common names are given. As for the chemical composition, most fishes have protein contents around 20%. There is a large variation in lipid contents, i.e. from less than 1% to more than 25%. The chemical composition is dealt with, in more detail, in the next four chapters.

Sharks

The members of this group of fishes are in general recognizable by their elongate body form. The tail fin is asymmetrical, the upper lobe being longer than the lower lobe. Fin rays are lacking in the fins, but some sharks have prominent spines in both dorsal fins. The jaws are well equipped with rows of sharply pointed teeth.

Carcharhinus plumbeus (Nardo). Sandbar shark (E); Requin gris (F); Grauhai (G)

Maximum length 2.5 m. Nose broad, round and shorter than the width of the mouth. First dorsal well developed, triangular; pectoral fins broad triangular and relatively long, upper lobe of tail well developed. Colour greyish-brown, white belly. Distribution: Atlantic coast from Portugal,

Spain, Morocco, Mediterranean. All in temperate and tropical seas; not in eastern Pacific and southern North Sea.

Squalus acanthias Linnaeus. Spurdog (E); Aiguillat (F); Dornhai (G)

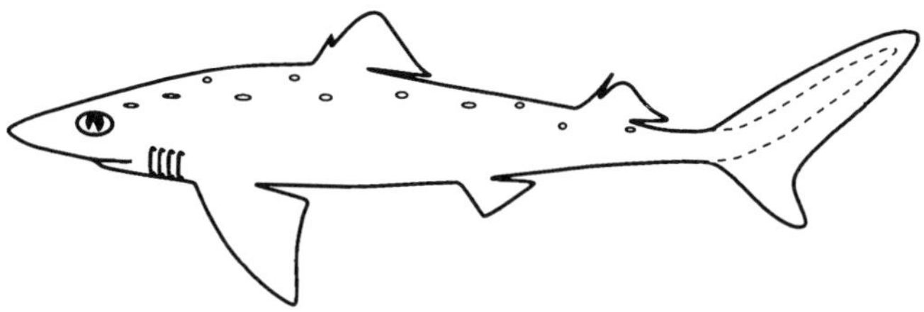

Maximum length 1.2 m. Colour grey with white patches, belly whitish. Prominent spines in front of both dorsal fins, no anal fin. The spines form a defence mechanism. The spurdog is a schooling species. They feed on sand eel, herring, haddock and squid. The species is ovoviviparous. Females have 2–11 juveniles with a length at birth of 30 cm. Distribution worldwide, but rare in Gulf of Mexico and Baltic. In 1989 37,570 t was caught by the fisheries (France 9892 t, Scotland (UK) 7463 t, Ireland 5336 t, Norway 5329 t, England (UK) 4369, Canada 2370 t).

Rays (skates)

Both names are more or less synonymous, but the larger species, i.e. the long-snouted ones, are referred to as skates, and the smaller species as rays. The species are flattened from above,' the pectoral fins are enormously developed and fused with the sides of the head and snout, forming a complete disc.

Raja clavata Linnaeus. Roker, Thornback ray (E); Raie bouclée (F); Nagelrochen (G)

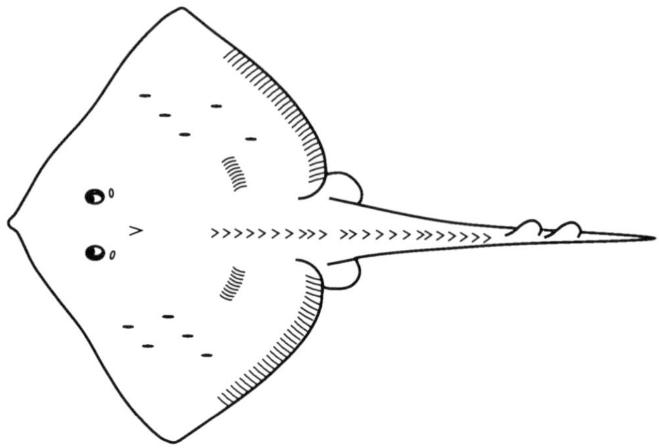

Maximum length 1 m. Dorsal side yellow-brown with irregularly shaped brown patches. Ventral side whitish, grey at the fringes. Dorsal side with coarse prickles. Adults usually have large spines with greatly swollen bases on the back in a median row on body and tail. Distribution: Atlantic coast of Europe and Africa, from Arctic Circle to South Africa, Mediterranean and western Black Sea. In 1989, French fisheries caught 2970 t. Rays, however, are not sorted by species but lumped together with related species such as *Raja batis, R. montagui, R. fullonica, R. naevus, R. oxyrinchus and R. senta*. The unspecified catches for 1989 were 25,613 t for Europe (France 5297 t, UK 10,081 t).

The ammonia smell of cooked rays (and shark) is not an indication that the fish is spoiled, but is the result of enzymic degradation of urea, which is present in relatively large amounts (see Chapter 3).

Sturgeons

Primitive fishes with only parts of the skeleton well ossified. They have elongate snouts, usually five rows of bony scutes along the body. The tail is heterocercal, the backbone continues in the upper lobe of the tail. Species of the Northern Hemisphere, most abundant in Asiatic and American continents. They are anadromous.

Acipenser sturio Linnaeus. Sturgeon (E), Esturgeon (F), Stör (G)

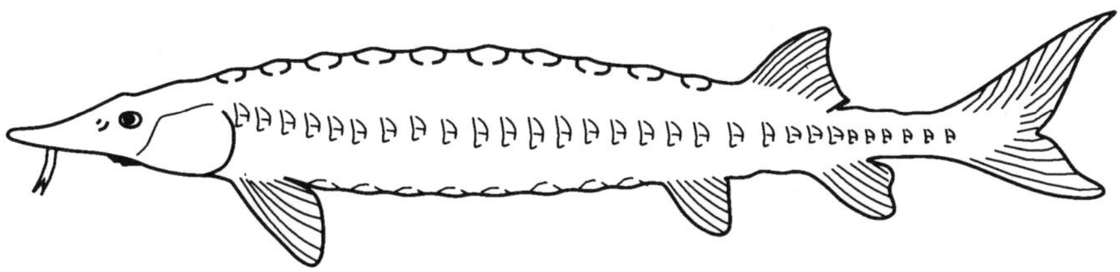

Maximum length 4 m. Back brownish-grey, sides greyish, belly yellow-white. The species has nearly vanished in Europe. At present an extensive programme in France has been set up to restock the large rivers (Gironde system). Distribution: Atlantic coasts from Morocco to Norway, North Sea, Baltic, Mediterranean and Black Sea. The meat is good to eat, but needs care in preparation. The eggs are better known. Several sturgeon species such as *A. gueldenstaedti* and *Huso huso* provide the true caviar. Caviar substitute is produced by the roe of the lumpsucker (*Cyclopterus lumpus*) after colouring to mimic the caviar of sturgeons.

Sturgeon catches for 1989 (all species) total 19,318 t – Russia 1880 t (European part 1312 t, Asian part 568 t); Iran 2051 t; USA 232 t; Canada 40 t.

Herring-like fishes

The species of the herring family are the most important fishes in economic terms. They are also important as a source of food for larger food fishes. They are pelagic fishes forming large shoals and perform periodic migrations. They have a fusiform body, well covered with scales except the head. There is a single dorsal fin usually situated near the midpoint of the body.

Clupea harengus Linnaeus. Herring (E); Hareng (F); Hering (G)

Maximum length 56 cm. Ventral and dorsal fins opposite each other. Back dark blue, side silvery. Inhabits the coastal zone to a depth of 200 m. Eggs are attached to the bottom on hard substrate, e.g. pebbles, dead shells, seaweeds. Distribution: Atlantic coasts, from Bay of Biscay to Greenland, Iceland, Spitsbergen, White Sea, the Baltic, the western Atlantic from South Carolina to Greenland. Total catches in 1989: eastern Atlantic 1,337,076 t; western Atlantic 275,110 t; Norway 269,106 t; Canada 220,484 t; Denmark 170,611 t; Sweden 157,213 t; Russia 137,377 t; Netherlands 90,000 t. The fat content may average from less than 5 to more than 25%.

Sardina pilchardus (Walbaum). Pilchard, Sardine (E); Sardine (F); Sardine (G)

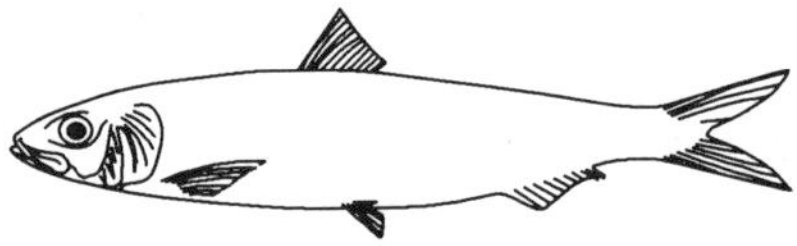

Maximum length 25 cm. Back olive-green, sides silvery, sometimes with a brownish hue. At eye level along the sides is a series of black dots, but these are not always present or only a few. Operculum (gill cover) with radiated grooves, scales rather loosely attached, hence scale loss after handling. Feeds as all other clupeids on zooplankton. Distribution: from southern Norway to Senegal, Mediterranean, Black Sea. World total catch in 1989 1,400,030 t (eastern North Atlantic 172,842 t, Mediterranean 253,545 t, Atlantic off northwestern Africa 973,643 t). About half of the total catch is made by Russian trawlers (609,539 t).

Sprattus sprattus (Linnaeus). Sprat (E); Sprat (F); Sprotte (G)

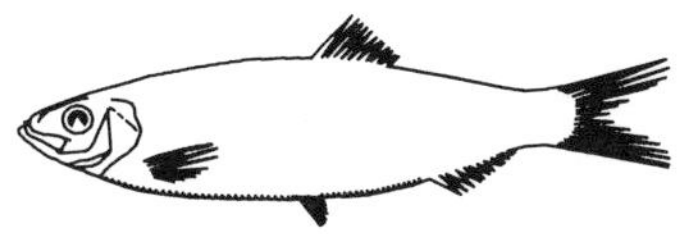

Maximum length 17 cm. Ventral fins before dorsal fin (side view). Back bluish-green, sides silvery. Schooling fish, often close to the shore, also in water with a low salinity. Migrates between foraging area and spawning area as all herring-like species do. Distribution: Atlantic coast from Morocco to British Isles, North Sea. Subspecies in Mediterranean and Black Sea *S.s. phalericus*, in Baltic *S.s. balticus*. Total catch in 1989 in the eastern North Atlantic 208,618 t and in the Mediterranean 106,208 t. Main landings by Denmark 113,554 t, UK 57,381 t and Poland 20,597 t. Fatty fish, containing 7% fat. Mostly eaten as a smoked product or in tomato sauce. Also source of fish oil.

Engraulis encrasicolus (Linnaeus). Anchovy (E); Anchois (F); Sardelle (G)

Maximum length 20 cm. Back light green to bluish-green, sides grey with a silvery horizontal stripe, the stripe dorsally with a dark rim. Only migrates into estuaries to spawn if water temperature in these areas is higher than at sea. Optimal spawning temperature 17–18 °C. Distribution: Atlantic coast from northwest Africa to south Norway, around British Isles, North Sea, Mediterranean, Black Sea, Sea of Azov. Product is mostly pickled in brine and stored for a long time before consumption. Anchovies are also marinated (Mediterranean).

Salmons

Freshwater or anadromous fishes. All possess a distinct rayless fleshy, adipose fin on the back between the dorsal fin and the tail. Pelvic fins are abdominal in position. Teeth are well developed in the jaws.

Salmo salar Linnaeus. Atlantic salmon (E); Saumon (F); Lachs (G)

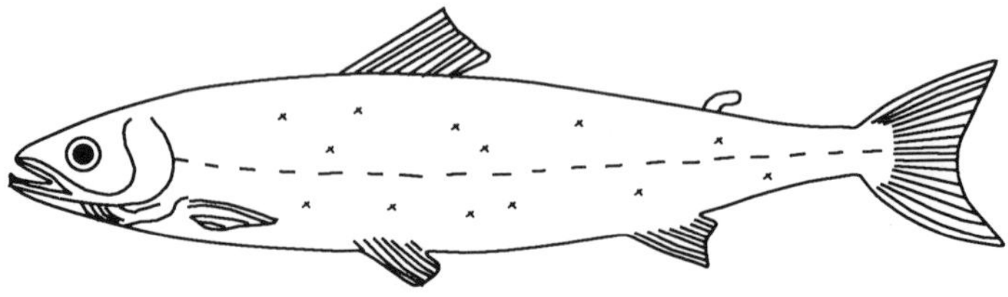

Maximum length 1.5 m (40 kg). Back blue–green, silvery sides, white belly. On the body are wide-apart, black to black-brownish dots, most of them dorsal of the lateral line. Upper jaw does not reach or only just reaches the posterior margin of the eye. The caudal peduncle is slender. Adult specimens live alone during their sea phase. Anadromous species spawn between October and December. Feeds at sea on shrimp-like animals and small fishes such as herring, sprat, sand eel and juvenile gadoids. The first juvenile stage is referred to as *alevin*. The second is the *parr* stage, i.e. 1 to 4 years in fresh water. Then the fish migrates to the open sea, where it stays for 1 to 3 years (*smolt*). Adults returning to the sea again after spawning are known as *kelts*. Distribution: North Atlantic, from the north of Spain to the north of Norway and Russia, around Iceland, and in the Western Hemisphere from Maine (USA) to the entrance of Hudson Bay. Has been introduced to the Pacific. Is also reared in large numbers, i.e. larger than the natural population, in Norway and also in Scotland, Ireland, Canada and the USA. Total production/catch in 1989 in North Atlantic 165,457 t, Norway 116,164 t, Scotland 28,999 t, Ireland 5500 t, Faeroe Islands 7901 t (mainly fish migrating to and from Scandinavia). Salmon are sold fresh or frozen. The smoked fish is a valuable product.

Oncorhynchus gorbuscha (Walbaum). Pink salmon, Humpback (E); Saumon rosé, Saumon à bosse (F)
Maximum length 60 cm (3.2 kg). Colour silvery. In spawning season black patches. Adult specimens during spawning season develop the characteristic 'hump'. Old animals develop a hooked beak. Anadromous species, lives like the Atlantic salmon. Sea phase 1 year. Distribution: northern Pacific. Three populations, i.e. British Columbia and Washington, Alaska and an Asiatic population. A few introductions of the species in northern Russia, north Sweden, Scotland and Iceland. In 1989, total catches were 363,092 t (Russia 144,748 t, USA 166,209 t, Canada 30,640 t, Japan 19,095 t).

Used as Atlantic salmon. Product is mainly light salted or smoked. In the Pacific (origin) there are also five other *Oncorhynchus* species, i.e.

O. keta (chum, keta, dog salmon; total catch 243,600 t in 1989); *O. masou* (masu-, cherry salmon; total catch in 1989 2845 t, by Japan); *O. nerka* (sockeye-, red salmon; total catch in 1989 168,879 t); *O. tshawytscha* (chinook-, spring-, king salmon; total catch in 1989 21,037 t); *O. kisutch* (coho-, silver salmon; total catch in 1989 59,964 t).

Oncorhynchus mykiss (Walbaum). Rainbow trout, Steelhead, Kamloops (E); Truite arc-en-ciel (F); Regenbogenforelle (G)

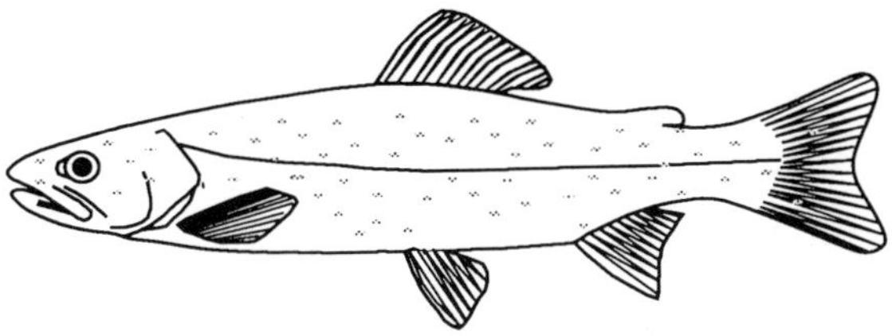

Maximum length 115 cm (20 kg). Back dark green to brown-green, sides lighter with a horizontal blue iridescent band. Heavily pigmented, black dots on the head, body, dorsal and tail fin. Anadromous species, but also landlocked (non-migratory). By far the most easy salmonid to cultivate, worldwide distribution. Name has been changed from the well-known name *Salmo gairdneri* into *Oncorhynchus mykiss* as original distribution area is the Pacific and systematic characteristics resemble more the genus *Oncorhynchus* than the genus *Salmo*. In fish farms the flesh is often coloured before slaughter with carotenoids to resemble the flesh of salmon. Total production in 1989 204,034 t: France 30,935 t; USA 27,725 t; Denmark 25,500 t; UK 19,362 t; Germany 18,584 t.

Carps

The carps (Cyprinidae) contain most of the freshwater species. They have a single dorsal fin, the first one or two fin rays spiny. The pelvic fins are placed near the middle of the body, well behind the pectoral fins. They have large swim bladders, divided into two lobes, and connected with the alimentary canal (Ostariophysi). The first four vertebrae are modified and form the Weberian apparatus, a device to improve hearing. In 1989 in total 4,974,044 t of cyprinids were caught.

Cyprinus carpio (Linnaeus). Carp (E); Carpe (F); Karpfen (G)

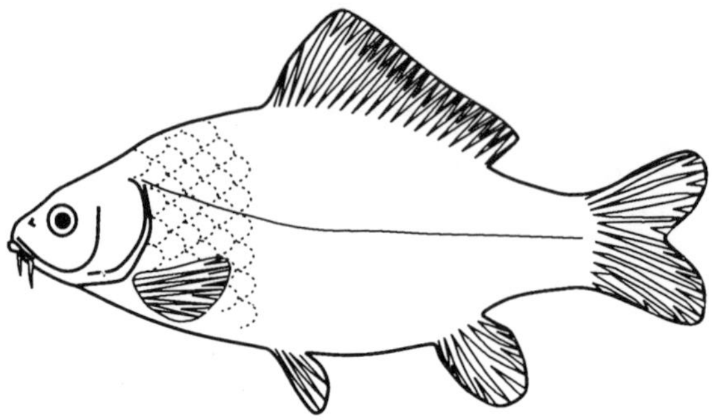

Maximum length 120 cm. Back brown to brownish-green; sides copper-coloured, dorsal fin dark, other fins reddish. Four barbels, two short ones on the upper lip, one large one in each corner of the mouth. Dorsal fin long, first fin ray well developed and spiny. Distribution originally around Black and Caspian Seas but now worldwide. Well-known species in aquaculture. Total production in 1989 1,085,341 t. Asia 679,315 t (China 460,532 t, Indonesia 107,050 t, Turkey 20,657 t); Europe 123,885 t (Poland 21,824 t, Germany 21,331 t, Czechoslovakia 17,628 t, Romania 16,722 t, Hungary 15,029 t); North America 25,050 t (Mexico 22,570 t).

Other important cyprinids are grass carp (*Ctenopharyngodon idella*), in 1989 950,910 t; silver carp (*Hypophtalmichthys molitrix*), in 1989 1,359,724 t; and bighead carp (*H. nobilis*), in 1989 653,566 t. All three species mainly in China.

Ictalurus punctatus (Rafinesque). Channel, Spotted, White and Lake catfish (E); Barbue d'Amérique (F); Wels (G)

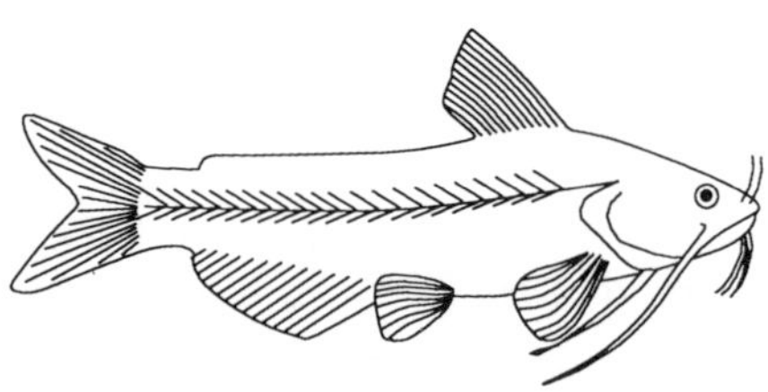

Maximum length about 55 cm, mostly 35–50 cm (1.4–2.3 kg). Juveniles pale blue to pale olive, with silvery overcast. Ventral surface silvery white to milky white, sides with varying number of olive to almost black spots. Adults steel-blue to grey. One dorsal fin, ahead of midpoint of body.

Adipose fin. Caudal fin very broad and deeply forked. Four pairs of barbels. One very long pair arises at the angle of the mouth and two pairs of barbels on the chin. Well-known aquaculture species. In 1989 165,111 t produced, nearly all in USA (165,000 t). Other important catfish species, Claridae and Siluridae, caught or produced in rearing facilities (also 1989 figures), are *Clarias* spp. 67,663 t (mainly Africa) and *Silurus glanis* (wels) 17,367 t (mainly Russia 16,713 t).

Eels

Eels have elongate bodies and a slimy skin. No ventral fins and minute or no scales at all. No separation between dorsal, anal and tail fins. Most species are marine, but some also enter fresh water (catadromous).

Anguilla anguilla (Linnaeus). Eel (E); Anguille (F); Aal (G)

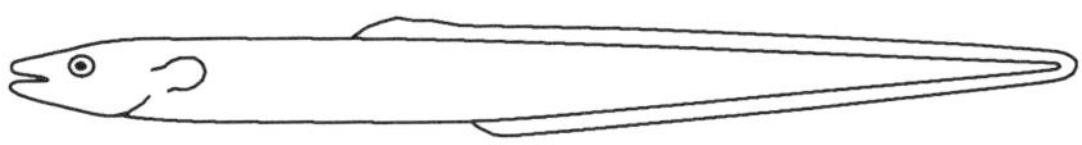

Maximum length 1.2 m. Large specimens are usually females. Males reach a length to 50 cm. In fresh water: olive-green–brown back, yellowish belly. The spawning takes place at sea (Sargasso Sea), the larvae are transported by the Gulf Stream towards the shores of Africa and Europe. They enter fresh water as elvers and stay there for about 10 years. Sexual differentiation starts after 4 to 5 years in fresh water. Before being sexually mature they leave the fresh water and swim back to the sea. Up till now no adult eel has ever been recaptured in the Sargasso Sea. Distribution: Europe, including Mediterranean countries; northern and western Africa. In 1989 15,352 t of eel was caught or produced in aquaculture facilities (Italy 3700 t, France 2716 t, Denmark 2316 t).

Conger conger Linnaeus. Conger (E); Congre (F); Meeraal (G)

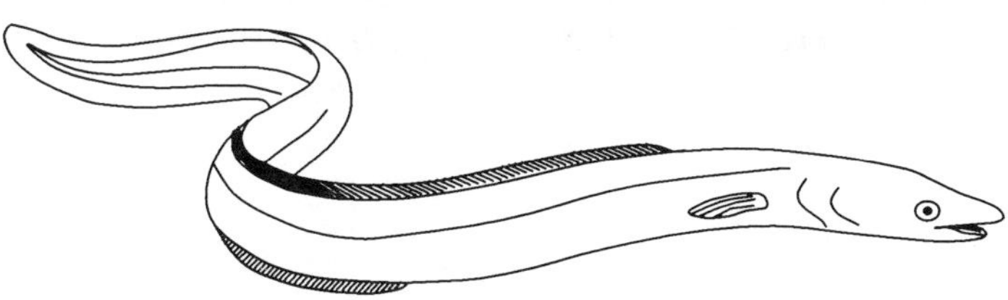

Maximum length 3 m. Greyish back, dorsal and anal fin with a black rim. No ventral fins. Upper jaw longer than lower jaw. Distribution: eastern North Atlantic, North Sea, Baltic. In 1989 13,689 t caught (France 5157 t, UK 3347 t, Spain 2080 t, Portugal 1655 t).

Cod-like fishes

The cod and hake families (Gadidae, Merluccidae) are of great economic importance (in 1989 12,830,536 t world catch). The Alaska pollack, caught in huge quantities, belongs to the *Gadus* family (see Chapter 1).

Members of these families are found in all the oceans, most commonly however in the cooler regions. Greatest abundance in the northern Atlantic, and in moderately shallow water on the continental shelf.

All species have soft rays in their fins and small scales. Most have a barbel on the chin, some have these on the upper part of the head. The ventral fins vary in number. Most cod-like species have three dorsal fins and two anal fins. The flesh is lean (about 0.8% lipids) but the liver contains much fat (more than 50%).

Gadus morhua Linnaeus. Cod (E); Cabillaud (F); Kabeljau, Dorsch (G)

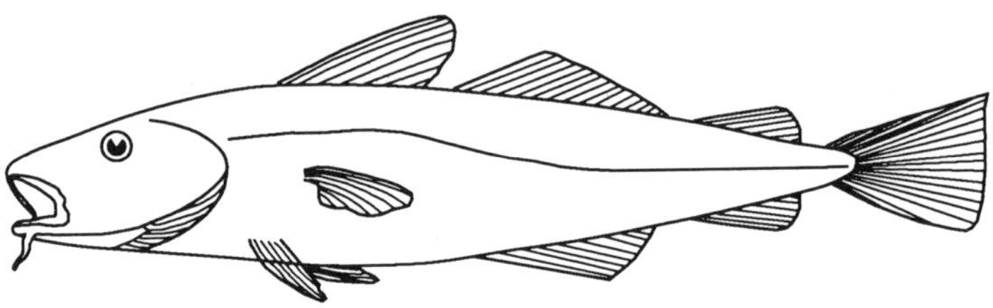

Maximum length 1.9 m, mostly 50–80 cm. Back and sides variable in colour, mostly greenish-brown with patches; belly white, white lateral line conspicuous, curved above the pectoral fins. Three back and two anal fins, barbel on chin. Distribution: northern Atlantic, Bay of Biscay to Greenland, arctic waters of Spitsbergen and Novaya Zemlya. In 1989 1,782,582 t landed (Canada 421,013 t; Iceland 353,985 t, Norway 186,489 t, Russia 149,315 t). Related species: Pacific cod, *G. macrocephalus*, with 1989 catches of 421,937 t.

Melanogrammus aeglefinus (Linnaeus). Haddock (E); Églefin (F); Schellfisch (G)

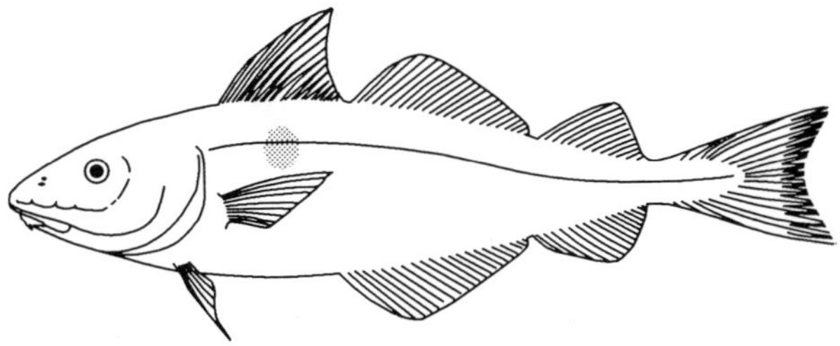

Maximum length 1.1 m, mostly 50–75 cm, back dark green–brown, sides silvery, belly white, large black dot between the pectoral base and lateral line, black lateral line. Three back and two anal fins, short barbel. Distribution: northern Atlantic, Portugal to Iceland, Greenland, White Sea, western Baltic. Catches in 1989: eastern Atlantic 239,683 t, western Atlantic 30,439 t (Scotland (UK) 77,774 t, Iceland 61,792 t, Norway 38,467 t, Canada 25,966 t).

Pollachius virens (Linnaeus). Saithe, Coalfish, Pollock (E); Lieu noir (F); Seelachs, Kohler (G)

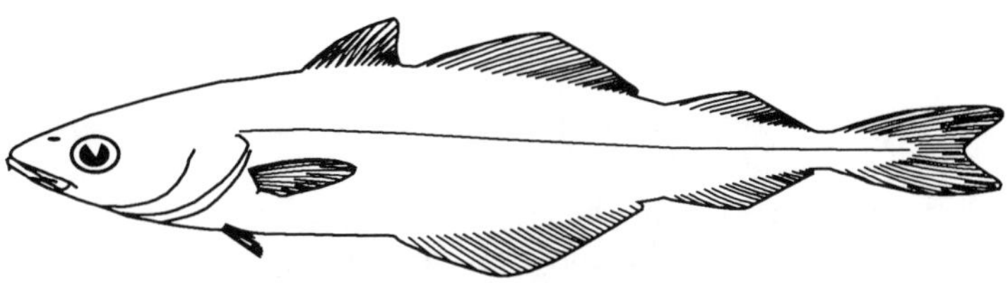

Maximum length 1.2 m, mostly 60–90 cm. Back dark green, sides and belly dull-silvery. Lateral line straight and whitish. Three dorsal fins and two anal fins. No barbel. Eastern North Atlantic from Bay of Biscay to Greenland, Iceland, White Sea, northern North Sea, western North Atlantic. In 1989 landed 375,752 t (Norway 144,484 t, Iceland 79,796 t, France 60,518 t, Faeroe Islands 44,746 t, Canada 44,591 t). Related species: pollack, *Pollachius pollachius*, in 1989 16,018 t.

Merlangius merlangus (Linnaeus). Whiting (E); Merlan (F); Wittling (G)

Maximum length 70 cm, mostly 30–40 cm. Back green-blue, sides silvery (when alive), white when dead, black dot at base of pectoral fin. Three dorsal and two anal fins. No barbel. Shallow water species. Distribution in eastern North Atlantic and western Mediterranean. In 1989 119,913 t landed (France 33,055 t, Scotland (UK) 32,325 t, Ireland 9,498 t).

Merluccius merluccius (Linnaeus). Hake (E); Merlu (F); Seehecht (G)

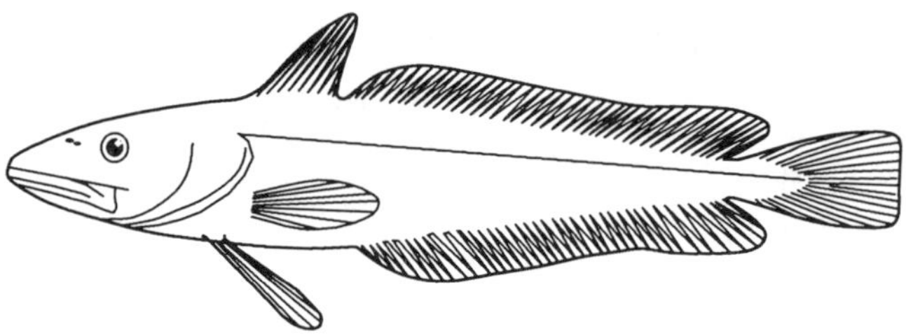

Maximum length 1.8 m. Back blue–grey, sides silvery, belly silvery white. Slender body, large head. Lower jaws longer than upper jaw. Sharp hinged teeth (which make it easy for the prey to enter when teeth fold back, but prevent their escape). Distribution: eastern Atlantic, on continental slopes, depth 100–300 m. Total catch in 1989 112,732 t (Spain 31,629 t, Italy 24,745 t, France 21,894 t).

Related species (with 1989 landings): Senegal hake, *M. senegalensis* (7077 t); silver hake, *M. bilinearis* (105,579 t), South Pacific hake, *M. gayi* (156,176 t); Argentine hake, *M. hubbsi* (404,254 t); North Pacific hake, *M. productus* (240,871 t); Cape hake, *M. capensis, M. paradox* (466,272 t).

Perch-like fishes

The perch-like fishes or cichlids (Cichlidae) are a family consisting of about 600 species. They resemble the perches (Percidae) in outer appearance. The body form is a disc or esoxiform. The dorsal fin consists of a part with spiny fin rays and one with soft fin rays. The anal fin often has three spines. Most species in South America, also in Africa, India and Sri Lanka. The name tilapia is applicable to many species. Several species are cultivated, many hybrid forms. In 1989, in total, 711,361 t produced or landed.

Oreochromis niloticus (Linnaeus). Tilapia (E), Tilapia (F), Tilapia (G)

Maximum length 61 cm (5.1 kg). Body dark brown to green, with dark blue-black vertical bars. Sometimes on each scale a little white dot. Throat and belly white, except in spawning season, when belly becomes deep red. Distribution: Israel, Ethiopia, river systems of Niger, Volta, Gambia and Senegal river, Lake Nasser (Egypt), Lake Tanganyika. Introduced in Lake Victoria. Has been transplanted to many other countries. Total production in 1989 153,659 t (China 89,473 t, Thailand 27,800 t, Kenya 16,892 t, Indonesia 10,650 t).

Sea perches (basses)

Marine fish with moderately elongated bodies. Two dorsal fins, the first with strong spines and connected to the second fin. The anal fin has three strong spines. Also spines and five rays in pelvic fins.

Dicentrarchus labrax (Linnaeus). Sea bass (E), Bar commun, Loup (F), Seebarsch, Wolfbarsch (G)

Maximum length 1 m. Back grey–green, silvery sides, white belly. Juveniles have black dots on the sides. Dark blotch on the gill cover. Two well-separated dorsal fins. Distribution: eastern Atlantic, from Senegal to south Norway, Mediterranean, Black Sea. In 1989 4115 t landed specified and 3551 t non-specified. Related species in North America: *Morone saxatilis* (striped bass), 134 t; and *M. americana*, 338 t.

Epinephelus aeneus (Geoffroy Saint-Hilaire). White grouper (E), Mérou blanc (F)

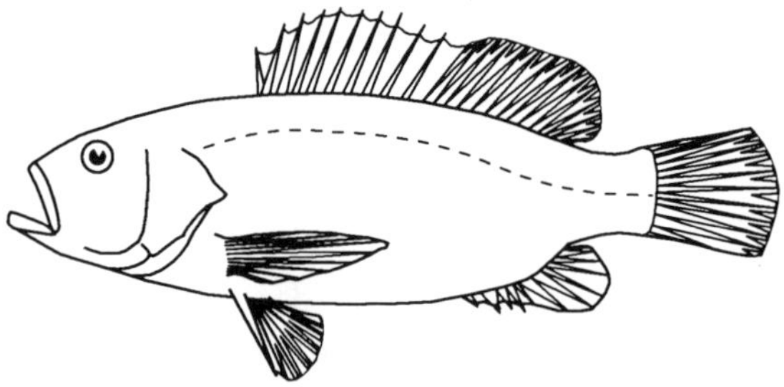

Maximum length 1.2 m. Variably greenish, sides with vertical stripes and black patches. Distribution: eastern Atlantic, from Angola to south England and Ireland, Mediterranean. Many species. In 1989 in total 74,866 t landed.

Pomatomus saltator (Linnaeus). Snapper, Blue fish (E), Tassergal (F)

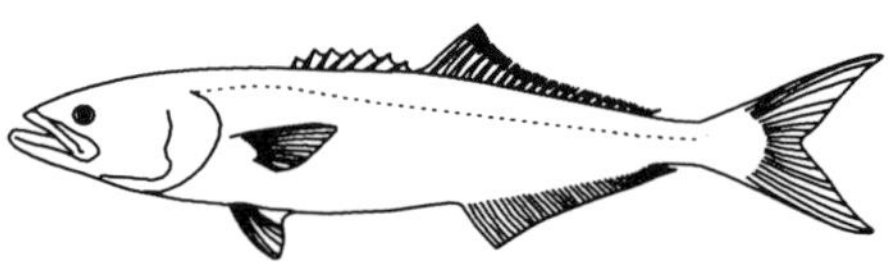

Maximum length 1.1 m, mostly 40–60 cm. Back silvery to grey-greenish. At the base of the pectoral fins a black dot. Two dorsal fins, sometimes connected. Anal fin with two spines, tail fin lightly forked. Large head, gill covers with short spine. Distribution: all tropical and subtropical seas, except central and western Pacific. Atlantic Ocean from South Africa to southern Portugal and Mediterranean. In 1989 23,023 t landed.

Horse mackerels

The Carangidae or horse mackerels have various body forms (long and elongate, flat or high). About 200 species. The caudal peduncle is slender, sometimes with keels. Species often live in large schools. Inhabit all seas. Many species grouped under one name, e.g. jacks, crevalles, scads, bumpers, runners, pompanos, leerfish, vadogo, amberjack, pilot fish.

Trachurus trachurus (Linnaeus). Scad, Horse mackerel (E); Chinchard (F), Stöcker, Bastardmakrele (G)

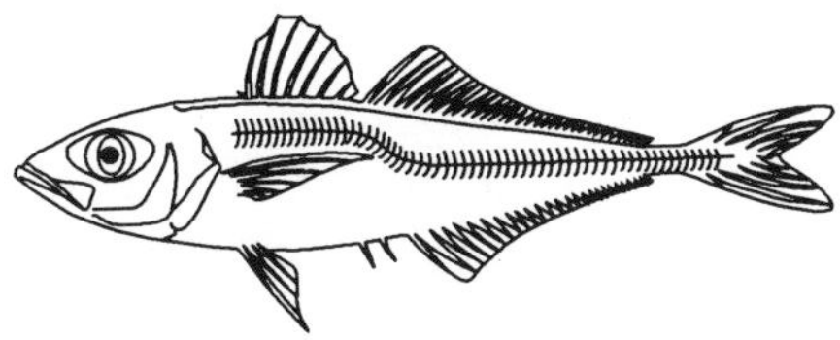

Maximum length 60 cm, mostly 30 cm. Back grey–black to blue–green, sides and ventral silvery white, small black dot on the gill cover. Shoaling species in water with depths of 100–200 m (maximum 500 m). Distribution: northeastern Atlantic, from Iceland to Cape Verde, Mediterranean. Along African coast subspecies (*T.t. capensis*), in Japan related species *T. japonicus*. In 1989 in total 375,984 t landed (Denmark 89,238 t, Norway 88,754 t, Japanese jack mackerel 229,306 t.

Sea breams

Most sea breams (Sparidae) have deep, compressed bodies, covered with scales. They have a long dorsal fin, the first part of it spiny. The anal fin is short, but always with the first three rays spiny. The pelvic fins have one spiny and five branched rays. Dentition is conspicuous, teeth are well developed in both jaws. Sea breams live in tropical to temperate water, most species on the continental shelves. Most species have commercial value.

Sparus aurata Linnaeus. Gilthead sea bream (E); Dorade (F); Goldbrasem (G)

Maximum length 70 cm, mostly 30–35 cm. Rather deep bodied. Silvery grey with a large black patch at the base of the pectoral fin, which spreads over the gill cover with a red tinge. A greenish-golden curved bar runs across the forehead. Adult specimens have a black stripe across the dorsal fin. Distribution: common in Mediterranean, Atlantic coasts from southwest England to northwest Africa. Also cultivated, but not yet in large numbers. In 1989 4713 t landed or produced, about half by Italy (2117 t) and 717 t by Turkey.

Mackerels and tunas

The mackerels and tunas (Scombridae) are a very homogeneous group, with 15 genera. All of them are fast swimmers with a fusiform body, optimally adapted to move through the water. All species have two dorsal fins and a row of dorsal and anal finlets along the tail behind the normal fins. Small keels are to be found on the base of the lobes of the tail fin. The tail fin is the most important part for propulsion. They often live in large shoals (e.g. mackerel) but also individually or in small shoals.

Many species have high protein contents (up to 25%). Most species belong to the fatty fish, but some (e.g. yellowfin tuna) are low in fat.

Katsuwonus pelamis (Linnaeus). Skipjack tuna (E); Bonite à ventre rayé, Listao (F); Echter Bonito (G)

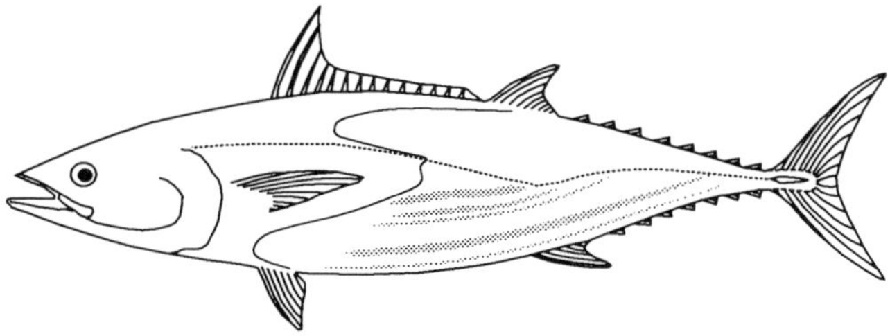

Maximum length 1.1 m (35 kg), mostly 80 cm. Back dark purple–blue, lower sides and belly silvery with conspicuous dark stripes. Pelagic species of warm and temperate waters. Migrates in shoals. Lives during daytime from near the surface to 260 m depth. Distribution: in all tropical and subtropical seas, but not in Black Sea and southern Mediterranean. The species contributes about 40% of all tuna catches. In 1989 in total 1,180,121 t was landed. Northeast Atlantic 5931 t; western Central Atlantic 4203 t; eastern Central Atlantic 84,851 t; southwest Atlantic 23,053 t; western Indian Ocean 217,323 t; eastern Indian Ocean 14,074 t, northwest Pacific 120,136 t; western Central Pacific 589,247 t; eastern Central Pacific 79,639 t; southwest Pacific 10,461 t; southeast Pacific 30,896 t. The flesh is of high quality, has a light brown colour and a taste of beef. The ventral parts of the fish provide the best meat.

Scomber scombrus Linnaeus. Atlantic mackerel (E); Maquereau commun (F); Makrele (G)

Maximum length 66 cm (1.8 kg), mostly 35–45 cm. Fusiform body. Back and upper sides blue–green with black irregular lines and patches, lower sides and belly white with a pink and golden tinge. The first dorsal fin has 11–13 spines. Spawns in water with a temperature of 11–14 °C. Related species: *S. japonicus*, spawns in water between 15 and 20 °C. Distribution: northeast Atlantic, Baltic, Mediterranean, Black Sea. In 1989 in total 626,041 t landed (UK 172,482 t, Norway 143,415 t, Ireland 84,216 t, Netherlands 34,670 t).

Thunnus albacares (Bonnaterre). Yellowfin tuna (E); Albacore, Thon à nageoires jaunes (F); Gelbflossenthun (G)

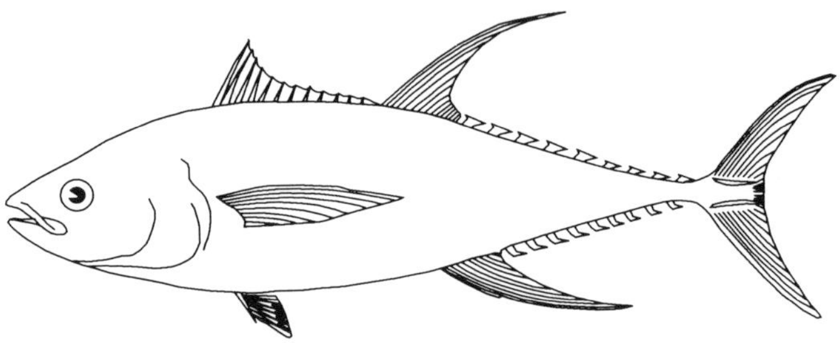

Maximum length 2 m (180 kg), mostly not exceeding 1.5 m. Adult specimens have a conspicuous long second dorsal fin and anal fin, both yellow. On the ventral side are about 20 vertical lines. Lives in shoals, sometimes migrating together with other tuna species. Distribution: in all tropical and temperate seas, but not in Mediterranean. In 1989 in total 905,493 t landed. Northwest Atlantic 1016 t, western Central Atlantic 26,339 t, eastern Central Atlantic 121,138 t, Western Indian Ocean 136,153 t, Eastern Indian Ocean 32,022 t, western Central Pacific 245,149 t, eastern Central Pacific 279,193 t, southeast Pacific 38,858 t. Pink-coloured firm flesh, taste of chicken. Dark-coloured meat unsuitable for consumption.

Thunnus thynnus (Linnaeus). (Northern) Bluefin tuna (E); Thon rouge (F); Roter Thun (G)

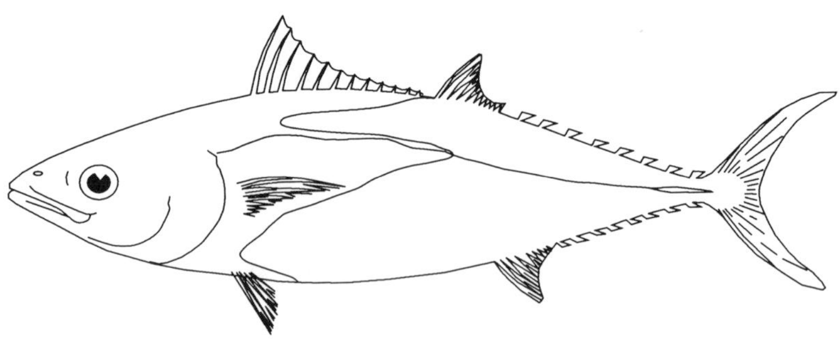

Maximum length 3.0 m (560 kg) mostly up to 2.0 m. Lower sides and belly silvery white with colourless patches. First dorsal fin yellow or bluish, second dorsal fin reddish-brown; anal fin and finlets yellow with a black rim. Keels black. Fast swimmers which cross the oceans. Shoaling species. Distribution in all tropical and temperate seas between 40°N and 40°S. Total catch in 1989 29,848 t (northeast Atlantic 4219 t; eastern Central Atlantic 1022 t; Mediterranean 12,694 t; northwest Pacific 6475 t; western Central Pacific 1931 t).

Swordfishes

The swordfish, spearfish and sailfish (Xiphiidae and Istiophoridae) are well known for the extreme prolongation of their snout and upper jaw into a spear or sword. They are large pelagic fishes, most inhabiting tropical and subtropical seas.

Xiphias gladius Linnaeus. Swordfish (E); Espadon (F); Schwertfisch (G)

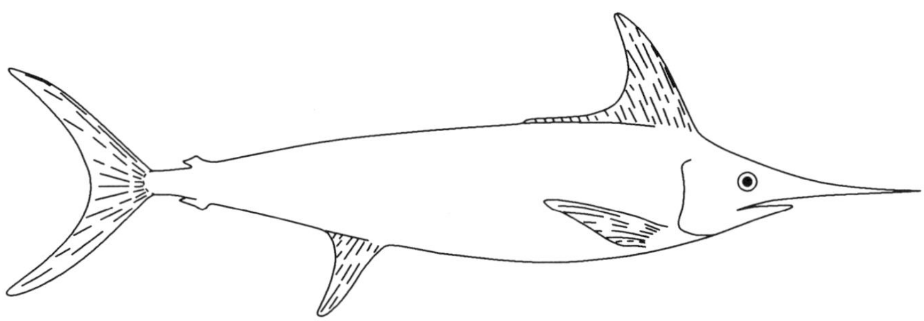

Maximum length 4.5 m (540 kg), mostly not exceeding 2 m. Females are larger than males. Specimens having a weight above 140 kg are mostly females. Back and sides black–brown, lower sides and belly light brown. First dorsal black–brown, other fins brown. Characteristic is the snout with 'sword'. Found in water above 13°C, mostly 18–20°C. Inhabits the surface layers to a depth of 650 m.

Main spawning area in the Mediterranean, near Sicily (Strait of Messina). Prey fish, feeding on pelagic fishes (tuna, dolphin fish, barracuda, flying fish and squid). Distribution over all seas between 50°N and 45°S. In 1989 76,737 t caught. Northeast Atlantic 3794 t, northeast Atlantic 4560 t (declining species), western Central Atlantic 5148 t, eastern Central Atlantic 3764 t, Mediterranean 15,532 t, southwest Atlantic 3216 t, southeast Atlantic 9081 t, Western Indian Ocean 2308 t, Eastern Indian Ocean 889 t, northwest Pacific 10,788 t, western Central Pacific 4968 t, eastern Central Pacific 4988 t, southwest Pacific 1328 t, southeast Pacific 6285 t.

Related species: Indo-Pacific sailfish (*Istiophorus platypterus*) 5469 t, Atlantic sailfish (*I. albicans*) 2292 t, Indo-Pacific blue marlin (*Makaira mazara*) 18,722 t, Atlantic blue marlin (*M. nigricans*) 2043 t, black marlin (*M. indica*) 8210 t, striped marlin (*Tetrapturus audax*) 16,765 t, Atlantic white marlin (*T. albidus*) 1406 t. Unspecified Istiophoridae 15,220 t.

Barracudas

The barracudas (Sphyraenidae) inhabit tropical seas, a few species also temperate waters. All 18 species are of commercial interest. Barracudas live in shoals and can be aggressive. The largest and most dangerous species is *Sphyraena barracuda* (Walbaum), which can reach a length of 3 m. In total in 1989 58,470 t of barracuda-like species were caught.

Sphyraena sphyraena (Linnaeus). Barracuda (E); Brochet de mer (F); Barracuda (G)

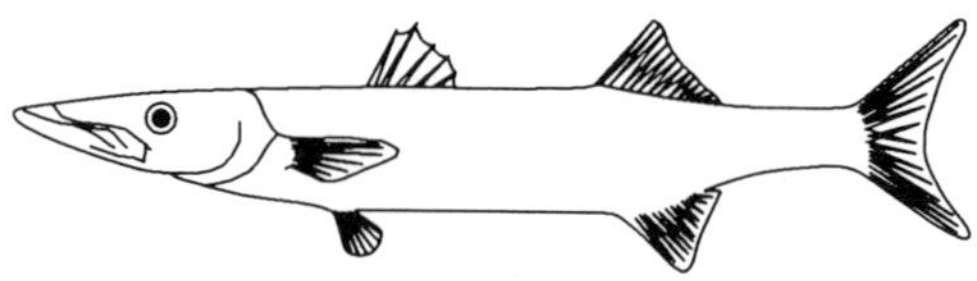

Maximum length 1.6 m, mostly 30–60 cm. Elongated body, large head, lower jaw longer than upper jaw, well-developed incisors. Back dark, sides silvery, parallel to lateral line a yellow band (only to be seen when fish is alive). Two dorsal fins, wide apart, tail forked. Lives solitary or in shoals. Feeds on fish, squid and crustaceans. Distribution: Atlantic coast, from the Bay of Biscay to Angola; Mediterranean and Black Sea; Brazil to Bermuda.

Mullets

The mullets (Mugilidae) are torpedo-shaped fishes inhabiting all tropical, subtropical and not too cold seas. Some species swim into fresh water. They are difficult to distinguish from each other as they look very much alike. They have large scales and a small mouth, and two short, wide-apart dorsals. Mullets live in shoals, and often in very shallow water. They feed on organic matter and graze objects overgrown with algae.

Mugil cephalus Linnaeus. Flathead grey mullet (E); Mulet à grosse tête (F); Meeräsche (G)

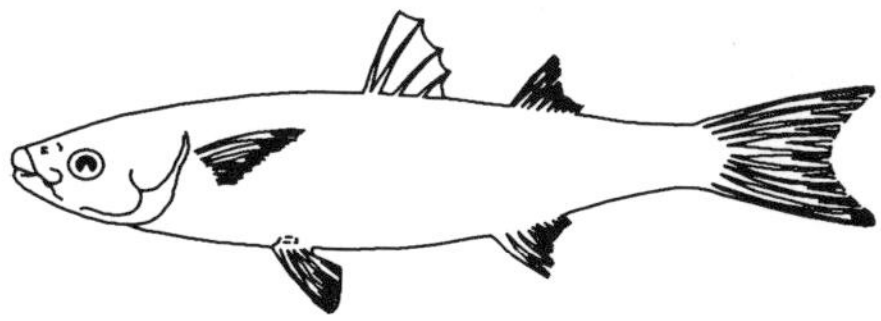

Maximum length 50 cm. Back greyish-blue, silvery sides with vague greyish length stripes, belly white. Broad head, distance between eyes more or less the same as the width of the mouth. Pectorals long. Distribution: all tropical seas of the world. Atlantic coasts from Bay of Biscay to Angola, Mediterranean and Black Sea. Most widely distributed mullet species. Collected fry is reared in aquaculture facilities. In 1989 38,613 t landed, together with 217,442 t unspecified mullet species.

Scorpaeniform fishes

Redfishes

The redfishes, including the scorpion fishes, belong to the family of the Scorpaenidae. They all have strong spines. The lower rays of the pectoral fins are unbranched.

Sebastes marinus (Linnaeus). Redfish, Norway haddock (E), Sebaste doré (F); Rotbarsch (G)

Maximum length 1 m, mostly 35–55 cm. Back and sides red, belly pink, gill covers grey. The lower three spines on the gill cover point to the belly (*S. viviparous* in the direction of the tail). Juvenile fish live near the coast, but adult specimens migrate into the open sea and live at depths ranging from 100 to 1000 m, where they shoal in small groups and feed on euphausids, herring and capelin. Distribution: northern Atlantic, North Sea to Spitsbergen, Barents Sea and Novaya Zemlya, Iceland, east and west Greenland, northern coasts of North America to Grand Banks and Gulf of St Lawrence. Statistics are only available for the northeast Atlantic. In 1989 12,501 t was landed. Also in 1989, 344,253 t of unspecified Atlantic redfish was landed (northwest Atlantic 163,630 t, northeast Atlantic 180,623 t).

Gurnards

Gurnards (Triglidae) are in most cases very colourful species, inhabiting tropical, subtropical and temperate waters. The average length is between 30 and 50 cm. Some species can attain 1 m. They are bottom-living, demersal species. Their depth range is to 500 m. Nearly all gurnards have the lowest three rays of each pectoral fin separate. These rays serve as a tactile organ to search the sea bottom for food. The gurnards have armoured heads with strong spines and rows of sharp spines along the back each side of the dorsal fins. In total in 1989 38,620 t of gurnards were landed.

Eutrigla gurnardus (Linnaeus). Grey gurnard (E); Grondin gris (F); Grauer Knurrhahn (G)

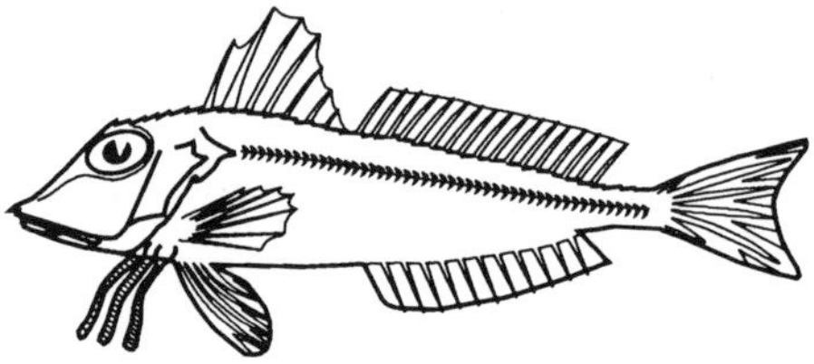

Maximum length 50 cm, mostly about 30 cm. Slender body, sharp snout, straight profile. Lateral line with spiny sharp scutes. Body brown–grey with white dots, belly cream-coloured. Sides sometimes with a red tinge. First dorsal fin with a rounded black mark. Lives on sandy bottom, from near the coast up to 140 m. Distribution: eastern Atlantic from Iceland and Norway to Morocco, Mediterranean and Black Sea. Most common gurnard in the North Sea. In 1989 26,862 t landed, mainly by Denmark (26,740 t). The annual catches of the related species *Aspitrigla cuculus* (red gurnard) amount to 253 t.

Flatfishes

The flatfishes (Pleuronectiformes) are divided into seven families, i.e. Psettodidae, Citharidae, Scophthalmidae, Bothidae, Pleuronectidae, Soleidae and Cynoglossidae. In total there are about 600 species, but the Psettodidae have only three species. A small number of species migrate into fresh water and an even smaller number have become true freshwater species; hence almost all of the flatfishes are sea fishes. Flatfish larvae emerging from the egg are symmetrical, but in due course the larvae are metamorphosed into 'flat' asymmetrical larvae, the pelagic phase of their life comes to an end and the demersal life begins.

Eyes of flatfish can be on the right side or the left side, depending on family (e.g. in Pleuronectidae the eyes usually turn to the right and in Bothidae to the left). Exceptions always occur in the families, e.g. 5–10% of North Sea flounder (*Platichthys flesus*) has eyes on the left. In some American flounders it is a 50–50 division. The length of flatfish can vary between 10 cm and 3 m (350 kg). The world flatfish catch in 1989 was 1,192,432 t and has been stable at this level at least since 1983.

Bothidae (turbot family)

Psetta maxima (Linnaeus) syn. *Scophthalmus*. Turbot (E); Turbot (F); Steinbutt (G)

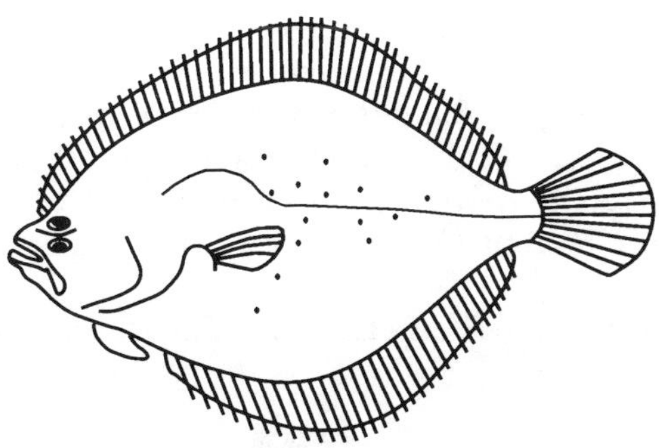

Maximum length 1 m. Pigmented side variable in colour, adapts to background, mostly sandy to brown-coloured with dark patches and dots. Almost square flatfish, large head and mouth (the related brill (*Scophthalmus rhombus*) is more ovoid in shape). Eyes on the left side. Feeds on fish, molluscs and crustaceans. Distribution: northeastern Atlantic from the Arctic Circle to northwest Africa and Mediterranean. In the Black Sea a subspecies (*P. m. maeotica*) is found. In 1989 in the eastern Atlantic 6581 t was caught, and in the Mediterranean and Black Sea 7259 t. The brill catches were 1573 t in the eastern Atlantic in the same year.

Pleuronectidae (plaice family)

Hippoglossus hippoglossus (Linnaeus). Halibut (E); Flétan (F); Heilbutt (G)

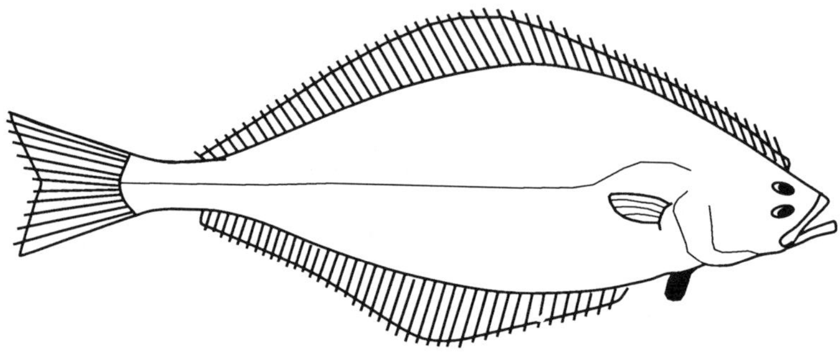

Maximum length 3 m, mostly 1.5–2 m. Pigmented side green–brown to black. Elongated, thick-set fleshy body. Large mouth, well-developed teeth on the jaws. Lower jaw protrudes. Tail rim almost straight. Demersal species in adult phase, at 50–2000 m. Occasionally leaves the bottom to feed on pelagic fishes. Fish from the open sea spawn at depths of 300–1000 m (at 5–7 °C from December to April). Distribution: eastern Atlantic from Spitsbergen, Barents Sea to Bay of Biscay; western Atlantic from Greenland to Cape Cod. In 1989 in total 6022 t was caught (western Atlantic 2718 t, eastern Atlantic 3304 t). In 1989 39,218 t of the related species, the Pacific halibut (*H. stenolepis*), was landed.

Pleuronectus platessa Linnaeus. Plaice (E); Plie, Carrelet (F); Scholle (G)

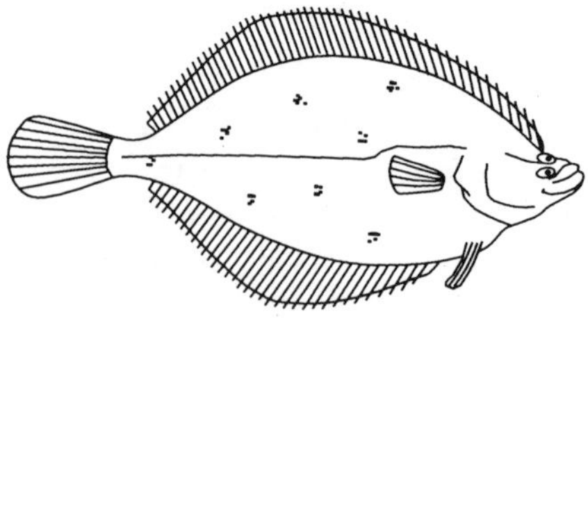

Maximum length 1 , mostly 35–40 cm. Dark brown to dark green with a number of orange dots over the whole body. Eyes nearly always on the right side. Head and mouth small. Skin smooth, on the head a row of four to seven bony tubercles from the eye region to the lateral line. Lives on sandy and stony bottoms, shallow to about 100 m. Older fishes tend to live deeper than younger ones. Feeds on molluscs and worms. Spawns from December to March. Distribution: northeast Atlantic, along the coasts of Europe and western Mediterranean. In 1989 in total 190,108 t was landed (Netherlands 79,620 t, UK 32,191 t, Denmark 30,805 t).

Soleidae (sole family)

Solea solea (Linnaeus). Common or Dover sole (E); Sole commune (F); Seezunge (G)

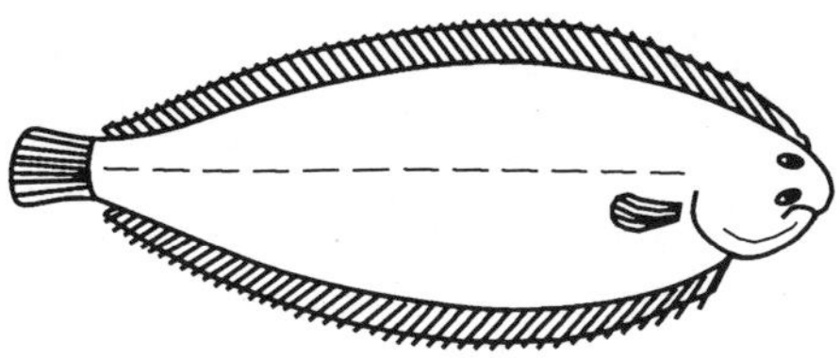

Maximum length 70 cm. Pigmented side brown to dark brown, with irregularly distributed patches and dots. Unpigmented side cream-white. Black oval dot on the end of pectoral fin. Demersal species. Feeds on worms and molluscs. Spawns in the North Sea from April to June, in the Mediterranean in February and in the Bay of Biscay from December to May. Distribution: eastern Atlantic from south of Norway to Senegal, Mediterranean. Grows faster in its northern distribution area. In 1989 43,314 t was landed (northeast Atlantic 29,766 t, (Netherlands 9920 t, France 7126 t, Belgium 4317 t); eastern Central Atlantic 5141 t; Mediterranean 8407 t).

Anglerfishes

The anglerfishes or Lophidae consist of a family with four genera and 12 species. Head and body are compressed. Large wide gaping mouth, strong jaws with large and curved teeth. Lower jaw longer than the upper jaw. Skin is loose and scaleless, many loose, fringed flaps, at the junction of the upper and lower surfaces. The gill opening is restricted and opens behind and above the pectoral fin. Lives in the coastal zone to 1600 m depth.

Lophius piscatorius Linnaeus. Anglerfish (E); Baudroie, Lotte (F); Seeteufel (G)

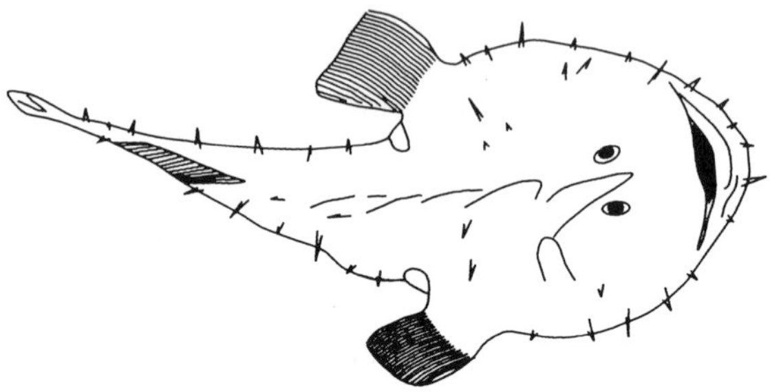

Maximum length 1.8 m, mostly up to about 1 m. Dorsal side reddish-brown to green-greyish with dark patches. Underside white, except hind parts of the pectorals and anal fin, which are black. The first dorsal fin rays without membrane, the first of these has a fleshy end, which is used as a lure to attract fish prey. The prey is sucked into the large gaping mouth, which opens as the attracted fish is just above it. Feeds on herring, whiting, smelt, even on seabirds. Distribution: northeast Atlantic, from Norway to the south of Spain, Mediterranean and Black Sea. In 1989 54,976 t was landed (France 18,177 t, UK 16,509 t, Spain 11,698 t).

For the trade the head is often cut off as it takes a lot of space and looks ugly. Related species: American angler (*L. americanus*) in 1989 11,645 t, Cape monk (*L. upsicephalus*) in 1989 7974 t (mainly South Africa). Also species in northwest Pacific, unspecified 17,525 t.

Molluscs

The Mollusca (molluscs) or shellfish form a whole phylum. The molluscs are built accordingly to a uniform plan, but show a wide diversity.

There is no standard molluscan shape. As a group they have a considerable commercial importance. Shellfish (bivalves and gastropods such as snails) and cephalopods are landed in large amounts, either collected, caught at sea or even reared in estuaries. The scaphopods and amphineurans have no commercial interest. Most species of commercial interest are marine species.

In 1989 in total 7,878,900 t of molluscs were landed, or about 8% of the total world fish catch in inland and marine waters. For 1989: freshwater molluscs 268,951 t, abalones, winkles, conchs 85,295 t; oysters 1,019,953 t; mussels 1,240,174 t; scallops, pecten 840,408 t; clams, cockles, arkshells

1,407,386 t; squids, cuttlefishes, octopuses 2,536,974 t; miscellaneous marine molluscs 479,708 t. About 20% of world landings consist of cultivated species. Most species form a cheap food supply. Exceptions are scallops, pecten and oysters, which fetch high prices.

Bivalves

Mussels

The mussels (Mytilidae) are a well-known family of bivalves and, from a commercial viewpoint, one of the most important. The shell has the typical mussel shape and varies in colour from green to deep blue or sometimes transparent yellow. In their juvenile stages they still are able to move, crawl over the bottom and sometimes bury themselves as a protection against storms when water movement may wash them away. Most species have so-called byssus threads, produced by a gland at the base of the molluscan foot, which act as mooring lines for the mussels. The byssus threads make the family very suited for cultivation.

Mussels are low in protein (12%) and in fat (1%).

Mytilus edulis Linnaeus. Blue mussel (E); Moule commune (F); Miesmuschel (G)

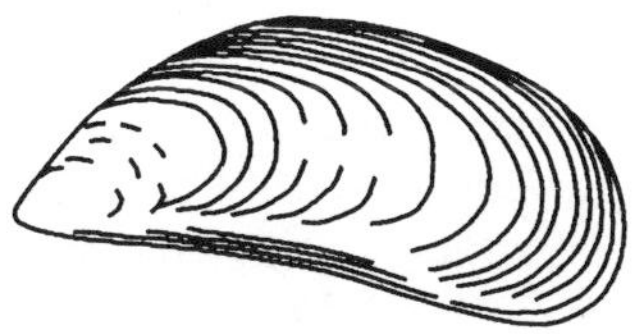

Maximum length about 10 cm, mostly 3–4.5 cm (wild), 4–6 cm (cultivated). Colour blue–black with a shiny periostracum, juveniles transparent yellow, inner side mother-of-pearl. The flesh is yellow–white, the edges of the mantle dark violet. Long and asymmetrical. Attaches itself with byssus threads to the bottom, rocks, etc. from the tidal zone to a depth of 20 m. Distribution: eastern and western Atlantic. In Europe from Norway to the south of Spain, also in Mediterranean. European production (Atlantic, Mediterranean) in 1989 495,616 t (Spain 193,010 t, Netherlands 107,000 t). Other species are *M. crassitesta* (Korean mussel), *M. chilensis* (Chilean mussel), *M. platensis* (River Plata mussel), *M. smaragdinus* (green mussel), *M. viridis* (brown mussel).

Mytilus galloprovincialis (Lamarck). Mediterranean mussel (E); Moule mediterranéenne, Moule de Provence (F)

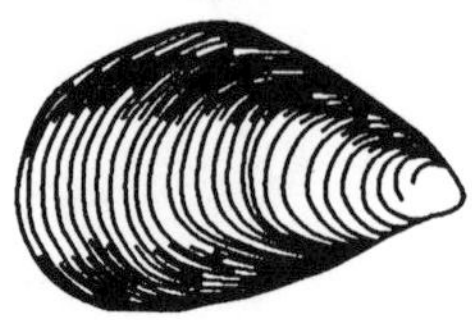

Maximum length 11 cm, mostly 7–8 cm. Colour of the shell dark violet, flesh of large specimen orange–red, small ones creamy white. Larger and broader than *M. edulis*. Has pointed umbo. Inhabits the waters just below the tidal zone. Attaches itself with byssus threads on hard substrate. Distribution: eastern Atlantic, south of UK, Mediterranean, Black Sea. Cultivated on large scale, e.g. in Spain. In 1989 in total landed 116,625 t (Italy 100,538 t).

Oysters

The oysters (Ostreidae) form the best known family of bivalves. They are highly modified for sedentary life. They have a rough outer shell, the byssal stage is short, and the spat cement themselves by the right valve immediately on settling. Adult oysters may be attached to objects on muddy ground, or may be zone-forming on rocks. They do not swim, but with a powerful muscle movement they can expel sediment from the inner shell. At temperatures of 5–10°C they can be stored for a week or longer, which makes them very well suited for transport alive. Oyster cultivation came into being in the middle of the 19th century. At present most oysters are produced in large cultures. World production 1989 1,019,953 t. Oysters contain not more than 8% of protein and are low in fat as well.

Ostrea edulis Linnaeus. European flat oyster (E); Huître plate européenne (F); Auster (G)

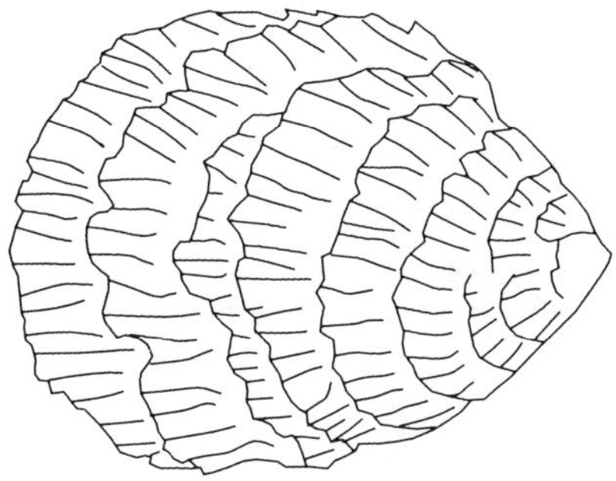

Length about 13 cm, mostly 6–10 cm. Colour variable, periostracum brown. Basic form round, sometimes asymmetrical 'ears' left and right of the hinge. The form varies between localities. The left valve is cupped, the right valve flat. The shell is built up from horny scales. One large central closing muscle. Lives in calm waters attached to rocks or objects. Maximum depth about 80 m. Distribution limited to waters where summer temperature exceeds 16 °C. Atlantic Ocean, east coastal Canada and USA, North Sea and Atlantic coasts of Europe, Mediterranean. European production in 1989 15,656 t (Atlantic 8769 t – Spain 4897 t, France 1975 t; Mediterranean 6887 t – Italy 5000 t, Greece 1410 t).

Related species; *O. lutaria* (New Zealand dredge oyster), *O. chilensis* (Chilean flat oyster).

Crassostrea gigas Thunberg. Pacific cupped oyster (E); Huître creuse japonaise (F); Japanische Auster (G)

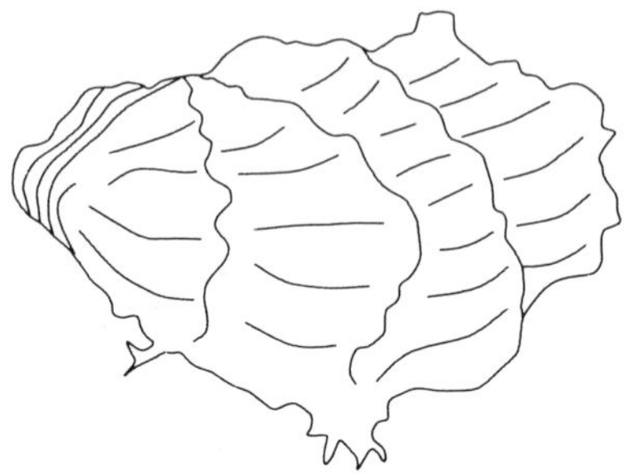

Maximum length 20 cm, mostly 7.5–15 cm. Colour greyish-white, juveniles purple bands. Form variable, mostly elongated and irregular. The horny scales are sharp and curled. The left valve is cupped, the right valve flat. One large closing muscle. Inhabits the tidal zone. Distribution all over the world since the 1950s when the species was transplanted for cultivation. In 1989 world production 636,156 t (northeast Atlantic 285 t; northwest Pacific 601,740 t – Japan 256,313 t, South Korea 242,956 t; northeast Pacific 30,371 t; eastern Central Pacific 1887 t; southwest Pacific 1664 t; southeast Pacific 209 t).

Crassostrea virginica (Gmelin). American cupped oyster, Eastern oyster (E); Huître creuse américaine, Huître de Virginie (F); Amerikanische Auster (G)
Maximum length 25 cm, mostly 8–15 cm. Form intermediate between flat oyster and Pacific oyster, a bit triangular, with a pointed umbo. Flesh in taste and colour similar to flat oyster. Inhabits shallow and brackish waters of the east coast of Canada, USA and Mexico. In 1989 total production 182,899 t (USA 128,589 t, Mexico 52,017 t). Related species: *C. rhizophorae* (mangrove cupped oyster), *C. angulata* (Portugese cupped oyster), *C. cucullata* (rock cupped oyster).

Scallops

The scallops (Pectinidae) belong to a large family. One of the valves is cupped, the other flat. The closing muscle (adductor muscle) is well developed. Pectinids can swim well, one species with a speed of 4 km hr^{-1}. Their pallial tentacles are highly developed, forming sensitive fringes of tactile organs or guarding tentacles for the exclusion of sediment. The tentacles of many species are tipped with small eyes, with a metallic blue sheen from the light-reflecting layer behind the retina. In most cases only the large adductor muscle is eaten, but some smaller species are eaten whole. Scallops are caught with heavy dredges. Artificial propagation is still in its infancy. In 1989 the total catch was 840,408 t.

Unlike the other bivalves described in this chapter, scallops have protein contents of 17–18% and fat contents of the order of 5%.

Pecten maximum (Linnaeus). Great scallop, Escallop (E); Coquille Saint-Jacques de l'Atlantique, Grande vanne (F); Grosse Kammuschel (G)

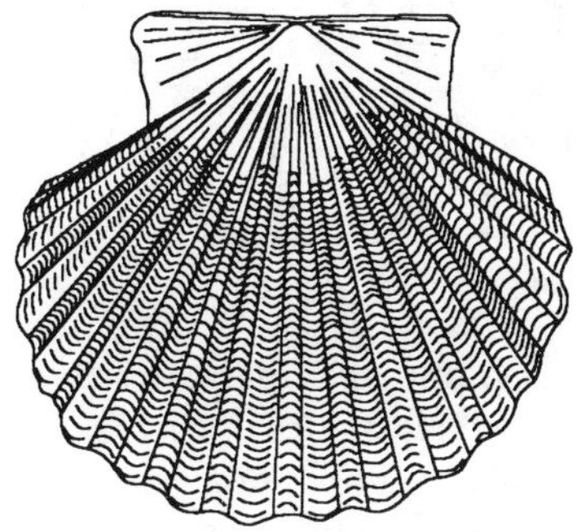

Maximum length 24 cm, mostly 10–12 cm. Largest European representative. The upper valve (left) flat, brown–pink-coloured, the lower valve (right) cupped, white to pinkish, with sculpture of 15–17 rounded ribs. On the mantle the 'eyes' can clearly be seen. Lives on hard sandy bottom or clay bottom, to about 100 m depth. The well-developed adductor muscle and the gonads (coral) are eaten. Distribution: eastern Atlantic from south Norway to south Spain, also in North Sea. At present the species is being overfished. In 1989 total landed 16,231 t (UK 9129 t, France 6505 t). Related species: *P. yessoensis* (Japanese scallop), in 1989 505,165 t caught (Japan 369,373 t, China 129,461 t).

Chlamys opercularis (Linnaeus). Queen's scallop, Queenie (E); Vanneau (F); Gedeckelte Kammuschel (G)

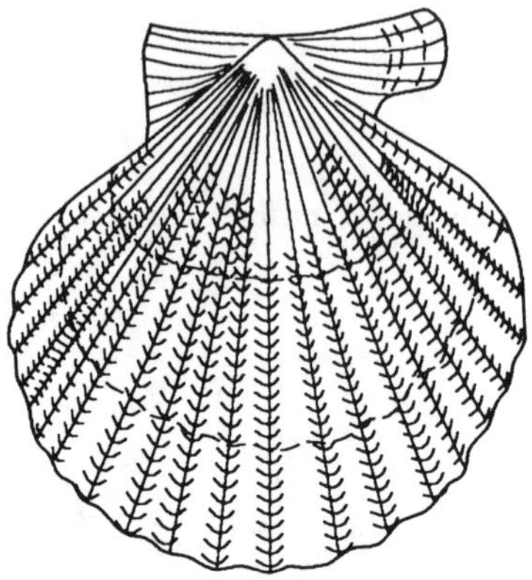

Maximum length 9 cm. Colour of upper valve very variable, orange, red, pink to brownish-purple and all shades of each gathered into a great variety of irregular patterns of streaks and blotches, very rarely white. Sculpture of 19–22 rounded radiating ribs and rows of fine transverse corrugations. Ears with fine ribs, byssal notch with small teeth. Margin strongly crenulate. In its first year it lives attached with byssus threads on stones and rocks, afterwards it lives free on sandy and gravelly bottom to depths of about 200 m.

Distribution: eastern North Atlantic, North Sea and Mediterranean. In 1989 total landed from the Atlantic 14,244 t (UK 8616 t (mainly Scotland, Isle of Man), Faeroe Islands 4757 t). Adductor muscle and gonads are eaten. When cultivated sold as Princess scallop.

Cockles

The cockles (Cardiacea) are among the best known of all bivalves. The edible species has long been known as food. The largest species have a maximum length of 11 cm. Shell inflated, heart-shaped with radiating ribs, often with scales or spines. Lives burrowed in the upper layers of the bottom, often in large quantities. The siphon sticking out of the sand is a food source for fishes and birds.

Cerastoderma edule (Linnaeus). Common edible cockle (E); Coque commune, Rigadot (F); Herzmuschel (G)

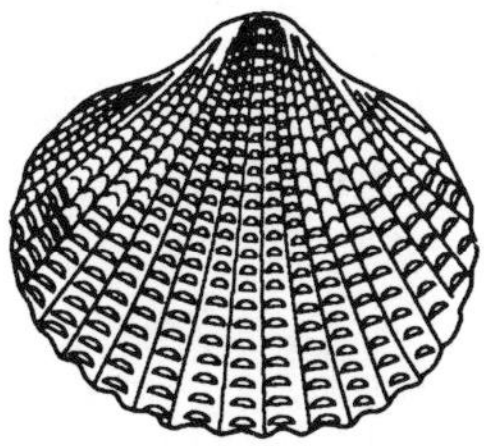

Maximum length 5 cm, mostly 2–3 cm. Shell solid, equivalved, somewhat broadly oval in outline. Colour dirty white, pale yellow or brown. Periostracum yellowish or greenish-brown. Sculptured with 22–28 radiating ribs, each with numerous scale-like spines and very fine irregular concentric lines. The cockle inhabits clean sand, muddy sand or muddy gravel to a depth of 40 m, from mid-tide level to just below low water mark. Often in densities of 10,000 specimens per square metre. Distribution: eastern Atlantic, from Barents Sea, the Baltic and the Iberian peninsula into the Mediterranean, the Black Sea and Caspian Sea, along the coast of west Africa to Senegal. In 1989 total landed 28,156 t (Atlantic), UK 14,748 t, France 5268 t, Denmark 4454 t.

Gastropods

The gastropods (Gastropoda) are very rich in species; about 40,000 are known. The group contains the spirally coiled snails, flat-shelled limpets, shell-less nudibranchs and terrestrial snails and slugs. Most gastropods have the shell and the visceral mass coiled in a right-handed spiral which they carry dorsally. Often the shell can be completely closed with the operculum. They crawl primitively by a flattened foot and with their radula (tongue rasp) graze algae growing on hard bottom or rocks. They reproduce via eggs from which a miniature animal emerges. Only a few species have free-swimming larvae.

Abalones

A characteristic of the species belonging to the abalones (Haliotidae) is that they lack an operculum. The aperture and foot are very wide and the animal lives in the shelter of rocks and in quiet pools. The shell is flat, the body whorl is enormously enlarged and the spire is reduced. The shell is perforated with a series of holes, providing an exit for the exhalant current of the mantle cavity. Most species have six of these holes.

The males spawn first, but only if they are close to a female. The female only sheds eggs when induced by the presence of sperm.

Haliotis tuberculatus Linnaeus. Abalone, Green ormer (E); Ormeau commun, Ormet, Oreille de Mer, Silieux (F); Seeohr (G)

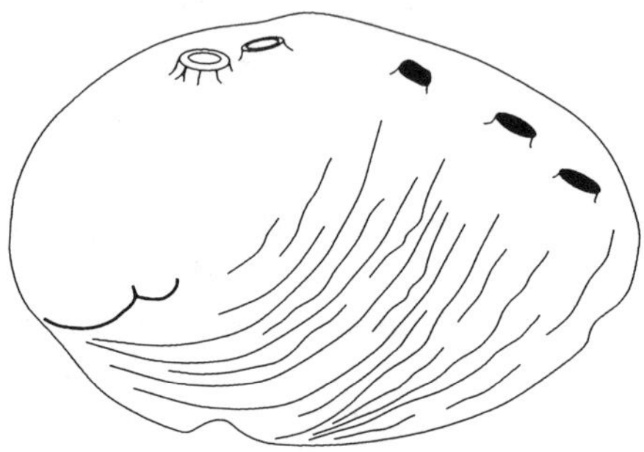

Maximum length 20 cm, mostly 6–8 cm. Colour greenish-brown–red, inner side covered with a layer of mother-of-pearl. Distribution: eastern North Atlantic, south of the Channel Islands (UK), Mediterranean. The total European production is about 10 t. The species is heavily overfished. Related species: *H. gigantea* (Giant abalone), Japan 3751 t in 1989, *H. midae* (Perlemoen abalone), South Africa 562 t in 1989, *H. ruber* (Blacklip abalone) 5531 t in 1989 (Australia 5500 t). Also unspecified catches of *Haliotis* spp. in 1989 of 3994 t (South Korea, Canada, Philippines, Mexico, USA, New Zealand).

Cephalopods

The cephalopods (Cephalopoda) form the most active and specialized class of molluscs. They have a chambered cell, e.g. *Nautilus*, an internal shell as in squids and cuttlefish, or no shell as in octopods. They are almost all fast-swimming carnivores and live pelagically. The close union between head and foot gave the name to this class. Out of this union two new organs developed, a series of prehensile tentacles around the mouth (eight or ten), the other organ being a muscular spout or funnel, lying behind the head where the mantle cavity is situated. The funnel controls the exit of water from the mantle cavity and produces a strong jet which is used in swimming. Cephalopods have some characteristics more in common with the more highly developed vertebrates, e.g. the eye with lens, pupil and eyelid, a well-developed nervous system and the capability to learn.

In all, the class has about 5000 species, belonging mainly to the cuttlefishes, squids and octopuses. In 1989 in total the world catch was 2,536,974 t. The squids (*Loligo, Ilex, Todarodes, Nototodarus*) are economically more important as compared to the cuttlefishes (*Sepia, Sepiola*) and octopuses (*Octopus*).

Squids

The most important families belonging to this order are the Loliginidae and Ommatostrephidae. They have a slender fusiform body and ten tentacles. Two of the tentacles are very long and retractile. They have well-developed paired fins enabling them to swim very well. Some species (flying squids) leave the water for a while and glide on their fins through the air, escaping in that way from dangerous situations. Squids live pelagically, in shoals in the open sea. The protein content averages 15%. The fat content is low.

Loligo vulgaris Lamarck. European squid, Calamary (E); Encornet, Calmar (F); Gewohner Kalmar (G)

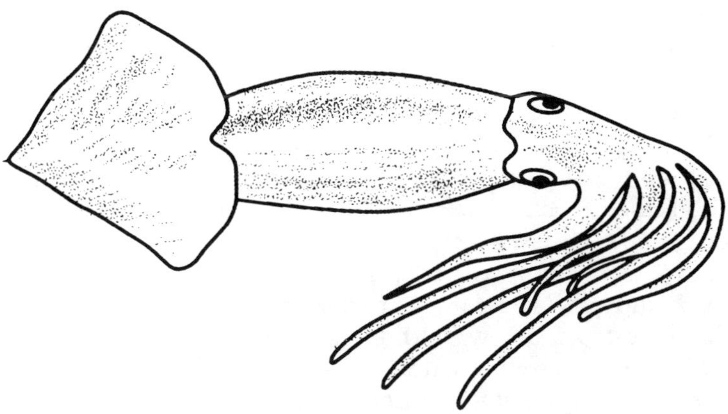

Mantle length 30–50 cm (1.5 kg). When alive the animal is of a transparent pinkish colour, when dead milky white. Fusiform body, the bases of the fins form about two thirds of the mantle length. Suckers in four rows on the tentacles. Lives in large shoals mostly between 60 and 130 m depth. Distributed in the Atlantic Ocean, the English Channel, Mediterranean Sea and the coast of northwest Africa.

In catch statistics often many types of squid are lumped together. In 1989 in total 177,203 t of *Loligo* spp. were landed worldwide, of which

5795 t came from European waters. Presumably the European catches were *Loligo vulgaris*.

The total world catch of squids was 1,254,506 t for 1989.

Crustaceans

The class crustaceans (Crustacea) is part of the large phylum of the Arthropoda, to which nearly a million species belong, i.e. probably more than 80% of all animal species on earth.

Crustaceans characteristically have two pairs of antennae, a pair of mandibles, and two pairs of maxillae. The rest of the appendages vary greatly from group to group. In fact, crustaceans are an enormously diverse class of animals that can hardly be characterized in any simple way. Some have a cephalothorax and an abdomen while others have a head and a trunk, or a head, thorax, and abdomen, or even a unified body. Most species are free living, but some are parasitic. Most of them are active swimmers, but some are sessile (such as barnacles). The majority are marine, but there are also many freshwater species, though only a few terrestrial.

Most of the economically important crustaceans belong to the decapods (Decapoda). These have a cephalothorax under an unsegmented carapax. The abdomen is segmented.

The world landings of crustaceans amounted in 1989 to 4,129,100 t, i.e. about 4% of landings by fisheries. The total landings of freshwater crustaceans for 1989 were 241,498 t (177,100 t from Asia, 42,728 m from North America); of sea spider crabs 1,146,408 t (northwest Pacific 750,380 t, northeast Pacific 106,059 t, northwest Atlantic 85,754 t, western Central Pacific 57,320 t, western Central Atlantic 52,480 t, northeast Atlantic 32,688 t); of lobster, spiny rock lobsters 202,869 t (northwest Atlantic 67,092 t, northeast Atlantic 51,932 t); of squat lobsters 4517 t; of shrimps and prawns 2,443,275 t (northwest Pacific 703,794 t, western Central Pacific 443,695 t, Western Indian Ocean 238,543 t, Eastern Indian Ocean 176,229 t, western Central Atlantic 172,414 t, northeast Atlantic 146,034 t); of krill (planktonic crustaceans) 394,129 t (Antarctica 393,111 t) and of miscellaneous marine crustaceans 69,761 t.

The protein contents are comparable to those of fish. The fat contents are usually between 1 and 2%.

Shrimps (Natantia)

Two groups of this suborder of the crustaceans are commercially important, i.e. the Penaeidae (penaeid shrimps) and Caridae (palaemonid and pandalid shrimps). The Natantia comprise about 1950 species, of which 300 are exploited or have the potential to be exploited.

Penaeid shrimps

More than 200 species belong to the family of the Penaeidae, of which at least 80 are of commercial interest. They are medium to large shrimps (20–30 cm) and live in the tropics and subtropics. The rostrum is well developed, the first three pereopods having a small claw. Most giant shrimps (such as king shrimps) are exploited in the coastal waters of the Indian subcontinent, southeast Asia, the Far East, Gulf of Mexico, Atlantic coasts of Central America, South America and West Africa. About 20% of their weight is edible.

Penaeus japonicus Bate. Kuruma prawn/shrimp, Japanese prawn/shrimp (E); Crevette kuruma, Crevette du Japon, Crevette impériale (F); Radgarnele (G)

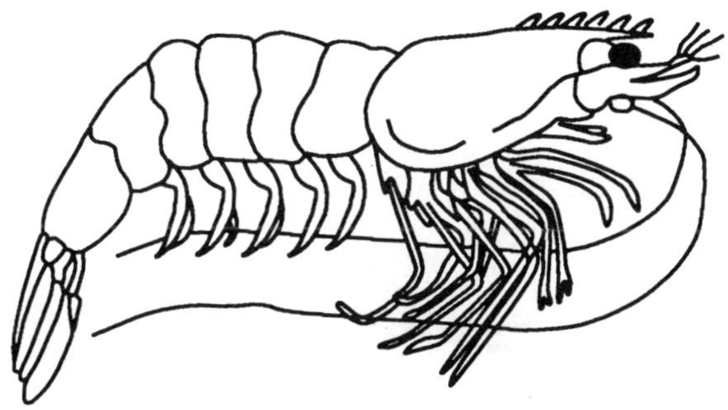

Maximum length 24 cm, mostly 16–29 cm. Colour yellowish to pink, dark brown bandings on rostrum, carapax, abdomen and uropods. Carapax smooth, rostrum with ridges on both sides, telson (tail part) with three pairs of flexible spines. Lives on muddy bottom, mostly between 20–40 m (extreme 90 m). Distribution: Indo-Western Pacific origin, southeastern Africa, Red Sea to South Korea, Japan and Malay archipelago. Entered the Mediterranean through the Suez Canal. Now important species of Turkish and Israeli waters (with *P. semisulcatus*, green tiger prawn). Total catch in 1989: 22,265 t (in northwest and western Central Pacific). Other penaeid catches for 1989: *P. aztecus* (northern brown shrimp, USA) 71,938 t; *P. merguiensis* (banana shrimp, Indonesia, Thailand) 82,581 t; *P. duorarum* (northern pink shrimp, Cuba, USA) 11,762 t; *P. monodon*

(giant tiger prawn, Indonesia, Thailand) 61,440 t; *P. chinensis* (fleshy prawn, China, South Korea) 212,231 t; *P. kerathurus* (caramote prawn, Mediterranean) 2399 t; *P. brevirostris* (crystal shrimp, Costa Rica, El Salvador, Panama) 4252 t; *P. occidentalis* (western white shrimp, Ecuador, Colombia) 81,106 t; *P. notialis* (southern pink shrimp, Gabon, Nigeria, Senegal) 12,813 t; *Metapenaeus* spp. (southeast Asia) 58,105 t; *Parapenaeus longirostris* (deepwater pink shrimp, northwest Africa) 4415 t.

Pandalid shrimps

A family with small to medium-sized shrimps with a characteristic rostrum, and no claws. Second most important shrimp family after the penaeid shrimps. Well-known genera are *Pandalus, Heterocarpus, Chlorotocus, Parapandalus, Plesionica.*

Pandalus borealis Krøyer. Northern shrimp, Deepwater prawn/shrimp, Pink shrimp (E); Crevette nordique, Crevette profonde (F); Grönlandgarnele/Krabbe (G)

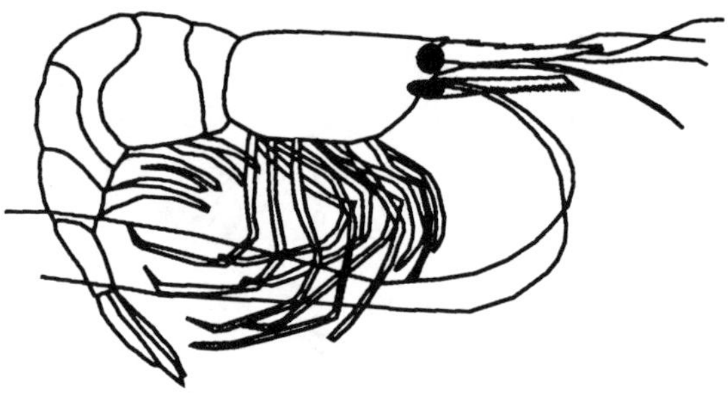

Maximum length 16.5 cm, mostly smaller. Uniform dark red. Lives on clay and muddy bottoms, depth from 20 to 1400 m, mostly between 50 and 450 m. Distribution: Northern Atlantic, from Greenland to Massachusetts (USA), Iceland, Spitsbergen to central North Sea, also in Pacific from Bering Sea to Japan and Oregon (USA). Total catch in 1989 188,421 t (Greenland 59,107 t, Norway 55,838 t, Iceland 26,785 t).

Crangonid shrimps

A small family with species in the cold and temperate waters of the Northern Hemisphere. Nearly all species live on sandy bottoms.

Crangon crangon (Linnaeus). Common or brown shrimp (E); Crevette grise (F); Nordseegarnele/krabbe, Sandgarnele (G)

Maximum length 9 cm, mostly 4.5–8 cm. Grey–brown, sometimes a dark band on fourth segment of the abdomen. Lives on sandy bottoms in the shallow coastal zone. Migrates in wintertime to deeper water. Distribution: eastern Atlantic, from Norway, North Sea to Portugal and south Spain, Baltic, Mediterranean, Black Sea. 25–35% edible. Total catch in 1989 19,857 t (Germany 13,268 t, Netherlands 2570 t, Denmark 1467 t).

Lobsters (Macrura)

In Macrura the power of swimming is reduced (early life stages) or extinct. The pereopods (walking legs) are well developed; hence they are well adapted to live on the bottom. About 700 species belong to the Macrura, of which 40 belong to the prime commercial species, e.g. lobsters, spiny lobsters and squat lobsters. In total in 1989 of this category 207,386 t was landed. Most of these were caught at sea, but also in fresh water (e.g. crayfish).

Homarid lobsters (Homaridae/Nephropidae)

Head and thorax are continued into a cephalothorax, which incorporates the first 14 segments. The abdomen consists of seven distinct segments, which are clearly marked externally. The jaws (mandibles) originating from the fourth segment are well developed for crushing their food, and are supported by the maxillas to bring the food to the mouth.

The true lobsters (*Homarus* spp.) live on rocky bottom, in shallow water in cold and temperate waters. During the day they hide under rocks and crevices, during the night they search for food (molluscs, worms and organic detritus).

Homarus americanus H. Milne Edwards. American lobster (E); Homard américain (F); Amerikanischer Hummer (G).
Maximum length 70 cm (9 kg), mostly smaller. Carapax broader than *H. gammarus*. Colour green–black, marbled. Lives on rocky bottom

to about 700 m depth. Distribution: Atlantic coasts of Canada, USA, Newfoundland and Labrador to North Carolina. Important species also cultivated on experimental scale. Total catch in 1989 67,092 t (Canada 43,085 t, USA 24,007 t).

Homarus gammarus (Linnaeus). Common or European lobster (E); Homard européen (F); Europäischer Hummer (G)

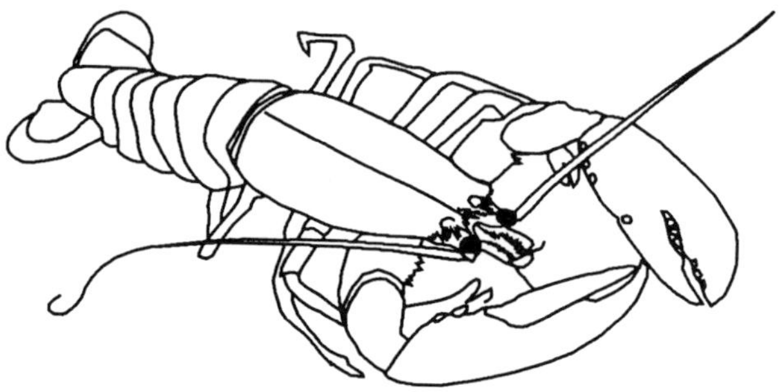

Maximum length 60 cm (5–6 kg), mostly 25–50 cm. Black to blue-black, marbled, sometimes reddish, belly yellowish. The carapax is smooth, with two pair of spines behind the eyes, long and sharp rostrum, with on both sides four to five teeth. Abdomen smooth. Left jaw in most cases a bit smaller than the right jaw. Lives on rocky bottom up to 70 m. Distribution: eastern Atlantic, shores of north and western Europe, from northern Norway to Morocco, western Mediterranean, not in Baltic and Black Sea. Around British Isles and Ireland. In 1989 2072 t landed (UK 1360 t, Ireland 361 t, Denmark 271 t, Morocco 14 t).

Nephrops norvegicus (Linnaeus). Norway lobster, Dublin Bay prawn, Prawn (E); Langoustine (F); Kaisergranat, Tiefseekrebs (G)

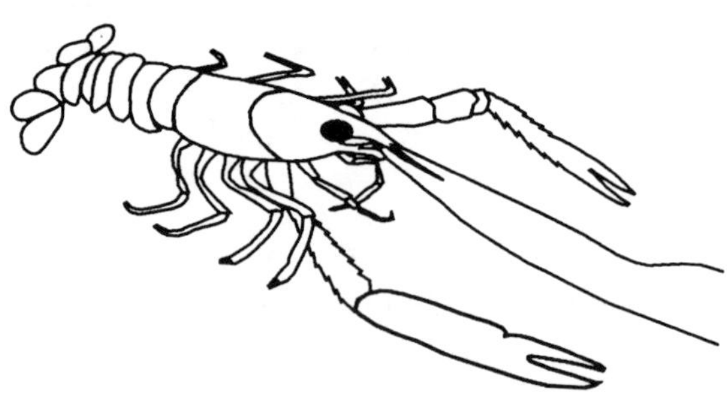

Maximum length 25 cm. Colour red–orange to orange, dark spots on the jaws and abdomen. Eggs whitish to black (when boiled red). Carapax smooth, long sharp eyes. First pereopods with long slender jaws. Lives on muddy bottom to a depth of 800 m (mostly between 50 and 300 m), lives in burrows. Distribution: eastern Atlantic, Iceland, from Norway and Faeroe Islands to northwest Africa, western Mediterranean, not in Baltic and Black Sea. In 1989 in total 55,105 t landed (Atlantic 49,423 t, Mediterranean 5617 t; UK 25,532 t; France 9828 t; Ireland 4332 t; Denmark 3530 t; Spain 2291 t).

Spider crabs, hermit crabs, king crabs (Anomura)

The suborder Anomura covers a wide collection of species when judged by their outer appearance. The last pair of pereopods is strongly reduced, hence it looks as if they have only eight walking legs.

King crabs (Lithodidae)

A family of large, strongly built crabs (up to 10 kg in weight), living in polar waters (Arctic, Antarctic). The flesh of these crabs is sold under trade names such as king crab, Kamchatka crab, Pacific king crab. *Paralithodes camschatica*, the Alaska king crab or red king crab (E) or crabe royal (F), is found in the North Pacific and *Lithodes antarcticus*, southern king crab (E) or crabe royal du sud (F), in the South Pacific and Atlantic, living under similar circumstances on hard rocky and stony bottoms. Their flesh is frozen into blocks and exported from the USA, Japan, South Korea, Russia, Argentina and Chile. Related species, with a similar economic interest, are *P. brevipes* (Hanasaki crab), *P. platypus* (blue king crab), *L. aequispina* (golden king crab), *L. murrayi* (from Antarctic regions). The total production in 1989 was 54,544 t for *Paralithodes* spp. (Russia 41,940 t, USA 11,971 t, Japan 623 t) and 2417 t for *Lithodes antarcticus* (mainly by Chile 2297 t).

Crabs (Canceridae)

This family of crabs from the Northern Hemisphere lives in cold and temperate waters. The carapax is almost oval in shape. The jaws are well developed, and almost the same size. The walking legs (pereiopods) are strongly built.

Cancer pagurus Linnaeus. Brown crab, Edible crab, European rock crab (E); Tourteau, Poupard, Dormeur, Poing-clos (F); Taschenkrebs (G)

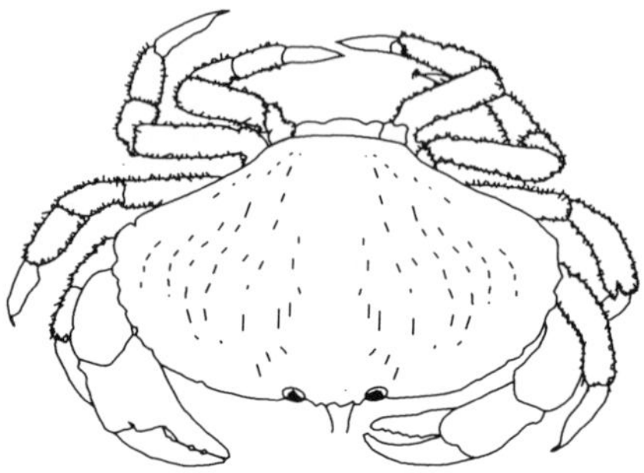

Maximum width 30 cm (6 kg), mostly up to 20 cm, carapax and legs brownish-red. The fingers of the jaws dark to almost black; belly white. Lives on rocky, stony bottoms, from shallow depths to 300 m; mostly between 20 and 200 m. Distribution: eastern Atlantic, from northern Norway to Portugal. Not in Mediterranean. In 1989 in total 26,365 t landed (UK 11,902 t, France 8302 t, Norway 1479 t). Per 100 g (flesh from jaws and legs) there is 22 g protein and 0.2 g fat; per 100 g (flesh from inside of carapax) there is 13 g protein and 7.5 g fat.

Callinectes sapidus Rathburn. Blue crab (E); Crabe bleu (F); Blaukrabbe (G)

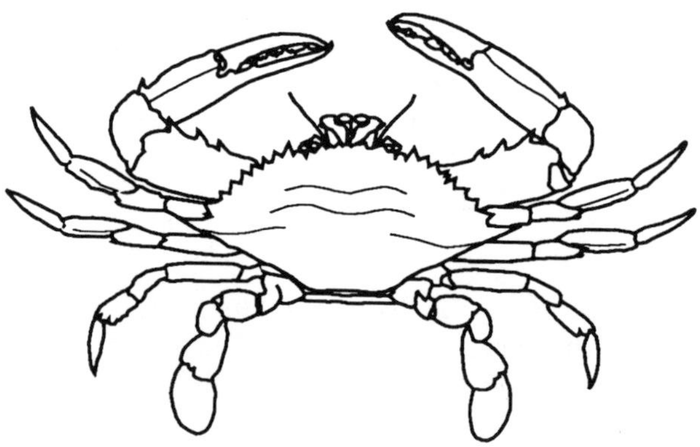

Maximum width 21 cm, mostly 10–16 cm. Colour grey–, blue– or brown–green, with shadowy patches on the carapax, belly white, legs blue and white, sometimes partly red. Adult males have blue jaws with a red tip, adult females orange jaws with a purple tip. Lives on muddy bottoms in the coastal zone and estuaries. Distribution: western Atlantic from Nova Scotia to Argentina, as introductions in Baltic and Mediterranean. In 1989 in total 102,105 t landed. Northwest Atlantic 56,740 t (USA) and western Central Atlantic 45,365 t (USA 37,684 t; Mexico 6126 t; Cuba 1555 t).

Related species: *C. danae* (Dana swimcrab) 4119 t (Brazil).

Spider crabs (Maiidae)

Family of strongly built species, with an almost triangular carapax and long legs. Meat is difficult to remove, but techniques are developed to overcome this problem. This explains the increase in landings of queen and snow crabs (*Chionocetes opilio* and related Pacific species) from 7000 t in the 1970s to nearly 100,000 t today (Canada (*C. opilio*) 22,865 t; USA 74,682 t; Japan 8500 t).

Maia squinado (Herbst). Spider crab, Spinous crab, Spinous spider crab (E); Araignée (F); Grosse Seespinne, Meerspinne (G)

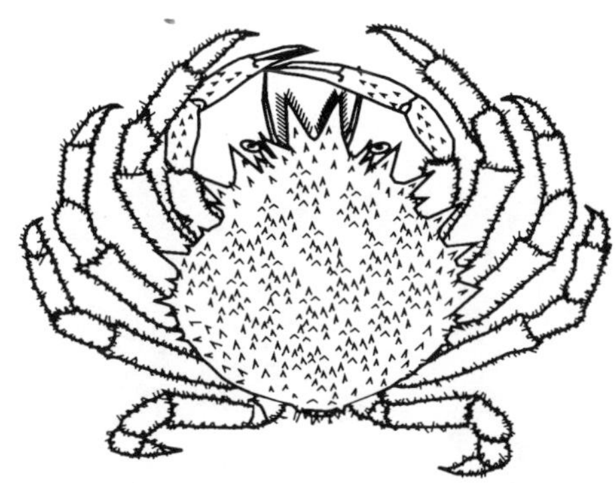

Maximum width 18 cm, mostly up to 10 cm, colour yellow–brown to red–brown. Carapax almost round, arched and full of sharp spines. The jaws are slender. Lives on rocky bottom, with seaweeds, from shallow to 150 m depth. Distribution: eastern Atlantic, from British Isles to northwest Africa, Azores, Madeira, Canary Islands, Mediterranean. Total landings in 1989: 3717 t (France 2376 t, UK 954 t, Ireland 181 t, Mediterranean 53 t).

References and Further Reading

Cohen, D.M. (1970) How many recent fishes are there? *Proceedings California Academy of Sciences*, 4th Series 38(17), 341–346.

FAO (1991) *FAO Yearbook Fishery Statistics, Catches and Landings*, Vol. 68. FAO, Rome, 516 pp.

Fischer, W., Schneider, M. and Bauchot, M.L. (1987) Fiches FAO d'identification des espèces pour les besoins de la pêche. *Mediterranée et Mer Noire, zone de pêche 37*, 2nd edn, Vols 1 and 2. FAO, Rome, 1529 pp.

Greenwood, P.H., Rosen, D.E., Weitzman, S.H. and Meyers, G.S. (1966) Phyletic studies of teleostean fishes with a provisional classification of living forms. *Bulletin of the American Museum of Natural History* 131, 341–455.

Holthuis, L.B. (1980) *FAO Species Catalogue*, Vol. 1: Shrimps and prawns of the world. An annotated catalogue of species of interest to fisheries. FAO Fisheries Synopsis 125. FAO, Rome, 271 pp.

Scott, W.B. and Crossman, E.J. (1973) Freshwater fishes of Canada. *Bulletin of the Fisheries Research Board of Canada* 84, 1–966.

Smith, M.M. and Heemstra, P.C. (eds) (1986) *Smith's Sea Fishes*, 6th edn. Springer-Verlag, Berlin, 1047 pp.

Souci, S.W., Fachmann, W. and Kraut, H. (1981) *Food Composition and Nutrition Tables 1981/82*, 2nd edn. H. Scheiz and G. Kloos (eds). Wissenschaftliche Verlagsgesellschaft MbH, Stuttgart, 1352 pp.

Whitehead, P.J.P., Bauchot, M.L., Hureau, J.C., Nielsen, J. and Tortonese, E. (1984–1986). *Fishes of the North-Eastern Atlantic and Mediterranean*, Vols. 1–3. UNESCO, Paris, 1473 pp.

Young, C.M. and Thompson, T.E. (1976) *Living Marine Molluscs*. Collins, Glasgow, 288 pp.

3

Composition and Nutritive Value of Fish Proteins and Other Nitrogen Compounds

NORMAN F. HAARD
Institute of Marine Resources, College of Agricultural and Environmental Sciences, University of California, Davis, California 95616, USA.

Introduction

Seafood products are derived from various tissues including the whole fish body, muscle, roe, stomach, kidney, liver, skin and fins. The meat is normally defined as the muscle or flesh of animals used as food, but its definition may also include other edible tissues. Muscle tissue contains hundreds of different proteins along with other nitrogen-containing molecules referred to as non-protein nitrogen (NPN). Three groups of muscle protein are: (i) sarcoplasmic proteins that are from the muscle sarcoplasm and are soluble in water or low ionic strength salt solution; (ii) myofibrillar proteins that are from the contractile apparatus and soluble in high ionic strength salt solution (e.g. 0.6 M KCl); and (iii) stroma or connective tissue proteins that are mostly from the extracellular matrix and are not soluble in water, alkaline solution, or high ionic strength salt solution. Other proteins are associated with cellular membranes.

NPN compounds are mostly from the sarcoplasm and include peptides, amino acids, amines, amine oxides, guanidine compounds, quaternary ammonium compounds, purines and urea. Protein and NPN together are called 'crude protein'. The percentage distribution of crude protein components varies to some extent in different food myosystems. Estimates of sarcoplasmic protein concentration based on nitrogen determination include NPN. It is, therefore, important to recognize that the NPN fraction may account for a relatively high percentage of total nitrogen in the muscle of some aquatic animals. Compared to food myosystems from land animals, finfish muscles contain less connective tissue protein (2–3% vs. 16–28%), more myofibrillar protein (70–80% vs. 39–68%), and more NPN (10–40% or more of the sarcoplasmic fraction in fishery products).

The cellular functions of nitrogenous compounds in the living animal include roles in enzyme catalysis, osmoregulation, antifreeze, intermediary metabolism, nitrogen storage, cell structure, muscle contraction and transport processes. Nitrogenous compounds contribute to the food value and utility of fishery products in many different ways. They collectively influence all the distinctive food attributes, i.e. colour, flavour, texture, nutrition, safety, and the postharvest deterioration of fish meat. Nitrogenous compounds in fish meat are also important because they contribute to chemical and physical changes during processing of fishery products.

The variable occurrence and properties of specific protein or NPN components in fish are a major determinant of why different species are more or less suitable for dehydration, freezing, thermoprocessing and fermentations. The purpose of this chapter is to briefly discuss our knowledge of proteins and NPN components in fishery products.

Interspecific factors

Despite the broad taxonomic diversity of animals that comprise seafood the basic composition of nitrogenous compounds in muscle from most aquatic animals is remarkably similar. The same metabolic patterns seem to be used throughout the animal kingdom (Dixon and Webb, 1979). Differences in metabolism are normally not the result of a change in an entire metabolic pathway. Genetic or phenotypic variation in metabolism can be explained by one or more of the following: (i) occasional 'all or none' addition or deletion of an enzyme to a basic pathway; (ii) differences in the amount of an enzyme; (iii) multiple molecular forms of an enzyme in the same tissue from the same organism; and (iv) homologous forms of an enzyme in the same tissue from different organisms. Exceptional enzymes that have been found only in specific sources of seafood include polyphenoloxidase in crustacea, carnosinase in eels and trimethylamine-*N*-oxide demethylase in gadoid fish. Similarly, a high concentration of other nitrogenous compound(s) is characteristic of muscle tissue from certain taxonomic groups, e.g. histidine in scombroid fish and urea in elasmobranchs.

Intraspecific factors

In contrast to white muscle tissue, dark muscles often contain less sarcoplasmic protein (Suzuki, 1981) and more NPN (Haard, 1992a). To further complicate matters, the quantity, physicochemical properties and catalytic properties of enzymes (Haard, 1990) and other constituents (Haard, 1992a) can vary with biological age, diet, exercise, habitat temperature, water depth and other ontogenic and environmental factors. For example, a proteinase from winter and spring burbot differs in thermal stability, kinetic constants and inhibitor sensitivity (Wiggs, 1974). The importance

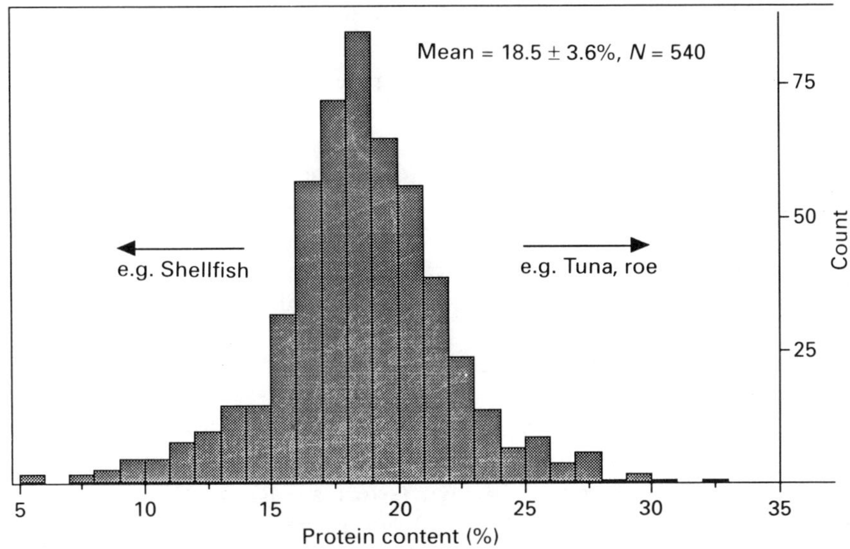

Fig. 3.1. Frequency distrubution for protein content in seafood. Data for flesh and roe of finfish and shellfish from 540 sources. Values for specific species are compiled from Iwasaki and Harada (1985), Nettleton (1985), and Wheaton and Lawson (1985). Statistical analysis was performed with JMP version 2.0 software (SAS Institute).

of habitat conditions is also reflected in findings that the amounts of free amino acids and other NPN compounds differ considerably in cultured and wild fish of the same species (reviewed by Haard, 1992a).

Total Content of Protein

The crude protein content of seafoods ranges from less than 8 to more than 25% of fresh weight (Fig. 3.1). However, most finfish muscle tissue contains about 18–22% protein (Sidwell, 1981) and the average for the 540 analyses summarized in Fig. 3.1 is 18.5%. Crustacean flesh (crabs, lobsters, shrimps) tends to contain more crude protein than that of molluscs (e.g. clams, oysters, mussels, squids). The protein composition of roe ranges from 12% (anglerfish) to 30% (crab), and within species is greater than that of muscle (Iwasaki and Harada, 1985).

Muscle type

Muscle is a heterogeneous tissue, normally being composed of mixtures of fibre types, an extracellular matrix and containing macrophages and residual blood. Within species, white flesh has more protein than dark flesh (Suzuki, 1981). These differences can be quite large in animals that contain large amounts of fat in dark muscle (e.g. *Sardinops melanostica*, 23.1% vs. 15.9%; *Scomber scombrus*, 21.2% vs. 14.9%; *Clupea harengus pallasi*, 22.0% vs. 15.5%). Protein content can differ by several per cent at the nape and tail end of fillets because of the varying proportion of dark and white muscle tissue. Fillets may contain more (e.g. Atlantic cod) or less (e.g. Atlantic salmon) protein at the nape end depending on the distribution of dark and white muscle. Therefore, care should be taken in the way fillets are sampled for determination of total protein.

Seasonal variation and rearing conditions

Although the fat content of muscle may vary considerably at different times of the year, the total protein content does not generally undergo large changes with fishing season. In ayu (sweet smelt), the protein content of white muscle decreases slightly from summer to autumn with no appreciable differences between wild and cultured fish (Hirano *et al.*, 1980). Differences in the protein content of white muscle from wild and cultured fish have been reported (Haard, 1992a). Of 23 species examined, the protein content was significantly higher in the muscle of wild fish for 13 species and was significantly higher in the muscle of cultured fish for 8 species. In most cases, the differences in protein content are less than 2%. However, wild striped jack contained about 5% more protein than cultured fish (Aoki *et al.*, 1991).

Protein depletion

Maturation of gonads and/or long periods of feed deprivation can result in tissue depletion with marked decreases in the content of protein in muscle from Pacific salmon (Bilinski *et al.*, 1984), Atlantic cod (Love, 1988), Dover sole (Hendrickson *et al.*, 1986) and American plaice (Haard, 1987). In severe cases of 'jellied' flounder the protein concentration of the muscle is less than half that of normal fish. Protein depletion is normally associated with increased water in the muscle.

Protein Groups

Sarcoplasmic proteins

The sarcoplasmic or 'myogen' fraction of muscle is a large family of proteins that share the property of being soluble in water and dilute salt solution. This group may inadvertantly include some proteins that are not strictly

Table 3.1. Sarcoplasmic protein content in muscle of different fish. Adapted from Suzuki (1981).

Fish	Species	Sarcoplasmic protein (g 100 g^{-1})
Mackerel	*Scomber japonicus*	6.4-8.8
Yellowtail	*Seriola quinqueradiata*	8.4-8.6
Red barracuda	*Sphyraena pinguis*	7.1
Horse mackerel	*Trachurus japonicus*	5.0-5.6
Sea bass	*Stereolepsis ischinagi*	4.2-5.7
Lizard fish	*Saurida undosquamis*	5.6
Grub fish	*Neopercis sexfasciata*	4.9
Flatfish	*Kareius bicoloratus*	4.4
Carp	*Cyrinus carpio*	3.9
Anchovy	*Engraulis japonica*	3.8
Cod	*Gadus morhua*	3.4

soluble in the sarcoplasm, notably membrane proteins. Fish sarcoplasmic proteins are more or less like those from land animals, i.e. they include myoglobin, hundreds of enzymes, and other albumins. The muscles of fish and lower vertebrates differ from those of land animals in that they contain large quantities of Ca^{2+}-binding proteins called parvalbumins (reviewed by Gazzaz and Rasco, 1993).

Content

Sarcoplasmic proteins normally comprise about 15–25% of the total protein in fish muscle. The content of sarcoplasmic protein (Table 3.1) is generally higher in pelagic fish such as sardine and mackerel, and lower in demersal fish like plaice and snapper.

Species identification

When separated by electrophoresis or isoelectric focusing, sarcoplasmic proteins can be used to 'fingerprint' or identify fish species (Toom *et al.*, 1982; Whitmore, 1986). Disc-gel electrophoresis of sarcoplasmic proteins from 16 species, in the absence of sodium dodecylsulfate (SDS), gives species-specific electrophoretic patterns which differ widely for saltwater and freshwater fishes (Nakagawa *et al.*, 1988a). Two-dimensional SDS-gel electrophoresis of sarcoplasmic proteins has been used to categorize fish into three groups: marine white-fleshed demersals, marine red-fleshed pelagics, and freshwater (Table 3.2). The distribution of a 43 kDa (creatine kinase), a 40 kDa (aldolase) and a 35 kDa (glyceraldehyde-3-phosphate dehydrogenase) protein was used to categorize these three groups of

Table 3.2. Classification of fish groups based on molecular mass distribution of sarcoplasmic proteins. Data provided for groups are: marine white, red sea bream; marine red, Pacific mackerel; freshwater, carp. Data from Nakagawa *et al.* (1988a, b). Components noted in bold type and with * are used to classify fish in groups.

Component		Relative amount (%)		
(kDa)	Identity	Marine white	Marine red	Freshwater
94		8.9	10.4	7.6
60		7.4	7.5	10.2
49-50		11.9	12.7	10.4
43	Creatine kinase	**12.6** *	**13.7** *	**17.1** *
40	Aldolase	13.7	**18.7** *	11.3
35	Glyceraldehyde-3-phosphate dehydrogenase	**13.5** *	**11.6** *	11.2
33		7.0	5.7	5.1
26		3.1	2.6	5.4
25		3.0	3.1	4.3
23		1.2	1.0	4.1
Others		17.7	13.0	16.7

fish (Nakagawa *et al.* 1988b). High performance liquid chromatography of sarcoplasmic protein has also been used for species identification (Armstrong *et al.*, 1992). Sarcoplasmic proteins can also be used to identify genetically distinct strains or stocks within the same species. In particular, enzyme polymorphism can be used to distinguish fish strains (Zajicek, 1991).

Heat coagulation

Most sarcoplasmic proteins of pelagic fishes are coagulated when heated in water at above 50 °C, e.g. 90 °C for 10 min. Only 65–75% of the sarcoplasmic protein from demersal fishes is heat coagulated (Suzuki, 1981). The low molecular mass (12 kDa) parvalbumins remain soluble when heated at 70 °C or above (Kawai *et al.*, 1992). Proteins from deep sea fishes are less heat stable than those from warmer water (Taguchi *et al.*, 1981). Small quantities of sarcoplasmic proteins have an adverse affect on the strength and deformability of myofibril protein gels. These proteins may interfere with myosin cross-linking during gel matrix formation as they do not form gels and have poor water-holding capacity (Smith, 1991). Sarcoplasmic proteins appear to be responsible for the low gel strength of products made from mackerel or sardine. In surimi manufacture, water washing of fish meat achieves several purposes, one of which is to remove the water-soluble proteins that interfere with gel formation. Small amounts of minerals in wash water can influence selective removal of sarcoplasmic proteins from minced fish (Martinez *et al.*, 1992). For this reason, hardness of water used to wash mince can influence the quality of surimi.

Enzymes

Fish sarcoplasmic proteins have, for the most part, the same basic, functional classes of enzymes common in mammals, birds and reptiles. These include the sarcoplasmic enzymes involved with physiological events of respiration, intracellular digestion, cell growth, cell division and secondary metabolism. On this basis, we can surmise that all six classes of enzymes (oxidoreductases, transferases, hydrolases, lyases, isomerases and lyases) are present in fish muscle. The comparative aspects of muscle enzymes in poultry, beef, lamb, pork and seafood were reviewed by Haard (1990). The main enzyme groups known to affect the edible qualities of fish are hydrolases, oxidoreductases and transferases. Marine enzymes are reviewed by Simpson *et al.* (1991).

HYDROLASES

Hydrolytic enzymes of importance in postharvest fish include proteinases and peptidases, lipases and phospholipases, and glycogen hydrolases.

Constitutive proteinases important in seafood include those present in: (i) muscle cells, (ii) the extracellular matrix and connective tissue surrounding muscle cells, and (iii) digestive and other organs.

Fish proteinases involved with the digestive process were first studied more than 100 years ago (Stirling, 1884). Some digestive enzymes, such as salmon pepsin, were purified to homogeneity more than 50 years ago (e.g. Norris and Elam, 1940). In more recent years we have gained better understanding of multiple molecular forms of digestive enzymes (Squires *et al.*, 1986; Simpson and Haard, 1987b) and their comparative biochemistry (Simpson and Haard, 1987a; Gildberg, 1988). The presence of lysosomal enzymes in fish muscle has been known for almost 30 years (Siebert, 1962) although new catheptic enzymes continue to be identified in fish and shellfish (e.g. Hara *et al.*, 1988; Ueno *et al.*, 1988; Zeef and Dennison, 1988). The high molecular mass, heat-stable, alkaline muscle proteinases were identified in fish muscle nearly 20 years ago (Iwata *et al.*, 1973) and have been extensively studied in connection with surimi gel weakening. More recently identified proteinases from seafood tissues include Ca^{2+}-activated neutral proteases (e.g. Jiang *et al.*, 1990), modori-inducing neutral proteinases (e.g. Kinoshita *et al.*, 1990), multicatalytic proteinases or proteosomes (e.g. Folco *et al.*, 1989), and metallo-proteinases with collagen-degrading activity (Bracho and Haard, 1991). Several groups of fish proteases are not strictly sarcoplasmic proteins being associated with organelles (e.g. cathepsins), connective tissue (collagenases) and myofibrils (e.g. alkaline proteinases). The literature on endogenous proteinases from seafood has been reviewed in detail by Haard (1994). Examples of specific proteinases in fish tissues are summarized in Table 3.3.

Lipolytic enzymes from fish include those from digestive organs and from the muscle (Table 3.4). Lipolysis is especially important in frozen fish where hydrolysis may occur in the lipid phase at low water activity (Haard, 1992b). Although soluble phospholipase has been isolated from pollack muscle (Audley *et al.*, 1978), most investigators have identified muscle phospholipase with the microsomal fraction. Muscle lipase has been identified with lysosomes (Geromel and Montgomery, 1980). Lipolysis in fish muscle was reviewed by Shewfelt (1981).

In addition to phosphorylase, carbohydrate hydrolases appear to contribute to glycogen degradation in postharvest fish muscle (Eskin, 1990). Fish muscle appears to differ from that of terrestrial animals in its capacity to mobilize glycogen via a hydrolytic (amylase-like) pathway as well as by the better known phosphorylase pathway. Muscle lysosomes contain proteoglycan-degrading enzymes like β-glucuronidase and β-*N*-acetylhexosaminidase that can contribute to postharvest softening of muscle texture (Kim and Haard, 1992). Marine organisms also possess various digestive polysaccharide-degrading enzymes (Anzai *et al.*, 1991a; 1991b; 1992; Yamada *et al.*, 1991).

Table 3.3. Proteinases identified in fish and shellfish. From Simpson *et al.* (1991).

Proteinase	Source	Tissue	Reference
Cathepsins	Albacore	Muscle	1
	Carp	Muscle	2
	Cod	Muscle	3
	Salmon	Muscle	4
	Pacific sole	Muscle	5
	Rockfish	Muscle	6
	Tilapia	Muscle	7
	Squid	Hepatopancreas	8
Alkaline proteases	Filefish	Muscle	9
	White croaker	Muscle	9
	Anchovy	Muscle	9
	Rockfish	Muscle	9
	Yellowtail	Muscle	9
	Tiger shrimp	Muscle	9
	Garfish	Muscle	9
Collagenase	Rockfish	Muscle	10
	Crab	Hepatopancreas	11
	Trout	Pyloric caeca	12
Pepsins	Tuna	Gastric mucosa	13
	Salmon	Gastric mucosa	14
	Dogfish	Gastric mucosa	15
	Bonito	Gastric mucosa	16
	Trout	Gastric mucosa	17
	Sardine	Gastric mucosa	18
	Hake	Gastric mucosa	19
	Cod	Gastric mucosa	20
	Smelt	Gastric mucosa	21
Trypsin	Cod	Pyloric caeca	22
	Sardine	Pyloric caeca	23
	Goldfish	Hepatopancreas	24
	Capelin	Intestine	25
	Catfish	Intestine	26
	Rainbow trout	Intestine	27
	Chinook salmon	Intestine	28
	African lungfish	Pancreas	29
Chymotrypsin	Spiny dogfish	Pancreas	30
	Herring	Intestine	31
	Capelin	Intestine	31

References: (1) Groninger, 1964; (2) Hara *et al.*, 1988; (3) Schmitt and Siebert, 1967; (4) Ting *et al.*, 1968; (5) Geist and Crawford, 1974; (6) Kim and Haard, 1992; (7) Sherekar *et al.*, 1988; (8) Hameed and Haard, 1985; (9) Iwata *et al.*, 1974; (10) Bracho and Haard, 1991; (11) Eisen and Jeffrey, 1969; (12) Yoshinaka *et al.*, 1978; (13) Tanji *et al.*, 1988; (14) Fruton and Bergmann, 1940; (15) Merrett *et al.*, 1969; (16) Kubota and Ohnuma, 1970; (17) Owen and Wiggs, 1971; (18) Noda and Murakami, 1981; (19) Sanchez-Chiang and Ponce, 1981; (20) Squires *et al.*, 1986; (21) Haard *et al.*, 1982; (22) Simpson and Haard, 1984; (23) Murakami and Noda, 1981; (24) Jany, 1976; (25) Hjelmeland and Raa, 1982; (26) Yoshinaka *et al.*, 1984; (27) Kitamikado and Tachino, 1960; (28) Croston, 1965; (29) De Haen *et al.*, 1977; (30) Ramakrishna *et al.*, 1987; (31) Kalac, 1978.

Table 3.4. Lipolytic enzymes in tissues of fish and shellfish. From Simpson *et al.* (1991).

Enzyme	Source	Tissue	Reference
Lipase	Skate	Pancreas	1
	Trout	Pancreas	2
	Cod	Digestive tract	3
	Crayfish	Digestive tract	4
	Lobster	Gastric juice	5
	Trout	Muscle	6
	Cod	Muscle	7
Phospholipase	Cod	Muscle	8
	Trout	Muscle	9
	Pollack	Muscle	10
	Flounder	Muscle	11

References: (1) Brockerhoff and Hoyle, 1965; (2) Brockerhof, 1966; (3) Leger, 1972; (4) Berner and Hammond, 1970; (5) Brockerhoff, 1967; (6) Geromel and Montgomery, 1980; (7) Olley *et al.*, 1962; (8) Chawla and Ablett, 1987; (9) Han and Liston, 1987; (10) Audley *et al.*, 1978; (11) Shewfelt *et al.*, 1981.

Oxidoreductases

Oxidoreductases known to occur in fish tissues are summarized in Table 3.5. Polyphenol oxidases are particularly important in Crustacea because they can cause postharvest discoloration. Lipoxygenases have been identified in fish skin, gill and muscle tissue. These enzymes have been implicated in skin carotenoid bleaching (Tsukuda, 1970) and fresh fish aroma (Josephson and Lindsey, 1986). Some fish lipoxygenase activity is associated with microsomal membranes in addition to the soluble sarcoplasmic proteins (Haard, 1992b). Other oxidoreductases that have been found in fish are peroxidases (Ilgner and Woods, 1985), catalase and superoxide dismutase (Scott and Harrington, 1990). Oxidoreductases are also important components of the glycolytic pathway. Although these enzymes are normally considered to be sarcoplasmic proteins, they may be particulate bound in fish muscle (Brooks and Storey, 1988a; Lushchak, 1991). Lactic dehydrogenases from marine fish are particularly well studied, although they are absent in the muscle of some invertebrates like squid.

Other enzymes

Among the many other soluble enzymes known to occur in fish are the transferases, phosphorylase (Assaf and Graves, 1969), phosphofructokinase (Brooks and Storey, 1988b), transglutaminase (Kishi *et al.*, 1991) and glutathione *S*-transferase (Nimmo and Spalding, 1985). Fish muscle contains three interconvertible forms of glycogen phosphorylase (Schmidt and Wagener, 1990). Transglutaminase catalyses acyl transfer reactions between

Table 3.5. Oxidoreductase enzymes from fish and shellfish tissues. From Simpson *et al.* (1991).

Enzyme	Source	Tissue	Reference
Phenol oxidase	Lobster	Cuticle	1
	Crab	Cuticle	2
	Lobster	Shell	3
	Shrimp	Shell	4
	Shrimp	Head	4
	Mussel	Digestive tract	5
	Goldfish	Skin	6
	Octopus	Ink gland	7
	Cuttlefish	Ink gland	7
	Squid	Ink gland	7
Lipoxygenase	Various finfish	Skin	8
	Trout	Skin	9
	Herring	Muscle	10
	Trout	Muscle	11
	Flounder	Muscle	12
9-Lipoxygenase	Trout	Gill	13
12-Lipoxygenase	Trout	Gill	14
15-Lipoxygenase	Trout	Gill	15
Peroxidase	Crayfish	Hepatopancreas	16
Superoxide dismutase	Salmon	Erythrocytes	17
Catalase	Salmon	Erythrocytes	17
Lactate dehydrogenase	Tuna	Muscle	18
	Halibut	Muscle	18
	Shrimp	Muscle	19
	Trout	Muscle	20
	Skate	Muscle	21
D-Glyceraldehyde-3-*P*-dehydrogenase	Lobster	Muscle	22
	Cod	Muscle	22
Glutamate dehydrogenase	Muscle	Hepatopancreas	23
	Molluscs	Muscle	24
Uricase	Various finfish	Liver	25

References: (1) Ferrer *et al.*, 1989; (2) Summers, 1967; (3) Savagaon and Sreenivasan, 1978; (4) Simpson *et al.*, 1988; (5) Waite, 1985; (6) Kim and Tchen, 1962; (7) Prota *et al.*, 1981; (8) Tsukuda, 1970; (9) Hsieh and Kinsella, 1986; (10) Slabyi and Hultin, 1982; (11) Han and Liston, 1987; (12) Macdonald and Hultin, 1987; (13) Winkler *et al.*, 1992; (14) German and Berger, 1990; (15) German and Crevling, 1990; (16) Ilgner and Woods, 1985; (17) Scott and Harrington, 1990; (18) Low *et al.*, 1973; (19) Thebault *et al.*, 1981; (20) Henry and Ferguson, 1985; (21) Lushchak, 1991; (22) Cowey, 1967; (23) Ruano *et al.*, 1985; (24) Wickes and Morgan, 1976; (25) Kinsella *et al.*, 1985.

γ-carboxamide groups of glutamine residues within proteins and suitable acceptors, usually primary amines. When lysine residues in protein are acyl acceptors, the reaction results in the formation of an ε-(γ-glutamyl)lysine isopeptide bond that serves to cross-link proteins. This reaction appears to be responsible for myosin heavy chain intermolecular cross-linking during the 'suwari' process of surimi paste (Numakura *et al.*, 1985).

Trimethylamine-*N*-oxide (TMAO) demethylase contributes to texture deterioration of frozen gadoid fish. It has been isolated from various fish tissues including pyloric caeca, liver, kidney, hepatopancreas and muscle (reviewed by Haard, 1992b). Although there are reports of soluble TMAO demethylase from fish muscle (Tokunaga, 1965), most research has shown the enzyme to be particulate bound (Parkin and Hultin, 1981; Lundstrom *et al.*, 1982).

Pigments

Several nitrogenous compounds contribute to the colour of fish and shellfish. Other than myoglobin and haemoglobin, most of these compounds are associated with epithelial tissue and are reviewed elsewhere (Haard, 1992c). Nitrogen-containing pigments include the blue–green to red carotenoproteins, blue–purple indigoids, brown–black melanins and melanoproteins, variously coloured ommatids, which are polycyclic, aromatic compounds from cephalopods, red–green tetrapyrroles such as biliverdin, flavins associated with the skin of scaleless fishes, purines responsible for white iridescence in leucophores of fish skin and pterins, which contribute to blue iridescence in fish scales. The haem proteins are particularly important because of their ubiquity, discoloration and influence on lipid oxidation in fish meat.

HAEM PROTEINS

The pigments associated with the bright red surface colour of fish meat include oxymyoglobin and oxyhaemoglobin. Normally haemoglobin contributes less to the appearance of seafood because it is lost rather easily during handling and storage while myoglobin is retained by the cellular structure. The biochemistry of haem proteins in meats is reviewed by Bandman (1987).

Haemoglobin is absent from some crustacea and molluscs and from certain Antarctic fish with colourless blood (Love, 1980). The prosthetic group in haemoglobins, protohaem 1, is the same in all animals; however, the protein part of the molecule differs somewhat with species. Fish haemoglobins are normally tetrameric proteins and in this respect are similar to mammalian haemoglobins. However, invertebrate haemoglobins can contain more than four haem groups and primitive fishes (lampreys and hagfish) have monomer, myoglobin-like haemoglobin. The combination of swordfish haemoglobin with hydrogen sulfide is reported to cause green discoloration (Amano and Tomiya, 1953). During this reaction, iso-valerianic acid is formed and is responsible for an unpleasant odour (Amano, 1949).

The red and white fibres tend to be more segregated in fish than in land animal muscles (Hultin, 1985). Therefore, red muscle of fish tends to have a darker appearance than other food myosystems due to the high concentration of myoglobin. Myoglobin content of muscle increases with age and during migration season (Love, 1988). Myoglobins from fish have amino acid compositions that differ considerably from those of mammals (Brown, 1962). A special feature of some fish myoglobins (e.g. tuna) is a cysteine residue adjacent to the haem. Such fish myoglobins can react with thiol compounds in the presence of trimethylamine oxide to cause green discoloration of the meat (Yamagata *et al.*, 1971). In general, fish myoglobins are more readily oxidized to metmyoglobin than mammalian myoglobins (reviewed by Haard, 1992c). Fish muscle contains oxidoreductase enzyme(s) capable of reducing metmyoglobin back to myoglobin (Shimizu and Matsura, 1971; Matsui *et al.*, 1975).

HAEMOCYANINS

Shellfish have copper-containing proteins called haemocyanins, which, like haemoglobins, combine reversibly with oxygen. Haemocyanin binds one molecule of oxygen per two copper atoms and has a functional subunit of 50–75 kDa. Heat-denatured haemocyanin or oxyhaemocyanin turns white and in the presence of hydrogen sulfide forms a blue–green colour. Haemocyanin also has catalase-like activity (Ghiretti, 1956).

Parvalbumins

Parvalbumins are a group of heat-stable, water-soluble, acidic, low molecular mass (about 12 kDa) Ca^{2+}-binding proteins. Parvalbumin, and another calcium-binding protein calmodulin, occur in the muscle of aquatic lower vertebrates, including fish. In fish, these molecules may constitute 20–30% of the sarcoplasmic protein fraction (Schwimmer, 1981). The calcium-binding proteins appear to serve the same function as the troponin complex and tropomyosin in higher vertebrates (Grandjean *et al.*, 1977). However, the exact physiological role of parvalbumins is not clear at this time. They are associated with fast, nerve-impulse-activated skeletal muscle (Pechere, 1977). Reports of parvalbumin isolation from fish muscle are summarized in Table 3.6.

Interestingly, the primary structures of fish parvalbumins are highly homologous with cod allergen M, also classified as a parvalbumin (Elsayed and Bennich, 1975). Allergen M is one of the most potent food allergens known (Aas, 1987). The immunology of fish parvalbumins is reviewed by

Table 3.6. Parvalbumins isolated from skeletal muscle of fishes.

Source	Components	Reference
Carp	3	1
Hake	1 major	2
Pike	1 major	3
Tilapia	2	4
Lungfish	5	5
Raja sp.	1 major	6
Perch	2 major	7
Cod	2	8
Oryzias sp.	5	9
Lepomis sp.	5	10
Cyprinidea family	5	11
Toadfish	1	12
Eel	3	13
Salmonidae family	5	14
Pollack	5	15

References: (1) Konosu *et al.*, 1965; (2) Pechere *et al.*, 1971; (3) Frankenne *et al.*, 1973; (4) Piront *et al.*, 1968; (5) Elsayed and Bennich, 1975; (6) Gerday and Teuwis, 1972; (7) Lehky and Stein, 1979; (8) Bhushana *et al.*, 1969; (9) Sakaizumi, 1985; (10) Whitmore, 1986: (11) Bobak and Slechta, 1988; (12) Gerday *et al.*, 1989; (13) Dubois and Gerday, 1990; (14) Rehbein and van Lessen, 1989; (15) Woychik and Dower, 1990.

Gazzaz and Rasco (1993). Crustacean IgE-mediated hypersensitivities are similar to those of finfish; however the allergens (Antigen I and II in shrimp) are different proteins (Lehrer and McCants, 1987; Lehrer and Daul, 1992). Shrimp antigen I is a heat-stable protein of 20.5 kDa containing 0.5% carbohydrate. Antigen II is 38.3 kDa with 4% carbohydrate and stable in shrimp boiled for 10 min.

The parvalbumins are also of interest because they can be used to confirm species origin of processed fish products (Woychik and Dower, 1990). It is also possible that removal of parvalbumins during surimi manufacture contributes to the improved gelation properties of water-washed fish mince.

Myofibrillar proteins

The myofibrillar proteins are those which compose the contractile apparatus or myofibril within the muscle cell (Table 3.7). Myosin and actin are the myofibrillar proteins that directly function in the contraction–relaxation cycle. Other proteins that participate indirectly in the cycle are called regulatory proteins (Maruyama and Ebashi, 1970) or major and minor

Table 3.7. Components of the myofibrillar protein fraction of vertebrate skeletal muscle.

Subclass	Protein	Location	Size (kDa)	Fish reference
Contractile	Myosin	A-Band	500	1
	α-Actin	I-Band	42	2
Regulatory	Tropomyosin	I-Band	66	3
	Troponin C	I-Band	21	4
	Troponin I	I-Band	21	4
	Troponin T	I-Band	35	4
	M-Protein	M-Filament	165	
	Creatine kinase	M-Bridges	42	
	β-Protein	M-Line(?)	130	
	C Protein	Myosin fil.	140	
	F Protein	A-Band	121	
	I Protein	A-Band	50	
	H Protein	A-Band	69	
	X Protein	A-Band	152	
	Paratropomyosin	A-Band	35	
	α-Actinin	Z-Line	95	
	β-Actinin	I-Filament	37	
	γ-Actinin	I-Band(?)	35	
	Eu Actinin	Z-Line	42	
	Z Protein	Z-Line	55	
	Z Nin	Z-Line	300	
Scaffold	Connectin	A-I Junc.	>1000	5
	Nebulin	N-Line	550	
	Titin	A-I Junc., etc.	>1000	
	Filamin	Z-Line and I-Band	230	
	Desmin	Z-Line	550	
	Synemin	Z-Line	220	
	Vimentin	Z-Line	58	

References: (1) Connell (1954), Buttkus (1966), Chung *et al.* (1967), Odense *et al.* (1969), Takashi (1973), Takashi *et al.* (1974), Tomioka *et al.* (1974), Tsuchiya *et al.* (1978), Watabe *et al.* (1983), Kanoh *et al.* (1983), Chen *et al.* (1988). (2) Connell (1954), Brigden (1971), Seki *et al.* (1973), Shenouda and Pigott (1975), Tsuchiya *et al.* (1977). (3) Hamoir (1955), Kubo (1957), Saad *et al.* (1959), Tsuchiya *et al.* (1980b), Sano *et al.* (1989). (4) Konno (1978). (5) Kimura *et al.* (1978; 1979a; 1979b), Seki and Watanabe (1984).

regulatory proteins (Obinata *et al.*, 1981). A third group called scaffold proteins or filament proteins plays a structural role in the myofibril and the muscle cell. It is believed this third group of proteins provides structural continuity along the myofibril and muscle cell.

The proportion of myofibrillar protein to total muscle protein is higher in fish than in land animals. However, the relative amounts of individual components are similar. The general subject of vertebrate myofibrillar proteins has been reviewed elsewhere (Asghar *et al.*, 1985; Bandman, 1987; Sikorski *et al.*, 1990; Skaara and Regenstein, 1990).

Myosin

Myosin molecules form the thick filaments and make up about half of the proteins in the contractile apparatus. The myosin molecule consists of two subunits called heavy chains (200 kDa) and up to four subunits called light chains (16–30 kDa) and is a long structure (2 nm × 160 nm) with two globular heads (5 nm × 20 nm). In fish, several myosin types with different light chain distribution have been identified (Focant *et al.*, 1981). Myosin from mackerel white muscle has three light chain subunits resembling fast myosin of rabbit. Myosin from dark muscle has two light chains like rabbit slow myosin (Watabe and Hashimoto, 1980). The myosin ATPase activity is located in the head region of the molecule and is required for the interaction of myosin with actin. The amino acid composition and molecular weights of fish myosins are similar (reviewed by Skaara and Regenstein, 1990). The high content of thiol groups in myosin (30–40) play an important role in meat processing (Hofmann and Hamm, 1978) and is similar in many species of fish (Skaara and Regenstein, 1990). Specific extraction conditions that avoid actomyosin formation are needed to isolate fish myosins (Connell, 1954; Mackie and Connell, 1964; Martone *et al.*, 1986). In extracting myosin from squid, care must also be taken to inhibit endogenous proteases that can extensively hydrolyse the molecule during isolation (Tsuchiya *et al.*, 1978).

Some differences between fish and mammalian myosin have been identified. The Mg^{2+}- and Ca^{2+}-ATPase activities of carp and tilapia myosin are less than those of rabbit myosin (Takashi, 1973). Myosin from cold water fish is less stable than that from tropical fish (Johnston *et al.*, 1973; 1975). Even for the same species, the water temperature (10°C vs. 30°C) during rearing influences the properties of carp myosin (Misima *et al.*, 1990). Fish myosin denatures and aggregates faster than rabbit myosin at moderate and freezing temperatures (Connell, 1960; Suzuki, 1981) and is influenced by the formation of disulfide bonds (Jiang *et al.*, 1988).

Paramyosin

Paramyosin occurs in the smooth and striated muscles of invertebrates. This 200 kDa molecule forms the central core of the thick filaments in such muscles. It comprises 14% of the salt-soluble proteins of squid (Tsuchiya *et al.*, 1980a) and 65% of abalone foot muscle (Pyeun *et al.*, 1973). Paramyosin, at relatively low concentrations, appears to contribute to the actomyosin structure of fish meat gel products (Sano *et al.*, 1986).

Actin

Actin is a globular protein (G-actin) that polymerizes at physiological salt concentration to form filaments called F-actin. Early studies showed that cod actin is homologous with rabbit actin except for significantly higher proline and lower aspartic, glutamic and lysine residues (Connell and Howgate, 1959). Actin is the principal component of the thin filaments and is second only to myosin as a major myofibrillar protein. Methods have been developed for the isolation of fish actin (Ueda *et al.*, 1967; Shenouda and Pigott, 1975). The actins from squid, carp, tilapia and rockfish have a similar molecular mass to that of rabbit muscle actin (Seki *et al.*, 1973; Tsuchiya *et al.*, 1977).

Tropomyosin

Tropomyosin is a major regulatory protein consisting of two subunit chains of approximately 33 kDa. The proportion of the two subunits (a and b) varies with the type of muscle fibre. Alaska pollack skeletal muscle tropomyosin is mostly a dimer of $\alpha\alpha$ 34 kDa chains (Sano *et al.*, 1989). The molecular masses of squid subunits (35 and 37 kDa) are somewhat higher than those of vertebrate species (Tsuchiya *et al.*, 1980a). Tropomyosin binds stoichiometrically with actin (1:7, mol/mol) and troponin (1:1, mol/mol). Tropomyosin was first isolated from fish (carp muscle) by Hamoir (1955) and has since been isolated from several species of fish (reviewed by Skaara and Regenstein, 1990; Table 3.7). The amino acid composition of purified cod tropomyosin is similar to mammalian sources with the exception of higher amounts of aspartic residues. It has also been reported that the quantities of glycine and leucine are species-specific in cold-blooded animals (Saad *et al.*, 1959).

Troponins

Troponin is a regulatory protein of about 80 kDa which confers Ca^{2+} sensitivity to actomyosin ATPase. Three subunits function in Ca^{2+} binding (troponin C), relaxation of the actin myosin complex in the presence of ATP (troponin I) and binding of whole troponin with tropomyosin (troponin T). The molecular masses of carp and shark troponin subunits have been reported by Shimomura *et al.* (1976). Squid troponins have molecular masses of 58, 28 and 24 kDa (Konno, 1978).

Other regulatory proteins

Actinins and other regulatory proteins identified in skeletal muscle are listed in Table 3.7 and reviewed elsewhere (Asghar *et al.*, 1985;

Bandman, 1987). Our knowledge of these proteins in fish muscle is fragmentary.

Scaffold proteins

Filament or scaffold proteins are believed to strengthen the integrity of the myofibrillar system by providing intrafibrillar and interfibrillar bridges. Proteins currently classified in this group are listed in Table 3.7. Connectin isolated from carp dark and white muscle tissues represents 2.8% and 4.4% of total muscle protein, respectively (Kimura *et al.*, 1979a). Seki and Watanabe (1984) found no significant differences in the connectin content of myofibrils (13%) from seven fish species. The elasticity of carp connectin is primarily a consequence of naturally occurring reducible cross-links, lysinonorleucine and another unidentified cross-link (Kimura *et al.*, 1979b). The amount of connectin decreases with increasing post-mortem storage and may contribute to the loss of muscle elasticity characteristic of postrigor fish muscle. Although other scaffold proteins have been identified in skeletal muscles of land animals there does not appear to be information available on their occurrence in fish.

Stroma proteins

The insoluble matter remaining after removing sarcoplasmic and myofibrillar proteins from muscle contains the stroma proteins. The two major types of protein in this fraction are the connective tissue proteins called collagen and elastin. Connective tissues are located primarily in the extracellular matrix and contain few living cells in an amorphous ground substance composed of carbohydrates and lipids as well as protein. Fibroblasts are cells that synthesize and secrete collagen, elastin and other substances, such as proteoglycans, that occupy the interstitial space of muscle and other tissues.

The stroma fraction also contains small amounts of other proteins not derived from the connective tissues. Other components of the stroma fraction are membrane lipoproteins, and the scaffold proteins discussed with the myofibrillar proteins.

Collagen

Collagen is a family of proteins that normally constitutes 20–25% of the total protein of mammalian muscles. The content of collagen in fish muscle is lower than that of mammals, ranging from about 1 to 12% of total protein and 0.2 to 2.2% of muscle wet weight (Sato *et al.*, 1986b; Sikorski *et al.*, 1990). Since the myofibrillar proteins of fish may denature and become insoluble during preparation of collagen, it is difficult to exclude denatured myofibrillar proteins from fish collagen using methods developed for mammalian and avian muscles. Sato *et al.* (1986a) devised a simple method where fish muscle is fractionated into hot-water-soluble and acid-soluble fractions that contain all the collagen and negligible non-collagen proteins. The collagen content in the flexible body musculature used for propulsion is higher than that in other muscles (Yoshinaka *et al.*, 1988). For this reason, the muscle of anguilliforms, such as eel, contains much more collagen than thunniforms, such as tuna.

COLLAGEN TYPES

More than ten distinct types of collagen are found in animal tissues. Collagen molecules are composed of three polypeptide α chains. At least four types of collagen (I, III, IV and V) have been identified in mammalian and avian muscles (Bailey *et al.*, 1979; Light and Champion, 1984; Asghar *et al.*, 1985). In mammals, type I collagen is composed of two $\alpha1$(I) chains and one $\alpha2$(I) chain represented by the formula [$\alpha1$(I)$\alpha2$(I)]. Type I collagen in fish skin and muscle often contains a unique $\alpha3$ chain with the structure [$\alpha1$(I)$\alpha2$(I)$\alpha3$(I)] (Kimura and Tanaka, 1986; Kimura *et al.*, 1987, 1988; Ramshaw *et al.*, 1988; Masui *et al.*, 1989; Bracho and Haard, 1990; Miyauchi and Kimura, 1990). The $\alpha3$ chain from carp skin (Miyauchi and Kimura, 1990) and blue grenadier (Ramshaw *et al.*, 1988) shows more similarity to the $\alpha1$ than to the $\alpha2$ chain.

In fish muscles, type V collagens have also been identified (Sato *et al.*, 1988; 1989a; 1989b) while type III-like collagen was not detectable (Sato *et al.*, 1989a). Types IV and V collagens do not form typical collagen fibrils, but are in the basement membranes surrounding muscle cells. In fish, the solubility of muscle collagen increases during post-mortem chilled storage (Cepeda *et al.*, 1990) and this appears to be associated with the degradation of type V collagen (Sato *et al.*, 1991) and proteoglycans (Kim and Haard, 1992). On the other hand, type I collagen appears to be very stable in post-mortem fish muscle (Cepeda *et al.*, 1990; Sato *et al.*, 1991).

COLLAGEN AMINO ACID COMPOSITION

Fish muscle and skin collagens differ from bovine collagens in having significantly higher contents of essential amino acids and often contain a lower concentration of hydroxyproline (Sikorski *et al.*, 1990). It has been suggested that tryptophan residues partially replace the residues of imino acids in shrimp collagen (Thompson and Thompson, 1968). The collagen content of muscle is often calculated on the basis of the hydroxyproline content of hydrolysates. Caution should be exercised in applying this method to fish because of the wide variation of this imino acid in fish collagen. Sato *et al.* (1989a) examined acid-soluble collagen from 22 species of fish and observed that the hydroxyproline content ranged from 4.7 to 10.0%. In contrast, the hydroxyproline content of bovine muscle collagen

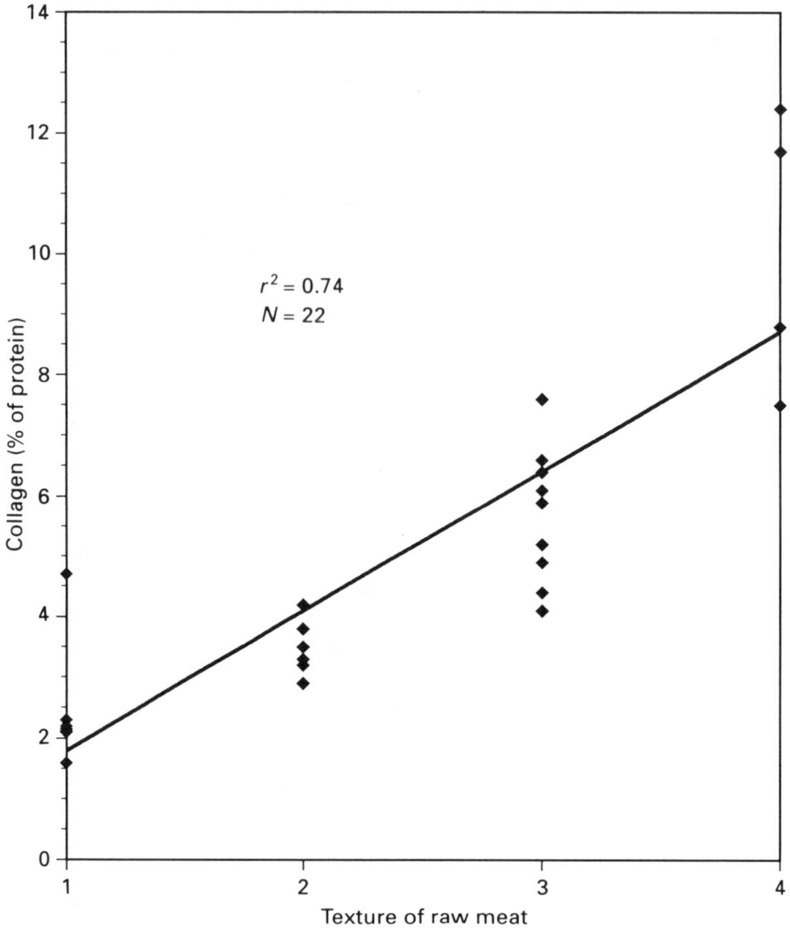

Fig. 3.2. Collagen content of the white muscle in relation to raw meat texture. Adapted from Sato *et al.* (1986b). Texture: 1 - tender; 2 - slightly tender; 3 - tough; 4 - very tough.

is 10.9% (McClain *et al.*, 1971). Hydroxyproline stabilizes the collagen molecule, and for this reason collagens of fish are normally denatured at relatively low temperatures.

COLLAGEN AND MEAT QUALITY

The importance of collagen to the seafood technologist has been reviewed by Sikorski *et al.* (1984).

Fish flesh that has strong mechanical strength has more collagen than flesh with a tender consistency (Fig. 3.2). During chill storage of some

Table 3.8. Percentage distribution of non-protein nitrogen compounds in finfish and shellfish. Adapted from Finne (1992).

Compound	Teleosts (mackerel)	Elasmobranchs (shark)	Crustacea (shrimp)	Molluscs (squid)
Amino acids	25	5	65	50
Peptides	5	5	15	15
Nucleotides	15	5	5	5
Creatine and creatinine	50	10	-	-
Trimethylamine oxide	15	20	5	15
Urea	-	55	-	-
Betaines	-	-	10	5
Ammonia and amides	5	-	-	-
Octopine		-	-	10

fish the myocommata may fail to hold the muscle cells together and thus cause gaping of the flesh (Hallett and Bremner, 1988). Collagen may also influence the water-holding capacity and gel-forming capacities of comminuted cooked products.

Other stroma proteins

Other stroma proteins, such as elastin and minor proteins of the extracellular matrix, do not appear to have been studied in fishes. Recently, Kim and Haard (1992) characterized the proteoglycan fraction of Pacific rockfish. The byssal threads of mussel contain a hydroxyproline-rich peptide that differs from collagen in its low content of glycine (Waite, 1983).

Non-protein Nitrogenous Compounds

The NPN content of seafood is significantly higher than that in other food myosystems, i.e. 9–18% of total nitrogen in teleosts and 33–38% in elasmobranchs (Belitz and Grosch, 1987). The major classes of sarcoplasmic components are free amino acids, peptides, guanidino compounds, urea, betaines, nucleotides and quaternary ammonium compounds. The distribution of these compounds varies with taxonomic family (Finne, 1992) as illustrated in Table 3.8. These compounds are important to the seafood technologist because (i) they influence the unique, delicate taste of seafood and (ii) they contribute to the spoilage of fishery products.

Free amino acids

The free amino acid content in muscles of aquatic organisms ranges from about 0.5% to 2% of muscle weight. Free amino acids apparently contribute to osmoregulation (Finne, 1992) and are depleted in fish muscle during starvation (Love, 1988). Crustacean muscles, e.g. from lobster, shrimp and crab, contain a higher content of free amino acids than finfish muscles. Cultured fish tend to contain less free amino acids than wild fish (reviewed by Simpson *et al.*, 1991; Haard, 1992a). Some unique (non-protein) free acids found in seafood are taurine, sarcosine, β-alanine, methylhistidine, and α-amino-*n*-butyric acid (Mackie and Ritchie, 1974). Taurine, a sulfonic acid, is common and abundant in marine invertebrates. In finfish, taurine is normally in greater concentration in the white than in the red muscle. Taurine can function in osmoregulation, but it may also serve as a food reserve. Taurine is very active in the Maillard browning reaction and contributes to the discoloration of dried squid (Haard and Arcilla, 1985). β-Alanine appears to be most common in cold water fish.

Red muscles tend to contain more histidine as a free amino acid than white muscles. Histidine is the predominant acid in fish of the Scombridae family, e.g. tuna, mackerel and mahi mahi. Decarboxylation of histidine by spoilage bacteria gives rise to biologically active histamine. Scombroid poisoning is a form of food intoxication with similar symptoms to those of food allergies. The allergic reaction is mediated by IgE hypersensitivity, which causes release of histamine from mast cells. Other biologically active amines that are formed in spoiled fish are putrescine and cadaverine (Mietz and Karmas, 1977).

Peptides

Small amounts of peptides have been identified in fish muscle extracts. These include carnosine (β-alanyl histidine), anserine (β-alanyl-1-methylhistidine) and balenine (β-alanyl-3-methylhistidine). Eel muscle is rich in carnosine (0.5–0.6%). Dark muscles tend to contain more of these peptides than white muscles. The ratio of carnosine to anserine is normally higher for freshwater than for marine fish. The metabolic role of these peptides is not clear at this time. Carnosine has antioxidant properties, probably due to inhibition of water-soluble lipid oxidation catalysts and/or active oxygen species (Decker and Crum, 1991).

Nucleotides

Normally, more than 90% of the nucleotides in fish muscle are purine derivatives resulting from ATP catabolism. At the moment of death, ATP is the main component of the nucleotide pool. Although the ATP catabolism to hypoxanthine that occurs in postharvest muscle is generally considered to be the consequence of endogenous enzymes, bacterial inosine nucleosidase can also influence the kinetics of this reaction (Surette *et al.*, 1988, 1990). ATP catabolism occurs at different rates in different fish species; e.g. ATP is depleted within an hour after death in free-living Pacific rockfish (Lee *et al.*, 1992) but its depletion may require almost a week in farmed sturgeon (Izquerdo-Pulido *et al.*, 1992). Crustacean muscle tends to have low AMP deaminase activity and for this reason AMP rather than IMP accumulation occurs immediately after death of the animal.

Other nucleotides in fish muscle include nicotinamide adenine nucleotides (NAD^+/NADH, 4–38 mg %; $NADP^+$/NADPH, 0.3–11 mg %). Normally, dark muscle contains about two times more of these compounds than white muscle. Ribose phosphates resulting from postharvest catabolism of these compounds contribute to Maillard browning in some seafood products (Haard and Arcilla, 1985).

Guanidino compounds

The phosphagens, creatine and creatine phosphate, serve as a high energy reservoir and may be present in fish muscle at 0.3–0.7%. White muscles tend to contain more of these compounds than red muscles (Sakaguchi *et al.*, 1982). Creatinine (6–35 mg %) is formed from creatine by cyclization and appears to be a metabolic end product for excretion. Invertebrates contain less creatine in their muscle than finfish. Cephalopods, such as squid or scallop, tend to accumulate the guanidino compounds arginine and octopine, intermediates of the arginine phosphate cycle.

Urea and quaternary ammonium compounds

Urea is common in all fish tissues but is generally present at less than 0.05% in muscle of teleost fish. Large amounts of urea are found in marine elasmobranchs (1–2.5%), where it functions in osmoregulation. Methylamines, like trimethylamine oxide, appear to stabilize intracellular proteins from denaturation by this protein denaturant (Somero, 1986). Decomposition of urea to ammonia and carbon dioxide is catalysed by bacterial urease (Finne, 1992). Since urea is water soluble and permeates cell membranes it can readily be washed from fillets by the processor.

Trimethylamine oxide (TMAO) is common in marine teleosts (15–250 mg %) and present at very high concentrations in marine elasmobranchs (1.0–1.8 mg %). Red muscles normally contain more TMAO than white muscles from the same species and concentration can vary with a variety of intraspecific factors. The occurrence and significance of TMAO in marine food products have been reviewed (Hebard *et al.*, 1982; Haard, 1992b). Betaines are especially abundant in mollusc and crustacea muscles

Table 3.9 Index of nutritional quality (INQ) for fish and other foodstuffs. Data from Snook (1984).

Food	Protein INQ[a]	Total INQ[b]	Total INQ from protein (%)
Fish	14	17	82
Chicken	11	17	65
Broccoli	5.3	180	3
Peas	4.6	12	38
Milk	3.8	17	22
Wholewheat bread	1.8	8	23

[a] INQ = percentage of nutrient requirement provided by food divided by percentage of kilocalorie requirement provided by food.
[b] Total INQ for seven major nutrients.

(0.4–1.0%), where they contribute to taste. The most common betaine is glycine betaine. Other betaines include β-alanine, 'betaine' and homarine. Betaine is an intermediate in choline metabolism in the living animal.

Nutritional Value

Like most animal foods, seafood contains protein having excellent nutritive value (Nettleton, 1985; Pigott and Tucker, 1990). Fish protein contains all essential amino acids, it is highly digestible, and the protein efficiency ratio ranges from 2.7 to 3.2 (Jhaveri *et al.*, 1984). The *in vitro* protein digestibility of fish is not influenced by the protein denaturation and aggregation that occurs during storage (Morey *et al.*, 1982). The protein INQ (index of nutritional quality) for fish also compares favourably with other foodstuffs although fish does not contain many other nutrients that significantly affect the total INQ (Snook, 1984; Table 3.9). In terms of nutritive value, fish protein ranks above casein.

Water washing of fish meat is a processing step for production of fillets, mince or surimi. This operation results in substantial amounts of protein in the waste stream. Methods are now available to recover water-soluble protein from the waste stream. Accordingly, the nutritional value of sarcoplasmic protein concentrates is of interest. Mamaeva *et al.* (1991) fed sarcoplasmic protein concentrates from three industrial fish (minthal, cod, putassu) to rats for 4 weeks and found that the biological value was equal to or greater than casein. Moreover, sarcoplasmic protein concentrates exhibited hypocholesterolaemic and hypolipidaemic effects in growing rats.

References and Further Reading

Aas, K. (1987) Fish allergy and the codfish allergen model. In: Brostoff, J. (ed.) *Food Allergy and Intolerance*. Tindall Baillière, London, pp. 356–366.

Amano, K. (1949) Studies on the green discoloration of frozen swordfish (*Xiphias gladius* L.). 1. Isolation of isovaleric acid like substance from green colored flesh. *Bulletin of the Japanese Society of Scientific Fisheries* 15, 467–474.

Amano, K. and Tomiya, F. (1953) Studies on green discoloration of frozen swordfish (*Xiphias gladius* L). II. Some experiments on the cause of discoloration. *Bulletin of the Japanese Society of Scientific Fisheries* 19, 671–687.

Anzai, H., Asada, H., Koshiba, A., Yoshida, S., Kobayashi, H., Uchida, N. and Nishide, E. (1991a) Distribution of polysaccharide digestive enzymes in a marine gastropod *Dolabella auricularia. Bulletin of the Japanese Society of Scientific Fisheries* 57, 2077–2081.

Anzai, H., Enami, Y., Chida, T., Okoshi, A., Omura, T., Uchida, N. and Nishide, E. (1991b) Polysaccharide digestive enzymes from midgut gland of abalone. *Bulletin of the College of Agriculture and Veterinary Medicine, Nihon University* 48, 119–127.

Anzai, H., Nanba, Y., Sawada, M., Uchida, N. and Nishide, E. (1992) Modes of action of cellulases from the gastric teeth of a gastropod *Dolabella auricularia. Bulletin of the Japanese Society of Scientific Fisheries* 58, 159–163.

Aoki, T., Takada, K. and Kunisaki, N. (1991) On the study of proximate composition, mineral, fatty acid, free amino acids, muscle hardness, and color difference of six species of wild and cultured fishes. *Bulletin of the Japanese Society of Scientific Fisheries* 57, 1927–1934.

Armstrong, S.G., Leach, D.N. and Wyllie, S.G. (1992) The use of HPLC protein profiles in fish species identification. *Food Chemistry* 44, 147–155.

Asghar, A., Samejima, K. and Yasui, T. (1985) Functionality of muscle proteins in gelation mechanisms of structured meat products. *Critical Reviews in Food Science and Nutrition* 22, 27–106.

Assaf, S.A. and Graves, D.J. (1969) Structural and catalytic properties of lobster muscle glycogen phosphorylase. *Journal of Biological Chemistry* 244, 5544–5555.

Audley, M.A., Shetty, K.J. and Kinsella, J.E. (1978) Isolation and properties of phospholipase A from pollock muscle. *Journal of Food Science* 43, 1771–1775.

Bailey, A.J., Shellswell, G.B. and Duance, V.C. (1979) Identification of the change of collagen types in differentiating myoblasts and developing chick muscle. *Nature (London)* 278, 67–68.

Bandman, E. (1987) Chemistry of animal tissues. Part 1: Proteins. In: Price, J.F. and Schweigert, B.S. (eds) *The Science of Meat and Meat Products*, 3rd edn. Food and Nutrition Press, Westport, Connecticut, pp. 73–77.

Belitz, H.-D. and Grosch, W. (1987) *Food Chemistry*. Springer Verlag, New York.

Berner, D.L. and Hammond, E.G. (1970) Phylogeny of lipase specificity. *Lipids* 5, 558–562.

Bhushana Rao, K.S.P., Focant, B., Gerday, Ch. and Hamoir, G. (1969) Low molecular weight albumins of cod white muscle *Gadus callarias* L. *Comparative Biochemistry and Physiology* 30, 33–48.

Bilinski, E., Jonas, R.E.E., Peters, M.D. and Choromanski, E.M. (1984) Effects of sexual maturation on the quality of coho salmon (*Oncorhynchus kisutch*) flesh. *Canadian Institute of Food Science and Technology Journal* 17, 271–273.

Bobak, P. and Slechta, V. (1988) Comparison of parvalbumins of some species of family Cyprinidae. *Comparative Biochemistry and Physiology* 91B, 697–699.

Bracho, G.E. and Haard, N.F. (1990) Determination of collagen crosslinks in rockfish skeletal muscle. *Journal of Food Biochemistry* 14, 435–451.

Bracho, G.E. and Haard, N.F. (1991) Characterization of alkaline metalloproteinases with collagenase activity from the muscle of Pacific rockfish (*Sebastes* sp.). In: Otwell, S. (ed.) *Proceedings of Joint Meeting of Atlantic Fisheries Technologists and Tropical/Subtropical Fisheries Technologists*. Florida Sea Grant, Gainesville, pp. 105–125.

Brigden, J. (1971) The amino acid sequence around four cysteine residues in trout actin. *Biochemical Journal* 123, 591–600.

Brockerhoff, H. (1966) Digestion of fat by cod. *Journal of the Fisheries Research Board of Canada* 23, 1835–1839.

Brockerhoff, H. (1967) Digestion of triglycerides by lobster. *Canadian Journal of Biochemistry* 45, 421–422.

Brockerhoff, H. and Hoyle, J.R. (1965) Hydrolysis of triglycerides by the pancreatic lipase of a skate. *Biochimica et Biophysica Acta* 98, 435–436.

Brooks, S.P. and Storey, K.B. (1988a) Subcellular enzyme binding in glycolytic control: *in vivo* studies with fish muscle. *American Journal of Physiology* 255, R289–R294.

Brooks, S.P. and Storey, K.B. (1988b) Re-evaluation of the 'glycolytic complex' in muscle: a multitechnique approach using trout white muscle. *Archives of Biochemistry and Biophysics* 267, 13–22.

Brown, W.D. (1962) The concentration of myoglobin and hemoglobin in tuna flesh. *Journal of Food Science* 27, 26–28.

Buttkus, H. (1966) Preparation and properties of trout myosin. *Journal of the Fisheries Research Board of Canada* 23, 563–573.

Cepeda, R., Chou, E., Bracho, G. and Haard, N.F. (1990) An immunological method for measuring collagen degradation in the muscle of fish. In: Voigt, M. (ed.) *Advances in Fisheries Technology and Biotechnology for Increased Profitability*. Technomic Publishing Co., Lancaster, Pennsylvania, pp. 487–506.

Chawla, P. and Ablett, R.F. (1987) Detection of microsomal phospholipase activity in myotomal tissue of Atlantic cod (*Gadus morhua*). *Journal of Food Science* 52, 1194–1197.

Chen, C.S., Hwang, D.C. and Jiang, S.T. (1988) Purification and characterization of milkfish (*Chano chanos*) myosin. *Bulletin of the Japanese Society of Scientific Fisheries* 54, 1423–1427.

Chung, C.S., Richards, E.G. and Olcott, H.S. (1967) Purification and properties of tuna myosin. *Biochemistry* 6, 3154–3160.

Connell, J.J. (1954) Studies on fish skeletal muscle. 3. Cod myosin and actin. *Biochemical Journal* 58, 360–367.

Connell, J.J. (1960) Studies of the proteins of fish muscle. 7. Denaturation and aggregation of cod myosin. *Biochemical Journal* 75, 530–538.

Connell, J.J. and Howgate, J.F. (1959) Studies on the proteins of fish skeletal muscle. 6. Amino acid composition of cod myofibrillar proteins. *Biochemical Journal* 71, 83–86.

Cowey, C.B. (1967) Comparative studies on the activity of D-glyceraldehyde-3-phosphate dehydrogenase from cold and warm blooded animals with reference to temperature. *Comparative Biochemistry and Physiology* 23, 969–976.

Croston, C.B. (1965) Endopeptidases of salmon ceca: chromatographic separation and some properties. *Archives of Biochemistry and Biophysics* 112, 218–223.

Decker, E.A. and Crum, A.D. (1991) Inhibition of oxidative rancidity in salted ground pork by carnosine. *Journal of Food Science* 56, 1179–1160, 1181.

De Haen, C., Walsh, K.A. and Neurath, H. (1977) Isolation and amino acid sequence analysis of a new pancreatic trypsinogen of the African lungfish *Protopterus aethiopicus*. *Biochemistry* 16, 4421–4425.

Dixon, M. and Webb, E.C. (1979) *Enzymes*, 3rd edn. Academic Press, New York, pp. 633–649.

Dubois, I. and Gerday, C. (1990) Soluble calcium-binding proteins: parvalbumins and calmodulin from eel skeletal muscle. *Comparative Biochemistry and Physiology* 95B, 381–385.

Eisen, A.Z. and Jeffrey, J.J. (1969) An extractable collagenase from crustacean hepatopancreas. *Biochimica et Biophysica Acta* 191, 517–520.

Elsayed, S. and Bennich, H. (1975) Primary structure of allergen M from cod. *Scandinavian Journal of Immunology* 4, 203–208.

Eskin, N.A.M. (1990) *Biochemistry of Foods*, 2nd edn. Academic Press, New York, pp. 24–26.

Feeney, R.E. (1988) Inhibition and promotion of freezing: Fish antifreeze proteins and ice-nucleating proteins. *Comments Agriculture and Food Chemistry* 1, 147–181.

Ferrer, O.J., Koburger, J.A., Simpson, B.K., Gleeson, R.A. and Marshall, M.R. (1989) Phenoloxidase levels in Florida spiny lobster (*Panulirus argos*): Relationship to season and molting stage. *Comparative Biochemistry and Physiology* 93B, 595–599.

Finne, G. (1992) Non-protein nitrogen compounds in fish and shellfish. In: Flick, G.J. and Martin, R.E. (eds) *Advances in Seafood Biochemistry, Composition and Quality*. Technomic Publishing, Lancaster, Pennsylvania, pp. 393–401.

Focant, B., Jacob, M.F. and Huriaux, F. (1981) Electrophoretic comparison of the proteins of some perch (*Perca fluviatilis* L.) head muscles. *Journal of Muscle Research and Cell Motility* 2, 295–305.

Folco, E.J., Busconi, L., Martone, C.B. and Sanchez, J.J. (1989) Fish skeletal muscle contains a novel serine proteinase with an unusual subunit composition. *Biochemical Journal* 263, 471–475.

Frankenne, F., Joassin, L. and Gerday, Ch. (1973) Amino acid sequence of the pike *Esox lucius* parvalbumin III. *Federation of European Biological Society Letters* 35, 145–147.

Fruton, J.S. and Bergmann, M. (1940) The specificity of salmon pepsin. *Journal of Biological Chemistry* 136, 559–560.

Gazzaz, S.S. and Rasco, B.A. (1993) Parvalbumins in fish and their role as food allergens. A review. *Reviews in Fisheries Science* 1(1), 1–26.

Geist, G.M. and Crawford, D.L. (1974) Muscle cathepsins in three species of Pacific sole. *Journal of Food Science* 39, 548–551.

Gerday, Ch. and Teuwis, J.-C. (1972) Isolation and characterization of the main parvalbumins from *Raja clavata* and *Raja montagui* white muscles. *Biochimica et Biophysica Acta* 271, 320–331.

Gerday, C., Collin, S. and Gerardin-Otthiers, N. (1989) Amino acid sequence of the parvalbumin from the very fast swimbladder muscle of the toadfish *Opsanus tau*. *Comparative Biochemistry and Physiology* 93B, 49–55.

German, J.B. and Berger, R. (1990) Formation of 8,15-dihydroxy eicosatetraenoic acid via 15-lipoxygenase and 12-lipoxygenase in fish gill. *Lipids* 25, 849–853.

German, J.B. and Crevling, R.K. (1990) Identification and characterization of a 15-lipoxygenase from fish gills. *Journal of Agricultural and Food Chemistry* 38, 2144–2147.

Geromel, E.J. and Montgomery, M.W. (1980) Lipase release from lysosomes of rainbow trout (*Salmo gairdneri*) muscle subjected to low temperature. *Journal of Food Science* 45, 412.

Ghiretti, F. (1956) The decomposition of hydrogen peroxide by hemocyanin and by its dissociation products. *Archives of Biochemistry and Biophysics* 63, 165–176.

Gildberg, A. (1988) Aspartic proteinases in fishes and aquatic invertebrates. *Comparative Biochemistry and Physiology* 91B, 425–435.

Gildberg, A. and Raa, J. (1983) Purification and characterization of pepsins from the Arctic fish capelin (*Mallotus villosus*). *Comparative Biochemistry and Physiology* 75A, 337–342.

Grandjean, J., Laszlo, P. and Gerday, C. (1977) Sodium complexation by the calcium binding site of parvalbumins. *Federation of European Biological Society Letters* 81, 376–380.

Groninger, H.S. (1964) Partial purification and some properties of a proteinase from albacore (*Germo alalunga*) muscle. *Archives of Biochemistry and Biophysics* 108, 175–182.

Haard, N.F. (1987) Protein and non-protein constituents in jellied American plaice, *Hippoglossoides platessoides*. *Canadian Institute of Food Science and Technology Journal* 20, 98–101.

Haard, N.F. (1990) Enzymes from food myosystems. *Journal of Muscle Foods* 1, 293–338.

Haard, N.F. (1992a) Control of chemical composition and food quality attributes of cultured fish. *Food Research International* 25, 289–307.

Haard, N.F. (1992b) Biochemical reactions in fish muscle during frozen storage. In: Bligh, E.G. (ed.) *Seafood Science and Technology*. Blackwell Scientific Publications, Boston, pp. 176–209.

Haard, N.F. (1992c) Biochemistry and chemistry of color and color change in seafoods. In: Flick, G.J. and Martin, R.E. (eds) *Advances in Seafood Biochemistry, Composition and Quality*. Technomic Publishing Co., Lancaster, Pennsylvania, pp. 305–360.

Haard, N.F. (1994) Protein hydrolysis in seafood, In: Shahidi, F. and Botta, R. (eds) *Seafoods: Chemistry, Processing Technology and Quality*. Chapman and Hall, London, pp. 10–33.

Haard, N.F. and Arcilla, R. (1985) Precursors of Maillard browning in dehydrated squid, *Illex illecebrosus*. *Canadian Institute of Food Science and Technology Journal* 18, 326–331.

Haard, N.F., Feltham, L.A.W., Helbig, N. and Squires, J. (1982) Modification of proteins with enzymes from the marine environment. In: Feeney, R. and Whitaker, J. (eds) *Modification of Proteins in Food, Pharmaceutical, and Nutritional Sciences*. Advances in Chemistry Series 198. American Chemical Society, Washington DC, pp. 223–244.

Hallett, I.C. and Bremner, H.A. (1988) Fine structure of the myocommata–muscle fibre junction in hoki (*Macruronus novaezelandiae*). *Journal of the Science of Food and Agriculture* 44, 245–261.

Hameed, K.S. and Haard, N.F. (1985) Isolation and characterization of cathepsin C from Atlantic short finned squid, *Illex illecebrosus*. *Comparative Biochemistry and Physiology* 82B, 241–246.

Hamoir, G. (1955) Fish proteins. *Advances in Protein Chemistry* 10, 227.

Han, T.-J. and Liston, J. (1987) Lipid peroxidation and phospholipid hydrolysis in fish muscle microsomes and frozen fish. *Journal of Food Science* 52, 294–296.

Hara, K., Suzumatsu, A. and Ishihara, T. (1988) Purification and characterization of cathepsin B from carp ordinary muscle. *Bulletin of the Japanese Society of Scientific Fisheries* 54, 1243–1252.

Hebard, C.E., Flick, G.J. and Martin, R.E. (1982) Occurrence and significance of trimethylamine oxide and its derivatives in fish and shellfish. In: Martin, R.E., Flick, G.J., Hebard, C.E. and Ward, D.R. (eds) *Chemistry and Biochemistry of Marine Food Products*. AVI Publishing Co., Westport, Connecticut, pp. 149–304.

Hendrickson, G.L., Fritzsche, R.A. and Fisher, R.A. (1986) Histology and ultrastructure of the 'jellied' condition in Dover sole (*Microstomus pacificus*). *California Sea Grant College Program*, University of California, A-032, La Jolla, pp. 230–234.

Henry, T. and Ferguson, A. (1985) Kinetic studies on the lactate dehydrogenase (LDH-5) isozymes of brown trout *Salmo trutta* L. *Comparative Biochemistry and Physiology* 82B, 95–98.

Hirano, T., Nakamura, H. and Suyama, M. (1980) Quality of wild and cultured ayu – II. Seasonal variation in proximate composition. *Bulletin of the Japanese Society of Scientific Fisheries* 46, 75–78.

Hjelmeland, K. and Raa, J. (1982) Characterization of two trypsin type isozymes isolated from the Arctic fish capelin (*Mallotus villosus*). *Comparative Biochemistry and Physiology* 71B, 557–562.

Hofmann, K. and Hamm, R. (1978) Sulfhydryl and disulfide groups in meats. *Advances in Food Research* 24, 1–111.

Hsieh, R.J. and Kinsella, J.E. (1986) Lipoxygenase catalyzed oxidation of the N-6 and N-3 polyunsaturated fatty acids: Relevance to and activity in fish tissue. *Journal of Food Science* 51, 940–945.

Hultin, H.O. (1985) Characteristics of muscle tissue. In: Fennema, O. (ed.) *Food Chemistry*. Marcel Dekker, New York, pp. 725–790.

Ilgner, R.H. and Woods, A.E. (1985) Purification, physical properties, and kinetics of peroxidases from freshwater crayfish (genus *Orconectes*). *Comparative Biochemistry and Physiology* 82B, 433–440.

Inaba, T., Shindo, N. and Fuji, M. (1976) Purification of cathepsin B from squid liver. *Agriculture and Biological Chemistry* 40, 1159–1165.

Iwasaki, M. and Harada, R. (1985) Proximate and amino acid composition of the roe and muscle of selected marine species. *Journal of Food Science* 50, 1585–1587.

Iwata, K., Kabashi, K. and Hase, J. (1973) Studies on muscle alkaline protease – I. Isolation, purification and some physicochemical properties of an alkaline protease from carp muscle. *Bulletin of the Japanese Society of Scientific Fisheries* 39, 1325–1337.

Iwata, K., Kabashi, K. and Hase, J. (1974) Studies on alkaline proteinases III. Distribution of alkaline proteinases in muscle of freshwater fish, marine fish and in internal organs of carp. *Bulletin of the Japanese Society of Scientific Fisheries* 40, 201–209.

Izquerdo-Pulido, M., Hatae, K. and Haard, N.F. (1992) Nucleotide catabolism and changes in texture during ice storage of cultured sturgeon, *Acipenser transmontanus*. *Journal of Food Biochemistry* 16, 173–192.

Jany, K.D. (1976) Studies on the digestive enzymes of the stomachless bonefish *Carassios auratus* Gibelio (Bloch): Endopeptidases. *Comparative Biochemistry and Physiology* 53B, 31–38.

Jhaveri, S.N., Karakoltsidis, P.A., Montecalvo, J. and Constantinides, S.M. (1984) Chemical composition and protein quality of some Southern New England marine species. *Journal of Food Science* 49, 110–113.

Jiang, S.T., Hwang, D.C. and Chen, C.S. (1988) Denaturation and change in SH group of actomyosin from milkfish (*Chanos chanos*) during storage at −20°C. *Journal of Agriculture and Food Chemistry* 36, 433–437.

Jiang, S.T., Tsao, C.Y., Wang, Y.T. and Chen, C.S. (1990) Purification and properties of proteases from milkfish muscle (*Chanos chanos*). *Journal of Agriculture and Food Chemistry* 38, 1458–1463.

Johnston, I.A., Frearson, N. and Goldspink, G. (1973) Effects of environmental temperature on the properties of myofibrillar adenosine triphosphate from various species of fish. *Biochemical Journal* 133, 735–738.

Johnston, I.A., Walesky, N.H., Davison, W. and Goldspink, G. (1975) Temperature adaption in myosin of antarctic fish. *Nature (London)* 254, 74–75.

Josephson, D.B. and Lindsey, R.C. (1986) Enzymic generation of fresh fish volatile aroma compounds. In: Parliment, T.H. and Croteau, R. (eds) *Biogenesis of Aromas*, ACS Symposium No. 317. American Chemical Society, Washington DC, pp. 201–219.

Kalac, J. (1978) Studies on herring (*Clupea harengus* L.) and capelin (*Mallotus villosus*, L.) pyloric ceca proteases III. Characterization of the anionic fractions of chymotrypsins. *Biologia* (Bratislava) 33, 939–945.

Kanoh, S., Watabe, S. and Hashimoto, K. (1983) Isolation and some physicochemical properties of requiem shark myosin. *Bulletin of the Japanese Society of Scientific Fisheries* 49, 757–763.

Kawai, Y., Uematsu, S. and Shinano, H. (1992) Effect of heat-treatment on some physicochemical properties and emulsifying activity of carp sarcoplasmic proteins. *Bulletin of the Japanese Society of Scientific Fisheries* 58, 1327–1331.

Kim, K. and Haard, N.F. (1992) Degradation of proteoglycans in the skeletal muscle of Pacific rockfish (*Sebastes* sp.) during ice storage. *Journal of Muscle Foods* 3, 103–121.

Kim, K.-H. and Tchen, T.T. (1962) Tyrosinase of the goldfish *Carassius auratus* L. I. Radioassay and properties of the enzyme. *Biochimica et Biophysica Acta* 59, 569–576.

Kimura, S., and Tanaka, H. (1986) Partial characterization of muscle collagens from prawns and lobsters. *Journal of Food Science* 51, 330–332, 339.

Kimura, S., Akashi, Y. and Kubota, M. (1978) Carp connectin: Amino acid composition. *Journal of Biochemistry* 83, 321–323.

Kimura, S., Akashi, Y. and Kubota, M. (1979a) Carp connectin: Its content in white and dark muscles. *Bulletin of the Japanese Society of Scientific Fisheries* 45, 237–240.

Kimura, S., Fujii, K., Kubota, M. and Maruyama, K. (1979b) Carp connectin: Reducible cross links in native fibrils. *Bulletin of the Japanese Society of Scientific Fisheries* 45, 241–243.

Kimura, S., Ohno, Y., Miyauchi, Y. and Uchida, N. (1987) Fish skin type I collagen: wide distribution of an $\alpha 3$ subunit in teleosts. *Comparative Biochemistry and Physiology* 88B, 27–34.

Kimura, S., Zhu, X.-P., Matsui, R., Shijoh, M. and Takamizawa, S. (1988) Characterization of fish muscle type I collagen. *Journal of Food Science* 53, 1315–1318.

Kinoshita, M., Toyohara, H. and Shimizu, Y. (1990) Diverse distribution of four distinct types of modori (gel degradation)-inducing proteinases among fish species. *Bulletin of the Japanese Society of Scientific Fisheries* 56, 1485–1492.

Kinsella, J.E., German, J.B. and Shetty, J. (1985) Uricase from fish liver: Isolation and some properties. *Comparative Biochemistry and Physiology* 82B, 621–624.

Kishi, H., Nozawa, H. and Seki, N. (1991) Reactivity of muscle transglutaminase on carp myofibrils and myosin B. *Bulletin of the Japanese Society of Scientific Fisheries* 57, 1203–1210.

Kitamikado, M. and Tachino, S. (1960) Studies on the digestive enzymes of rainbow trout II. Proteases. *Bulletin of the Japanese Society of Scientific Fisheries* 26, 685–694.

Konno, K. (1978) Two calcium regulation systems in squid (*Omnastrephes sloani pacificus*) muscle. *Journal of Biochemistry* 84, 1431–1440.

Konosu, S., Hamoir, G. and Pechere, J.-F. (1965) Carp myogens of white and red muscles: Properties and amino acid composition of the main low molecular weight components of white muscle. *Biochemical Journal* 96, 98–112.

Kubo, S. (1957) On fish tropomyosin. I. Isolation and amino acid composition. *Hokkaido Daigaku Faculty of Fisheries Bulletin* 8, 147–157.

Kubota, M. and Ohnuma, A. (1970) Studies on bonita pepsin. I. Purification of pepsin. *Bulletin of the Japanese Society of Scientific Fisheries* 36, 1147–1156.

Lee, Y.-Z., Singh, R.P. and Haard, N.F. (1992) Changes in freshness of Chilipepper rockfish (*Sebastes goodei*) during storage measured by biochemical biosensors. *Journal of Food Biochemistry* 16, 119–129.

Leger, C. (1972) Essai de purification de la lipase du tissu intercaecal de la truite

(*Salmo gairdneri* Rich). *Annales de Biologie Animale Biochimie et Biophysique* 12, 341–345.

Lehky, P. and Stein, E.A. (1979) Perch muscle parvalbumin: General characterization and magnesium binding properties. *Comparative Biochemistry and Physiology* 63B, 253–259.

Lehrer, S.B. and Daul, C.B. (1992) Allergenic reactions to seafood: Identification of allergens. In: Flick, G.J. and Martin, R.E. (eds) *Advances in Seafood Biochemistry, Composition and Quality*. Technomic Publishing Co., Lancaster, Pennsylvania, pp. 185–197.

Lehrer, S.B. and McCants, M.L. (1987) Reactivity of IgE antibodies with crustacea and oyster allergens: Evidence for common antigenic structures. *Journal of Allergy and Clinical Immunology* 80, 133–139.

Light, N. and Champion, A.E. (1984) Characterization of muscle epimysium, perimysium and endomysium collagens. *Biochemical Journal* 219, 1017–1026.

Love, R.M. (1980) *The Chemical Biology of Fishes*, Vol. 2. Academic Press, New York, pp. 277–278.

Love, R.M. (1988) *The Food Fishes: Their Intrinsic Variation and Practical Implications*. Farrand Press, London, p. 59.

Low, P.S., Bada, J.L. and Somero, G.N. (1973) Temperature adaptation of enzymes: roles of free energy, the enthalpy and entropy of activation. *Proceedings of the National Academy of Sciences of the USA* 70, 430–432.

Lundstrom, R.C., Correia, F.F. and Wilhelm, K.A. (1982) Enzymatic dimethylamine and formaldehyde production in minced American plaice and blackback flounder mixed with red hake TMAO-ase active fraction. *Journal of Food Science* 47, 1305.

Lushchak, V.I. (1991) Characteristics of microsome-bound lactate dehydrogenase from skate white muscles. *Biokhimiya* 56, 2173–2180.

Macdonald, R.E. and Hultin, H.O. (1987) Some characteristics of the enzymic lipid peroxidation system in the microsomal fraction of flounder skeletal muscle. *Journal of Food Science* 52, 15–21.

Mackie, I.M. and Connell, J.J. (1964) Preparation and properties of purified cod myosin. *Biochimica et Biophysica Acta* 93, 544–552.

Mackie, I.M. and Ritchie, A.H. (1974) Free amino acids of fish flesh. *Proceedings of the IVth International Congress of Food Science and Technology* 1, 29–38.

Mamaeva, E.M., Bikbov, T.M., Gyn, E.E., Safronova, A.M., Kminchenko, V.M. and Plashina, I.G. (1991) Biological value and assimilation of sarcoplasmic protein concentrates from commercial fish products and the effect of their consumption on rats. *Voprosy Pitaniya* 2, 52–54.

Martinez, I., Solberg, C., Lauritzen, K. and Ofstad, R. (1992) Two-dimensional electrophoretic analyses of cod (*Gadus morhua*, L.) whole muscle proteins, water-soluble fraction and surimi. Effect of the addition of $CaCl_2$ and $MgCl_2$ during the washing procedure. *Applied and Theoretical Electrophoresis* 2, 201–206.

Martone, C.B., Busconi, L., Folco, E.J., Trucco, R.E. and Sanchez, J.J. (1986) A simplified myosin preparation from marine fish species. *Journal of Food Science* 51, 1554–1555.

Maruyama, K. and Ebashi, S. (1970) Regulatory protein in muscle. In: Briskey, E.J.,

Cassers, R.G. and Marsh, B.B. (eds) *Physiology and Biochemistry of Muscle as Food, 2*. University of Wisconsin Press, Madison, pp. 373–382.

Masui, R., Yamauchi, H., Suganuma, H., Ishida, M. and Kimura, S. (1989) Subunit composition of type I collagen from fish in Salmonoidei. *Bulletin of the Japanese Society of Scientific Fisheries* 55, 1421–1426.

Matsui, T., Shimizu, C. and Matsuura, F. (1975) Studies on metmyoglobin reducing systems in the muscle of bluewhite dolphin. II. Purification and some physico-chemical properties of ferrimyoglobin reductase. *Bulletin of the Japanese Society of Scientific Fisheries* 41, 771.

McClain, P.E., Wiley, E.R. and McCague, K.E. (1971) Species variations in the cross-linking characteristics of collagen from the intramuscular connective tissues of striated muscles (*Bos taurus, Ovis aries, Sus scofra*). *International Journal of Biochemistry* 2, 167.

Merrett, T.G., Bar-Eli, E. and Van Vunakis, H. (1969) Pepsinogens A, C, and D from the smooth dogfish. *Biochemistry* 8, 3696–3702.

Mietz, J.L. and Karmas, E. (1977) Chemical quality index for canned tuna as determined by high pressure liquid chromatography. *Journal of Food Science* 42, 155.

Misima, T., Yokoyama, T., Yano, K. and Tsuchimoto, M. (1990) The influence of rearing water temperature on the properties of Ca^{+2}- and Mg^{+2}-ATPase activity on carp myofibril. *Bulletin of the Japanese Society of Scientific Fisheries* 56, 477–487.

Miyauchi, Y. and Kimura, S. (1990) Characterization of an $\alpha 3$ chain from carp skin type I collagen. *Bulletin of the Japanese Society of Scientific Fisheries* 56, 1509–1514.

Morey, K.S., Satterlee, L.D. and Brown, W.D. (1982). Protein quality of fish stored in modified atmospheres as predicted by the C-PER assays. *Journal of Food Science* 47, 1399–1400.

Murakami, K. and Noda, M. (1981) Studies on proteinases from the digestive organs of sardine. I. Purification and characterization of three alkaline proteinases from the pyloric ceca. *Biochimica et Biophysica Acta* 658, 17–26.

Nakagawa, T., Watabe, S. and Hashimoto, K. (1988a) Electrophoretic analysis of sarcoplasmic proteins from fish muscle on polyacrylamide gels. *Bulletin of the Japanese Society of Scientific Fisheries* 54, 993–998.

Nakagawa, T., Watabe, S. and Hashimoto, K. (1988b) Identification of three major components in fish sarcoplasmic proteins. *Bulletin of the Japanese Society of Scientific Fisheries* 54, 999–1004.

Nettleton, J.A. 1985. *Seafood Nutrition: Facts, Issues and Marketing of Nutrition in Fish and Shellfish*. Osprey Books, Huntington, New York, p. 26.

Nimmo, I.A. and Spalding, C.M. (1985) The glutathione S transferase activity in the kidney of rainbow trout (*Salmo gairdneri*). *Comparative Biochemistry and Physiology* 82B, 91–94.

Noda, M. and Murakami, K. (1981) Studies on proteinases from the digestive organs of sardine. II. Purification and characterization of two acid proteinases from the stomach. *Biochimica et Biophysica Acta* 658, 27–34.

Norris, E.R. and Elam, D.W. (1940) Preparation and properties and crystallization of tuna pepsin. *Journal of Biological Chemistry* 204, 673–680.

Numakura, T., Seki, N., Kimura, I., Toyoda, K., Fujita, T., Takama, K. and Arai, K. (1985) Cross-linking reaction of myosin in the fish paste during setting (Suwari). *Bulletin of the Japanese Society of Scientific Fisheries* 51, 1559–1565.

Obinata, T., Maruyama, K., Sugita, H., Kohama, K. and Ebashi, S. (1981) Dynamic aspects of structural proteins in vertebral skeletal muscle. *Muscle Nerve* 4, 456–488.

Odense, P.H., Leung, T.C., Green, W.A. and Dingle, J.R. (1969) The isolation of cod muscle tropomyosin by heat treatment. *Biochimica et Biophysica Acta* 188, 124–131.

Olley, J., Pirie, R. and Watson, H. (1962) Lipase and phospholipase activity in fish skeletal muscle and its relationship to protein denaturation. *Journal of the Science of Food and Agriculture* 13, 501.

Owen, T.G. and Wiggs, A.J. (1971) Thermal compensation in the stomach of brook trout (*Salvelinus fontinalis*). *Comparative Biochemistry and Physiology* 40B, 465–473.

Parkin, K.I. and Hultin, H.O. (1981) A membrane fraction from red hake muscle catalyzing the conversion of TMAO to dimethylamine and formaldehyde. *Science and Technology of Refrigeration* 4–6, 475.

Pechere, J.-F. (1977) The significance of parvalbumins among muscle calciproteins. In: Wasserman, R.H. (ed.) *Calcium-Binding Proteins and Calcium Function*. North Holland, New York, pp. 213–221.

Pechere, J.-F., Demaille, J. and Capony, J. (1971) Muscular parvalbumins: Preparative and analytical methods of general applicability. *Biochimica et Biophysica Acta* 236, 391–408.

Pigott, G.M. and Tucker, B.W. (1990) *Seafood. Effects of Technology on Nutrition*. Marcel Dekker, Inc., New York, pp. 41–42.

Piront, A., Hamoir, G. and Crocaert, R. (1968) Proteinic composition of the low ionic strength extracts of *Tilapia macrochir* Boulanger white muscle. *Archives of International Physiology and Biochemistry* 76, 1–25.

Prota, G., Ortonne, J.P., Voulet, C., Khatchadourian, C., Nardi, G. and Palumbo, A. (1981) Occurrence and properties of tyrosinase in the ejected link of cephalopods. *Comparative Biochemistry and Physiology* 68B, 415–419.

Pyeun, J.H., Hashimoto, K. and Matsuura, F. (1973) Isolation and characterization of abalone paramyosin. *Bulletin of the Japanese Society of Scientific Fisheries* 39, 395.

Ramakrishna, M., Hultin, H.O. and Atallah, M.T. (1987) A comparison of dogfish and bovine chymotrypsins in relation to protein hydrolysis. *Journal of Food Science* 52, 1198–1202.

Ramshaw, J.A., Werkmeister, J.A. and Bremner, H.A. (1988) Characterization of type I collagen from the skin of blue grenadier (*Macruronus novaezelandiae*). *Archives of Biochemistry and Biophysics* 267, 497–502.

Rehbein, H. and Lessen, E. van (1989) Parvalbumin in salmon and trout species. *Archiv für Fischereiwissenschaft* 39, 147–162.

Ruano, A.R., Riano, J.L.A., Amil, M.R. and Santos, M.J.H. (1985) Some enzymatic properties of NAD^+-dependent glutamate dehydrogenase of mussel hepatopancreas (*Mytilus edulis* L.) – Requirement of ADP. *Comparative Biochemistry and Physiology* 82B, 197–202.

Saad, F., Kominz, D.R. and Laki, K. (1959) A study of the tropomyosins of three cold-blooded vertebrates of different classes. *Journal of Biological Chemistry* 234, 551–555.

Sakaguchi, M., Murata, M. and Kawai, A. (1982) Changes in free amino acids and creatine contents in yellowtail (*Seriola quinqueradiata*) muscle during ice storage. *Journal of Food Science* 47, 1662–1666.

Sakaizumi, M. (1985) Species-specific expression of parvalbumins in the genus *Oryzias* and its related species. *Comparative Biochemistry and Physiology* 80B, 499–505.

Sanchez-Chiang, L. and Ponce, O. (1981) Gastricsinogens and gastricsins from *Merluccius gayi* – Purification and properties. *Comparative Biochemistry and Physiology* 68B, 251–257.

Sano, T., Noguchi, S.F., Tsuchiya, T. and Matsumoto, J.J. (1986) Contribution of paramyosin to marine meat gel characteristics. *Journal of Food Science* 51, 946–950.

Sano, T., Noguchi, S.F., Tsuchiya, T. and Matsumoto, J.J. (1989) Contribution of tropomyosin to fish gel characteristics. *Journal of Food Science* 54, 258–264.

Sato, K., Yoshinaka, R., Itoh, Y. and Sato, M. (1986a) Molecular species of collagen in the intramuscular connective tissue of fish. *Comparative Biochemistry and Physiology* 92B, 87–91.

Sato, K., Yoshinaka, R., Sato, M. and Ikeda, S. (1986b) A simplified method for determining collagen in fish muscle. *Bulletin of the Japanese Society of Scientific Fisheries* 52, 889–893.

Sato, K., Yoshinaka, R., Sato, M. and Shimizu, Y. (1986c) Collagen content in the muscle of fishes in association with their swimming movement and meat texture. *Bulletin of the Japanese Society of Scientific Fisheries* 52, 1595–1600.

Sato, K., Yoshinaka, R., Sato, M., Itoh, Y. and Shimizu, Y. (1988) Isolation of types I and V collagens from carp muscle. *Comparative Biochemistry and Physiology* 90B(1), 155–158.

Sato, K., Yoshinaka, R. and Sato, M. (1989a) Hydroxyproline content in the acid-soluble collagen from muscle of several fishes. *Bulletin of the Japanese Society of Scientific Fisheries* 55, 1467.

Sato, K., Yoshinaka, R., Sato, M. and Tomita, J. (1989b) Biochemical characterization of collagen in mycommata and endomysium fractions of carp and spotted mackerel muscle. *Journal of Food Science* 54, 1511–1514.

Sato, K., Ohashi, C., Ohtsuki, K. and Kawabata, M. (1991) Type V collagen in trout (*Salmo gairdneri*) muscle and its solubility change during chilled storage of muscle. *Journal of Agriculture and Food Chemistry* 39, 1222–1225.

Savagaon, K.A. and Sreenivasan, A. (1978) Activation mechanism of prephenyloxidase in lobster and shrimp. *Fishery Technology* 15, 49–55.

Schmidt, H. and Wagener, G. (1990) Glycogen phosphorylase in fish muscle: demonstration of three inter-convertible forms. *American Journal of Physiology* 258C, 344–351.

Schmitt, A. and Siebert, G. (1967) Differenzierung aliphatischer Dipeptidasen aus Dorschmuskel. *Hoppe-Seylers Zeitschrift für Physiologische Chemie* 348, 1009–1016.

Schwimmer, S. (1981) *Source Book of Food Enzymology*. AVI Publishing Co., Westport, Connecticut, 498 pp.

Scott, E.M. and Harrington, J.P. (1990) Comparative studies of catalase and superoxide dismutase activity within salmon erythrocytes. *Comparative Biochemistry and Physiology* 95B, 91–93.

Seki, N. and Watanabe, T. (1984) Connectin and its post-mortem changes in fish muscle. *Journal of Biochemistry (Tokyo)* 95, 1161–1167.

Seki, N., Kitao, M., Konno, K. and Arai, K. (1973) On the purification of actin from fish muscle. *Bulletin of the Japanese Society of Scientific Fisheries* 39, 1211–1219.

Shenouda, S.Y.K. and Pigott, G.M. (1975) Lipid–protein interaction during aqueous extraction of fish protein: Fish actin preparation and purification. *Journal of Food Science* 40, 520–522.

Sherekar, S.V., Gore, M.S. and Ninjoor, V. (1988) Purification and characterization of cathepsin B from the skeletal muscle of freshwater fish, *Tilapia mossambia. Journal of Food Science* 53, 1018–1023.

Shewfelt, R.L. (1981) Fish muscle lipolysis – A review. *Journal of Food Biochemistry* 5, 79–100.

Shewfelt, R.L., McDonald, R.E. and Hultin, H.O. (1981) Effect of phospholipid hydrolysis on lipid oxidation in flounder muscle microsomes. *Journal of Food Science* 46, 1297–1301.

Shimizu, C. and Matsura, F. (1971) Occurrence of new enzyme reducing metmyoglobin in dolphin muscle. *Agricultural and Biological Chemistry* 35, 465.

Sidwell, V. (1981) *Chemical and Nutritional Composition of Finfishes, Whales, Crustaceans, Mollusks, and their Products*, NOAA Technical Memorandum, NMFS F/SEC-11. United States Department of Commerce, Washington, DC, 432 pp.

Siebert, G. (1962) Enzymes in marine fish muscle and their role in fish spoilage. In: Heen, E. and Kreuzer, R. (eds) *Fish in Nutrition*. London Fishing News, London, p. 80.

Sikorski, Z.E., Scott, D.N. and Buisson, D.H. (1984) The role of collagen in the quality and processing of fish. *Critical Reviews in Food Science and Nutrition* 20, 301.

Sikorski, Z.E., Kolakowska, A. and Pan, B.S. (1990) The nutritive composition of the major groups of marine food organisms. In: Sikorski, Z.E. (ed.) *Seafood Resources, Nutrition, Composition and Preservation*. CRC Press Inc., Boca Raton, Florida, pp. 29–54.

Simpson, B.K. and Haard, N.F. (1984) Trypsin from Greenland cod *Gadus ogac*. Kinetic and thermodynamic characteristics. *Canadian Journal of Biochemistry and Cell Biology* 62, 894–900.

Simpson, B.K. and Haard, N.F. (1987a) Cold-adapted enzymes from fish. In: Knorr, D. (ed.) *Food Biotechnology*. Marcel Dekker, Inc., New York, pp. 495–527.

Simpson, B.K. and Haard, N.F. (1987b) Trypsin and trypsin-like enzymes from the stomachless cunner. Kinetic and other physical properties. *Journal of Agriculture and Food Chemistry* 35, 652–656.

Simpson, B.K., Marshall, M.R. and Otwell, W.S. (1988) Phenoloxidases from pink

and white shrimp: kinetic and other properties. *Journal of Food Biochemistry* 12, 205–217.
Simpson, B.K., Smith, J.P. and Haard, N.F. (1991) Marine enzymes. In: *Encyclopedia of Food Science and Technology*. J. Wiley & Sons, New York, pp. 1645–1653.
Skaara, T. and Regenstein, J.M. (1990) The structure and properties of myofibrillar proteins in beef, poultry, and fish. *Journal of Muscle Foods* 1, 269–291.
Slabyi, B.M. and Hultin, H.O. (1982) Lipid peroxidation by microsomal fraction isolated from light and dark muscle of herring (*Clupea harengus*). *Journal of Food Science* 47, 1395–1398.
Smith, D.M. (1991) Factors influencing heat-induced gelation of muscle proteins. In: Parris, N. and Barford, R. (eds) *Interactions of Food Proteins*, American Chemical Society Series 454. American Chemical Society, Washington DC, pp. 243–256.
Snook, J.T. (1984) *Nutrition, A Guide to Decision Making*, Prentice-Hall, Englewood Cliffs, New Jersey, 120 pp.
Somero, G. (1986) From dogfish to dogs: Trimethylamines protect proteins from urea. *News in Physiological Sciences* 1, 9–12.
Squires, J., Haard, N.F. and Feltham, L.A.W. (1986) Pepsin isozymes from Greenland cod, *Gadus ogac*. 1. Purification and physical properties. *Canadian Journal of Biochemistry and Cell Biology* 65, 205–209.
Stirling, W. (1884) On the ferments or enzymes of the digestive tract in fishes. *Journal of Anatomy and Physiology* 18, 426–435.
Summers, N.M. (1967) Cuticle sclerotization and blood phenoloxidase in the fiddler crab *Uca pugnax*. *Comparative Biochemistry and Physiology* 23, 129–138.
Surette, M.E., Gill, T.A. and LeBlanc, P.J. (1988) Biochemical basis of post-mortem nucleotide catabolism in cod (*Gadus morhua*) and its relationship to spoilage. *Journal of Agriculture and Food Chemistry* 36, 19.
Surette, M., Gill, T. and MacLean, S. (1990) Purification and characterization of purine nucleoside phosphorylase from *Proteus vulgaris*. *Applied and Environmental Microbiology* 56(5), 1435–1439.
Suzuki, T. (1981) *Fish and Krill Protein Processing Technology*. Applied Science Publishers, London, pp. 10–13.
Taguchi, T., Tanaka, M. and Suzuki, K. (1981) On heat coagulation of water soluble proteins from deep sea fishes. *Nippon Suisan Gakkaishi* 47(4), 551–554.
Takashi, R. (1973) Studies on muscular proteins of fish – VIII. Comparative studies on the biochemical properties of highly purified myosins from fish dorsal and rabbit skeletal muscle. *Bulletin of the Japanese Society of Scientific Fisheries* 39, 197–205.
Takashi, R., Morozuka, T. and Arai, K. (1974) Light chains from fish dorsal and rabbit skeletal muscles. *Bulletin of the Japanese Society of Scientific Fisheries* 40, 1155–1161.
Tanji, M., Kageyama, T. and Takahashi, K. (1988) Tuna pepsinogens and pepsins: Purification, characterization and amino-terminal sequences. *European Journal of Biochemistry* 177, 251–259.
Thebault, M.T., Bernicard, A. and Lennon, J.F. (1981) Lactate dehydrogenase

from the caudal muscle of the shrimp *Palaemon serratus*. Purification and characterization. *Comparative Biochemistry and Physiology* 68B, 65–70.

Thompson, H.C. and Thompson, M.H. (1968) Isolation and amino acid composition of collagens of the white shrimp *Penaeus setiferus*. *Comparative Biochemistry and Physiology* 27, 127.

Ting, C.-Y., Montgomery, M. and Anglemier, A.F. (1968) Partial purification of salmon muscle cathepsins. *Journal of Food Science* 33, 617–621.

Tokunaga, T. (1965) Studies on the development of dimethylamine and formaldehyde in Alaska pollack. II. Factors affecting the formation of dimethylamine and formaldehyde. *Bulletin of the Hokkaido Regional Fisheries Research Laboratory* 30, 90.

Tomioka, H., Yamaguchi, K., Hashimoto, K. and Matsuura, F. (1974) Studies on myosin of spiny lobster – I. Isolation and physico-chemical properties. *Bulletin of the Japanese Society of Scientific Fisheries* 40, 1269–1275.

Toom, P.M., Ward, C.F. and Weber, J.R. (1982) Identification of fish species by isoelectric focusing. In: Martin, R.E., Flick, G.F., Hebard, C.E. and Ward, D.R. (eds) *Chemistry and Biochemistry of Marine Food Products*. AVI Publishing Co., Westport, Connecticut, pp. 51–65.

Tsuchiya, T., Suzuki, H. and Matsumoto, J.J. (1977) Physicochemical and biochemical properties of squid actin. *Bulletin of the Japanese Society of Scientific Fisheries* 43, 1233–1240.

Tsuchiya, T., Yamada, N. and Matsumoto, J.J. (1978) Extraction and purification of squid myosin. *Bulletin of the Japanese Society of Scientific Fisheries* 44, 175–179.

Tsuchiya, T., Fukuhara, S. and Matsumoto, J.J. (1980a) Physico-chemical properties of squid paramyosin. *Bulletin of the Japanese Society of Scientific Fisheries* 46, 197–200.

Tsuchiya, T., Shinohara, T. and Matsumoto, J.J. (1980b) Physico-chemical properties of squid tropomyosin. *Bulletin of the Japanese Society of Scientific Fisheries* 46, 893–896.

Tsukuda, N. (1970) Studies on the discoloration of red fishes – VI. Partial purification and specificity of the lipoxygenase-like enzyme responsible for carotenoid discoloration in fish skin. *Bulletin of the Japanese Society of Scientific Fisheries* 36, 725.

Ueda, T., Shimizu, Y. and Simidu, W. (1967) Extractability of fish actin. *Bulletin of the Japanese Society of Scientific Fisheries* 33, 59–64.

Ueno, R., Ikeda, S., Sakanaka, K. and Horiguchi, Y. (1988) Characterization of pepstatin insensitive protease in mackerel muscle. *Bulletin of the Japanese Society of Scientific Fisheries* 54, 699–707.

Waite, J.H. (1983) Evidence for a repeating 3,4-dihydroxyphenylalanine and hydroxyproline containing decapeptide in the adhesive protein of the mussel, *Mytilus edulis* L. *Journal of Biochemistry* 258, 2911–2915.

Waite, J.H. (1985) Catechol oxidase in the byssus of the common muscle, *Mytilus edulis* L. *Journal of the Marine Biology Association, UK* 65, 359–371.

Watabe, S. and Hashimoto, K. (1980) Myosins from white and dark muscles of mackerel. Some physico-chemical and enzymatic properties. *Journal of Biochemistry* 87, 1491–1499.

Watabe, S., Dinh, T., Ochiai, Y. and Hashimoto, K. (1983) Immunochemical specificity of myosin light chains from mackerel ordinary and dark muscles. *Journal of Biochemistry* 94, 1409–1420.

Wheaton, F.W. and Lawson, T.B. (1985) *Processing Aquatic Food Products.* John Wiley and Sons, New York.

Whitmore, D.H. (1986) Identification of sunfish species by muscle protein isoelectric focusing. *Comparative Biochemistry and Physiology* 84B, 177–180.

Wickes, M.A. and Morgan, R.P. (1976) Effects of salinity on three enzymes involved in amino acid metabolism from the American oyster *Crassostrea virginica. Comparative Biochemistry and Physiology* 53B, 339–343.

Wiggs, A.J. (1974) Seasonal changes in the thyroid proteinase of a teleost fish, the burbot, *Lota lota* L. *Canadian Journal of Zoology* 52, 1071–1078.

Winkler, M., Pilhofer, G. and German, J.B. (1992) Stereochemical specificity of the *N*-9-lipoxygenase of fish gill. *Journal of Food Biochemistry* 15, 437–448.

Woychik, J.H. and Dower, H.J. (1990) Characterization of parvalbumins of Alaska and Atlantic pollock and Gulf and Atlantic menhaden. *IFT National Meeting*, Anaheim, California. Abstract No. 757. Institute of Food Technologists, Chicago, Illinois.

Wright, J.L.C. (1992) Shellfish toxins: a Canadian perspective. In: Bligh, E.G. (ed.) *Seafood Science and Technology*. Blackwell Scientific Publications, Cambridge, Massachusetts, pp. 366–375.

Yamada, A., Takano, K. and Kamoi, I. (1991) Purification and properties of amylases from tilapia intestine. *Bulletin of the Japanese Society of Scientific Fisheries* 57, 1903–1909.

Yamagata, M., Horimito, M. and Nagaoka, C. (1971) Accuracy of predicting occurrence of greening in tuna based on content of trimethylamine oxide. *Journal of Food Science* 36, 55–57.

Yoshinaka, R., Sato, M. and Ikeda, A. (1978) Distribution of collagenase in the digestive organs of some teleosts. *Bulletin of the Japanese Society of Scientific Fisheries* 40, 263–268.

Yoshinaka, R., Sato, M., Suzuki, T. and Ikeda, S. (1984) Enzymatic characterization of anionic trypsin of the catfish (*Parasilurus astus*). *Comparative Biochemistry and Physiology* 77B, 1–6.

Yoshinaka, R., Sato, K., Anbe, H., Sato, M. and Shimizu, Y. (1988) Distribution of collagen in body muscle of fishes with different swimming modes. *Comparative Biochemistry and Physiology* 89B, 147–151.

Zajicek, P. (1991) Enzyme polymorphism of freshwater trypanosomes and its use for strain identification. *Parasitology* 102, 221–224.

Zeef, A.H. and Dennison, C. (1988) A novel cathepsin from the mussel (*Perna perna*) (Linne). *Comparative Biochemistry and Physiology* 90B, 201–204.

4

Composition and Nutritive Value of Fish and Shellfish Lipids

R.G. Ackman

Canadian Institute of Fisheries Technology, Technical University of Nova Scotia, Halifax, Nova Scotia, Canada.

Introduction

The health benefits of marine foods are well known (Nettleton, 1985; 1987; Kinsella, 1986; 1987; Somogyi and Hötzel, 1990), but good data on their fatty acids are scattered in a large literature, mostly because the lipid chemist is as narrow-minded as any other type of scientist. Probably thousands of reports exist for recovery and analysis of fatty materials of marine origin. Unfortunately many are concerned only with fat as energy, some focus on sterols, and most are divorced from other nutritional elements such as protein or vitamins and minerals. There is a compensating factor, however. Edible fish muscle is nominally 18% protein (see below and Chapter 3) and usually contains 1 to 2% ash. 'Whole fish' data have more ash and a little less protein (Nunes *et al.*, 1992). The balance of 80% of the wet weight of muscle is made up of lipid and water, and in high-fat fish knowledge of the water content will give an approximation of the fat content by difference. Before tabulating lipids and fatty acids for their nutritional value it is useful to review the types of lipids in fish muscle.

Lipid Types and Their Variations

To most people animal fat is the visible triglyceride they see as a white solid on parts of uncooked or cooled, cooked meats. In fish the corresponding depot fat is always liquid at room temperature and is seldom even visible to the consumer. An exception is in the belly flaps of salmon steaks. These are often seen left on plates in restaurants because they are too high in fat (30–50%) to be palatable. The 'dark' muscle of some fish such as mackerel

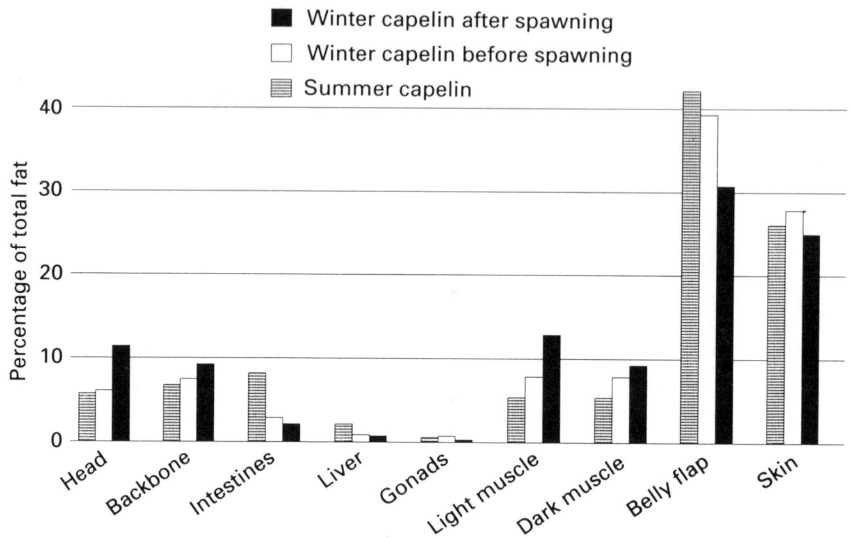

Fig. 4.1. Distribution of fats (essentially triglycerides) in the body of capelin *Mallotus villosus*. Adapted from Ackman (1980).

may at times be higher in triglyceride fat than the white muscle (Ackman and Eaton, 1971). This unfortunately has influenced consumers in the United Kingdom, who are used to eating white fish such as Atlantic cod *Gadus morhua* to the extent that they perceive fish with dark-coloured flesh as high in fat, thus hindering market development for such fish (Marshall and Currall, 1992).

Figure 4.1 shows the distribution of depot fat in Atlantic capelin *Mallotus villosus*, but it is rare to find such detail, even for commercially important fish. The Japanese sardine *Sardinops melanosticta* deposits subdermal fat in measurable layers in parallel with muscle fat content (Yamada, 1981). The Indian oil sardine *Sardinella longiceps* has been reported to have 27.4% or more lipid in the skin and only 6.0% in the muscle (Nair *et al.*, 1979). Polvi (1989) recorded 29–35% lipid in the subdermal fat of small (~600 g) Atlantic salmon *Salmo salar*. This skin fat is mostly triglyceride (Table 4.1). In the case of mackerel the subdermal triglycerides were very similar to the muscle fat triglycerides in fatty acid composition (Ackman *et al.*, 1991). In the instance of capelin (Fig. 4.1), where subdermal fat is plentiful, the fish has a remarkable resistance to oxidation in frozen storage (Ackman *et al.*, 1969; Botta *et al.*, 1983), whereas in the case of Atlantic mackerel *Scomber scombrus* this subdermal fat is especially prone to oxidation (Ke and Ackman, 1976; Ke *et al.*, 1977a, b). This susceptibility is also known in the Indian oil sardine

Table 4.1.. Lipid classes (g 100 g^{-1} tissue) of light and dark muscle and skin layer of Atlantic salmon (*Salmo salar*) fed special diets.[a]

		Supplemental diet lipid			
Muscle layer	Lipid classes[b]	Herring oil	Canola oil	EPA/DHA conc.[c]	Egg fatty acid[c]
Light	Total	6.6	6.3	4.7	4.5
	PL	0.7	0.6	0.8	0.9
	CH	0.01	0.03	0.01	0.1
	TG	5.8	5.7	3.9	3.6
	FFA	0.01	<0.01	0.1	0.02
Dark	Total	19.2	22.1	17.5	15.8
	PL	0.9	1.1	1.4	0.8
	CH	0.2	0.1	0.02	0.1
	TG	17.8	20.2	15.1	14.9
	FFA	0.2	0.6	1.0	0.04
Skin	Total	35.4	34.5	24.8	29.3
	PL	2.4	0.8	1.2	0.4
	CH	0.06	0.07	0.2	0.1
	TG	32.7	31.8	23.3	28.7
	FFA	0.04	0.2	0.3	0.03

[a] Adapted from Polvi (1989). All diets included a basic fish fatty acid mix from fish meal of about 4% of diet. Supplemental diet lipid was added at 10%.
[b] PL, phospholipids; CH, cholesterol; TG, triglycerides; FFA, free fatty acid.
[c] Ethyl esters.

S. longiceps (Nair *et al.*, 1979), the Japanese sardine *S. melanosticta* (Yamada, 1979), and various other species (Toyomizu *et al.*, 1980).

The triglyceride depot fat in edible fish muscle is subject to seasonal variations in all marine fish from cold northern (or southern) latitudes. Figure 4.2 shows how this can literally double the fat content of fillets from winter flounder (*Pseudopleuronectes americanus*) after the midsummer spawning period. This is a modest change in 'fat' compared to that of the Atlantic herring *Clupea harengus* (Fig. 4.3). Newfoundland herring fillet depot fat can cycle between 5% and just over 16% (Ackman and Eaton, 1976). The European sardine (*Sardina pilchardus*) fat cycles between 1.6% and 22.4% on a whole body basis (Nunes *et al.*, 1992).

Data for seasonal variations of fat in edible muscle of Atlantic mackerel have been published (Ackman, 1982), and Ackman and Eaton (1971) examined both lean (spring) and fat (autumn) muscle in detail. It was observed that the dark muscle seemed to consistently have a relatively high level of total lipid (9–18%), whereas the light muscle went from a low of 2.2% in spring to a high of 10% by October. The samples were restricted to muscle and did not include the subdermal fat. This could add up to

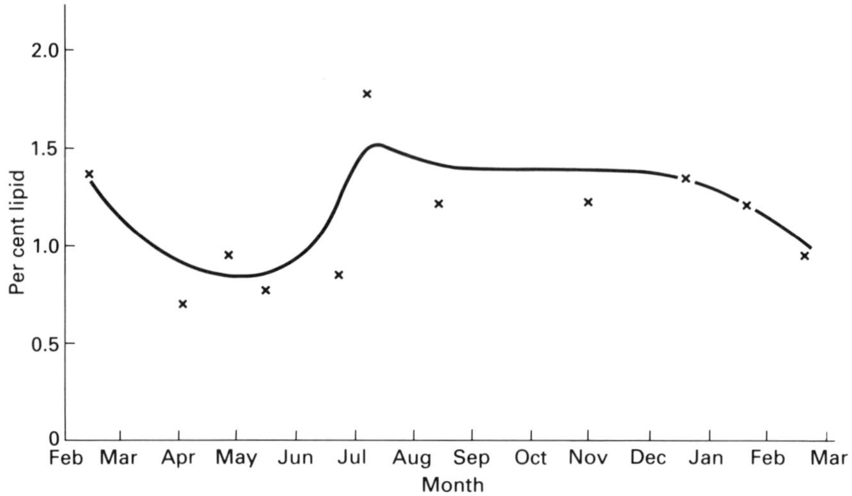

Fig. 4.2. Seasonal variation in total lipid recovered from fillets of *Pseudopleuronectes americanus*. (Ackman and Castell, unpublished results.)

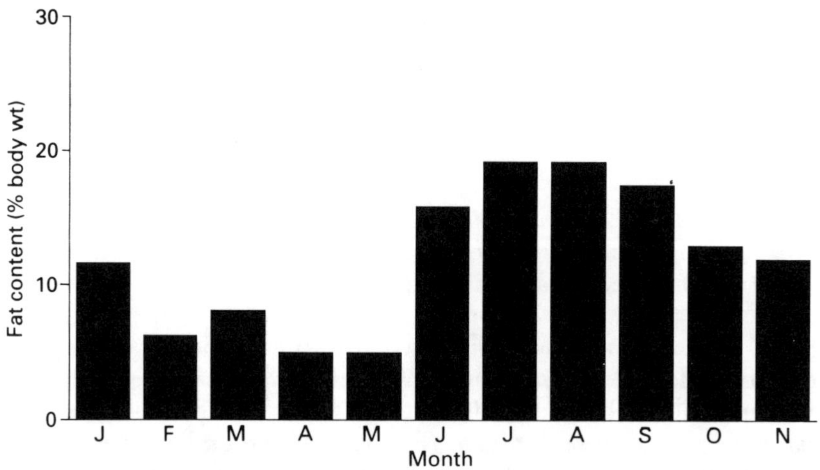

Fig. 4.3. Seasonal variation in total fat of early-autumn-spawning herring bodies from the Irish Sea, 1973-1974. Adapted from Wallace (1988).

100% extra fat to the total for a fillet of muscle plus skin (Ackman, 1980; Ackman and Gunnlaugsdottir, 1992). Unlike the capelin (Fig. 4.1) or salmon, the belly flap in mackerel has relatively little fat compared to the muscle or skin.

The basic lipid of fish muscle is cellular in nature. The muscle lipids of the Atlantic cod *G. morhua* have been much studied (Jangaard *et al.*,

1967a; 1967b; Addison *et al.*, 1968; Hardy *et al.*, 1979; Ackman, 1980; Hardy, 1980; Orlick *et al.*, 1991). There is very little variation in total lipid of cod muscle with season (Jangaard *et al.*, 1967a) or in fatty acids (Jangaard *et al.*, 1967b). Table 4.2 indicates that in the polar lipids of fresh cod phosphatidylcholine (PC) is the main lipid class and this is true for the phospholipids of the edible muscle of most vertebrate fish and of shellfish. Phosphatidylethanolamine (PE) is the second most important polar lipid class, and the two together total about 75% of cod light muscle lipids. In the report of Orlick *et al.* (1991) on cod lipids, considerable hydrolysis of polar lipids had taken place in a frozen stored sample prior to lipid class analysis, with a resulting formation of 16.4% free fatty acids. In fresh fish muscle this class would be of the order of 5% of lipids (Hardy *et al.*, 1979). However, the proportions of all other lipid classes were still approximately the same as had been reported by Addison *et al.* (1968) for fresh cod (Table 4.2). Orlick *et al.* (1991) reported, for muscle of Atlantic cod, 33.5% and 13.6% lipid, respectively, of these phospholipid classes and for saithe (*Pollachius virens*), 43.5% and 17.8% respectively. Obviously triglyceride is a minor part of the lipid in this case and, as shown by its fatty acid composition, is essentially similar to that of cod liver oil. It has been assumed that this triglyceride in cod muscle is 'cellular' and part of membranes, but it could also be present intercellularly as scattered adipocytes (see Yamada and Nakamura, 1964; Yamada, 1972; Shindo *et al.*, 1986). Total lipid of light cod muscle is invariably in the range of 0.60–0.70% of wet weight depending on the analyst, the extraction method and to some extent on any inclusion of the dark (also called red or lateral line) muscle. This muscle type (Obatake and Heya, 1985; Shindo *et al.*, 1986), which is more prominent near the tail in most fish, has been compared with light muscle for lipids of farmed rainbow trout *Oncorhynchus mykiss* by Ingemansson *et al.* (1991) as shown in Table 4.3. The ultrastructure of freshwater fish muscle is the same as that of marine fish (Shindo *et al.*, 1986) and the very fundamental nature of the cellular phospholipids in edible muscle of freshwater as well as of marine fish is confirmed by the higher proportion of phosphatidylcholine relative to phosphatidylethanolamine in both muscle types. In channel catfish (*Ictalurus punctatus*) muscle the proportions were also 3 : 1 (Erickson, 1992). The proportions among the polar lipids in the catfish flesh are the same as for Atlantic cod, but the trout light muscle (Table 4.3) report indicates more polar lipid (1.71%) than would be expected. Polvi (1989) found light muscle of Atlantic salmon *Salmo salar* to contain about 0.7% polar lipid (Table 4.1). Salmonids may require more investigation since lipids of Pacific sockeye salmon (*Oncorhynchus nerka*) are reported by Porter *et al.* (1992) to include the two major phospholipids in light muscle at 0.31% and in dark muscle at 0.76%. These proportions seem reasonable but the total lipid of 2.3% in light muscle indicates that these fish were

Table 4.2. Lipid classes (w/w%) and fatty acids (w/w%) of Atlantic cod *Gadus morhua* muscle and liver.

	Muscle						Liver[b,c]
		Phosphatides[a]					
	Total lipid[d]	Choline	Ethanolamine	Serine	Sterol esters[a]	Triglyceride[a]	Triglyceride
% in lipid[e]	—	60.0	16.4	4.3	2.2	4.1	—
Fatty acids							
14:0	1.3	1.4	0.5	0.5	1.5	4.4	3.2
16:0	15.3	23.9	8.8	8.0	8.4	14.1	13.5
18:0	2.8	2.6	4.4	18.9	1.6	2.6	2.7
Σ Saturated	19.4	27.9	13.7	27.4	11.5	21.1	19.4
16:1	2.6	4.2	1.7	1.4	4.6	12.0	9.8
18:1	13.0	12.6	14.4	14.4	10.0	22.4	23.7
20:1	3.2	1.7	4.0	5.6	3.9	12.1	7.4
22:1	—	0.4	0.3	0.5	3.6	11.4	5.1
Σ MUFA[f]	18.8	18.9	20.4	21.9	22.1	57.9	46.0

18:2n-6	0.3	0.9	0.8	0.7	0.7	0.9	1.4
20:4n-6	—	2.4	1.6	2.9	1.4	0.5	1.6
18:3n-3	1.0	0.2	0.3	0.2	0.4	0.2	0.6
18:4n-3	1.1	0.3	0.3	<0.1	0.2	1.2	0.9
20:5n-3	21.2	19.4	10.7	7.5	27.5	7.3	11.1
22:5n-3	2.0	1.6	1.9	1.7	2.2	0.7	1.7
22:6n-3	32.2	27.3	40.5	35.6	30.7	7.4	12.6
Σ PUFA[g]	57.8	52.1	56.1	48.6	63.1	18.2	29.9

[a] Adapted from Addison *et al.* (1968).
[b] Adapted from Jangaard *et al.* (1967b).
[c] Oil content 35%.
[d] Adapted from Orlick *et al.* (1991).
[e] Not including cholesterol at 6.3% of total muscle lipid.
[f] Monounsaturated fatty acids.
[g] Polyunsaturated fatty acids.

Table 4.3. Comparison of lipids of light and dark muscle of farmed rainbow trout *Oncorhynchus mykiss* and total lipids of farmed channel catfish *Ictalurus punctatus*.

	Trout muscle[a]		Catfish[b]
Lipids	Light	Dark	Whole muscle
Total lipids (%)	3.0	7.3	5.4
Neutral lipids (% of total lipid)	43	68	89.4
Triglyceride	24.8	40.8	88.3
Cholesterol	—	—	1.1
Cholesterol ester	14.7	22.4	—
Free fatty acid	3.3	3.8	—
1,2-Diacylglycerol	0.3	0.8	—
Polar lipids (% of total lipid)	57	32	10.5
Phosphatidylcholine	31	18.5	7.3
Lysophosphatidylcholine	2.3	1.0	—
Phosphatidylethanolamine	18	9.5	2.6
Lysophosphatidylethanolamine	1.8	0.8	—
Sphingomyelin	3.2	2.4	—
Phosphatidylserine	—	—	0.3
Phosphatidylinositol	—	—	0.3

[a] Adapted from Ingemansson *et al.* (1991).
[b] Adapted from Erickson (1992).

presumably taken after their migration, presumably pre-spawning, and therefore seriously depleted of lipid. Possibly this condition accounts for their report of PE > PC in dark muscle (19% total lipid), but the PC > PE reported in the light muscle is normal.

Although cholesterol is not a main focus of this review, it is safe to say that the cholesterol is typically in the 40–60 mg range for 100 g of fish edible muscle. Higher values are of course reported for comparable cooked products (Exler, 1987). Also higher levels are often found in the gonads (Shima *et al.*, 1979; Exler, 1987). Lobster (Exler, 1987) and crab meats (Gordon, 1982; King *et al.*, 1990) may have 0.06–0.10 g 100 g^{-1} but among popular seafoods shrimps are a special case, cholesterol normally being ≈0.15 g 100 g^{-1} meat (Krzynowek and Panunzio, 1989). It is thus extraordinarily difficult to present nutritional data on fish ‘fat’ in a way meaningful to the consumer or the nutritionist. Although Atlantic cod muscle lipid may be listed as 0.65%, the fact that this is mostly phospholipid means that by weight the fatty acids will be only about 0.5% of this muscle. As the proportions of triglycerides in edible muscle lipid increase there will be a higher percentage of fatty acid available per unit mass of extractable lipid.

Paul *et al.* (1980) supplemented McCance and Widdowson’s (1978) tabulations of foods with seafood fatty acids for northeast Atlantic fish and shellfish, and later Canadian tabulations appeared (Ackman and McLeod, 1988) covering many of the same Atlantic species. American data for similar species have been published (Krzynowek and Murphy, 1987; Krzynowek *et al.*, 1989). Ackman (1989a) reviews many of these data in one publication and includes a tabulation of fat in fish and fishery products analysed in Sweden.

A remarkably useful compilation of nutritional information is provided by Exler (1987) in United States Department of Agriculture Human Nutrition Information Service Agriculture Handbook Number 8-15 *Composition of Foods: Finfish and Shellfish Products. Raw, Processed, Prepared.* Figure 4.4 is a reproduction of part of one page, of lipid data only, for nutritional components of the popular Atlantic ocean perch *Sebastes marinus*, also known as redfish. It should be noted that the data are averaged from analyses of many samples, in this case as many as 208 for lipid, but about 20 for fatty acids. In Table 4.4 a comparison is made between basic raw and cooked lipid composition data of the Handbook for *Reinhardtius hippoglossoides*. It is important to record that loss of water on cooking enriches food fish lipids in most cases, since only with high-fat species is fat sometimes 'cooked out' of the fish. This effect depends also on the mode of cooking (Gall *et al.*, 1983; Hearn *et al.*, 1987; Ohta *et al.*, 1988).

Since so many fish lipid nutritional data have been recorded from species fished in New England, two species (snapper and pompano) have been selected from Handbook 8-15 to illustrate the similarity of fat contents and fatty acids for fish often available in the southern US (Table 4.5). Fatty acid details for other species of finfish in the south are available elsewhere (Gooch *et al.*, 1987; 1989). Although sharks are wide-ranging they have only recently become a popular food fish in North America. Because of the very large number of species eaten in the USA the vexing confusion of popular and regional names has been partly resolved by publication of an inexpensive reference book, *The Fish List* (Food and Drug Administration, 1989; also reproduced in Ackman, 1989a).

Table 4.6 presents a broad classification of marine and freshwater fishes, by both particular fish and groups of fish, in terms of fat content. Different publications may give figures falling into two classes among low, medium and high fat contents. If the consumer is in doubt about a particular but unfamiliar fish offered for sale, it is often reasonably safe to relate it by shape and appearance to one of the familiar species of this table.

n-3 Polyunsaturated Fatty Acids in Fish Lipids

It is already apparent from Tables 4.2, 4.4 and 4.5 that the polyunsaturated fatty acids of many fish lipids are dominated by two members of the

FINFISH
Ocean Perch, Atlantic, raw
Sebastes marinus **(L.)**

Nutrients and units		Amount in 100 grams, edible portion			Amount in edible portion of common measures of food		Amount in edible portion of 1 pound of food as purchased
		Mean	Standard error	Number of samples	Approximate measure and weight: 1 fillet[1] = 64 g	3 oz = 85 g	Refuse: 0
(A)		(B)	(C)	(D)	(E)	(F)	(G)
Proximate:							
Water	g	78.70	0.129	183	50.37	66.89	356.97
Food energy	kcal	94			60	80	427
	kJ	394			252	335	1,788
Protein (N X6.25)	g	18.62	0.137	135	11.92	15.83	84.48
Total lipid (fat)	g	1.63	0.092	208	1.04	1.39	7.40
Carbohydrate, total	g	0.00			0.00	0.00	0.00
Crude fiber	g	0.00			0.00	0.00	0.00
Ash	g	1.20	0.021	115	0.77	1.02	5.43
Lipids:							
Fatty acids:							
Saturated, total	g	0.244			0.156	0.207	1.107
4:0	g						
6:0	g						
8:0	g						
10:0	g						
12:0	g	0.003		1	0.002	0.003	0.014
14:0	g	0.063		21	0.040	0.054	0.286
16:0	g	0.139		21	0.089	0.118	0.631
18:0	g	0.034		21	0.022	0.029	0.154
Monounsaturated, total	g	0.625			0.400	0.531	2.835
16:1	g	0.086		21	0.055	0.073	0.390
18:1	g	0.207		21	0.132	0.176	0.939
20:1	g	0.100		20	0.064	0.085	0.454
22:1	g	0.229		21	0.147	0.195	1.039
Polyunsaturated, total	g	0.427			0.273	0.363	1.937
18:2	g	0.028		21	0.018	0.024	0.127
18:3	g	0.057		21	0.036	0.048	0.259
18:4	g	0.024		19	0.015	0.020	0.109
20:4	g	0.004		21	0.003	0.003	0.018
20:5	g	0.080		21	0.051	0.068	0.363
22:5	g	0.023		21	0.015	0.020	0.104
22:6	g	0.211		21	0.135	0.179	0.957
Cholesterol	mg	42	2.380	10	27	36	191
Phytosterols	mg						

Fig. 4.4. Reproduction of lipid parts of one page from US Department of Agriculture Handbook 8-15 (Exler, 1987).

Table 4.4. Composition and lipid components (g 100 g^{-1}, edible portion) of raw and cooked Greenland halibut, *Reinhardtius hippoglossoides* (also known as Newfoundland turbot).[a]

Nutrient	Raw (mean)	Cooked, dry heat (mean)
Proximate		
Water (g)	70.3	61.9
Food energy (kcal)	186	239
(kJ)	779	999
Protein (*n* × 6.25) (g)	14.4	18.4
Total lipid (fat) (g)	13.8	17.7
Carbohydrate, total (g)	0.0	0.0
Crude fibre (g)	0.0	0.0
Ash (g)	1.00	1.28
Fatty acids		
12:0 (g)	0.01	0.01
14:0 (g)	0.75	0.97
16:0 (g)	1.44	1.85
18:0 (g)	0.19	0.25
Σ Saturated (g)	2.42	3.10
16:1 (g)	1.76	2.26
18:1 (g)	2.39	3.07
20:1 (g)	2.24	2.87
22:1 (g)	1.97	2.52
Σ MUFA (g)	8.38	10.74
18:2 (g)	0.12	0.16
18:3 (g)	0.04	0.06
18:4 (g)	0.13	0.17
20:4 (g)	0.06	0.08
20:5 (g)	0.53	0.67
22:5 (g)	0.09	0.14
22:6 (g)	0.39	0.50
Σ PUFA (g)	1.37	1.75
Cholesterol (mg)	46	59

[a] Adapted from US Department of Agriculture Handbook No. 8-15 (Exler, 1987).

n-3 family, 20:5n-3 (eicosapentaenoic acid, popularly known as EPA) and 22:6n-3 (docosahexaenoic acid, popularly known as DHA). It seems strange to find lay-people discussing these 'omega-3' fatty acids and their health benefits, but this is a unique case where media exploitation of medical news allows an interpretation that fish or fish oils are good for the cardiovascular system (Fletcher, 1989; Saynor and Ryan, 1990). The popular Eskimo image and the mystique of 'omega-3' for the type of fatty acids in fish facilitated this process and after more than a decade it is useful to remember that the pioneering fieldwork on Eskimo health

Table 4.5. Lipid content, fatty acids and cholesterol details for several species of marine fish available in the USA.

	Snapper[a] (Lutjanidae)	Florida pompano[a] (*Trachinotus carolinus*)	Shark[a] (mixed species)	Halibut[a] (*Hippoglossus* sp.)	Herring[a] (*Clupea harengus*)	Ocean perch[a] (*Sebastes* sp.)	Salmon, Atlantic[b] Farmed (*Salmo salar*)	Wild
Total lipid (g 100 g^{-1})	1.3	9.5	4.5	2.29	9.04	1.63	6.2	5.8
Fatty acids (g 100 g^{-1})	0.99	7.24	3.93	1.81	7.91	1.30	5.7	5.3
14:0	0.04	0.41	0.08	0.06	0.55	0.06	0.45	0.27
16:0	0.16	2.22	0.73	0.21	1.35	0.14	0.90	0.84
18:0	0.08	0.82	0.12	0.06	0.11	0.03	0.11	0.13
Σ Saturated	0.28 (29%)	3.51 (48%)	0.92 (24%)	0.32 (18%)	2.04 (26%)	0.24 (19%)	1.46 (26%)	1.24 (23%)
16:1	0.06	0.56	0.25	0.16	0.62	0.09	0.43	0.33
18:1	0.16	2.02	0.98	0.36	1.52	0.21	0.78	1.13
20:1	0.02	–	0.21	0.12	0.72	0.10	0.57	0.56
22:1	–	–	0.31	0.10	0.84	0.23	0.52	0.46
Σ MUFA	0.25 (26%)	2.59 (36%)	1.81 (46%)	0.65 (36%)	3.74 (47%)	0.63 (48%)	2.30 (40%)	2.48 (47%)
18:2n-6	0.02	0.12	0.08	0.03	0.13	0.03	0.22	0.06
20:4n-6	0.05	0.25	0.11	0.14	0.06	0.01	0.02	0.03
18:3n-3	<0.01	–	0.03	0.06	0.10	0.06	0.05	0.03
18:4n-3	0.01	–	0.03	0.04	0.21	0.02	0.07	0.03
20:5n-3	0.05	0.18	0.32	0.07	0.71	0.08	0.32	0.24
22:5n-3	0.06	0.20	0.11	0.09	0.06	0.02	0.11	0.10
22:6n-3	0.26	0.39	0.53	0.29	0.86	0.21	0.51	0.63
Σ PUFA	0.46 (46%)	1.14 (16%)	1.20 (30%)	0.84 (46%)	2.13 (27%)	0.43 (33%)	1.30 (23%)	1.12 (21%)
Cholesterol (mg 100 g^{-1})	37	50	51	32	60	42	30	20

[a] Values compiled from US Department of Agriculture Handbook No. 8-15 (Exler, 1987).
[b] Values compiled from Polvi (1989).

and dietary lipids (Dyerberg and Bang, 1979; Dyerberg and Jørgensen, 1992; Dyerberg, 1992) was backed by sound pharmacology. In 1979 the Wellcome Foundation filed a UK patent application GB 2033745, 'Fatty acid and derivatives thereof in treatment or prophylaxis of thrombo-embolic conditions'. Their experience in prostaglandin research had shown that EPA could be converted to the prostacyclin PGI_3, a compound with presumably potent benefits in the vascular wall (Bruckner, 1992; Morita *et al.*, 1993). DHA by itself has physiological effects (Yazawa and Kageyama, 1991), but has attracted less attention. It can be converted to EPA in the body (von Schacky and Weber, 1985), but it has recently become controversial in its own right in the context of supplements for full brain development in late pregnancy or in pre-term infants (Carlson *et al.*, 1992; Innis, 1992; Neuringer *et al.*, 1992; Nettleton, 1993). It is also probably beneficial in reducing cardiac arrhythmia (Charnock *et al.*, 1991).

The original epidemiological evidence for Greenland Eskimos eating a diet rich in omega-3 fatty acids has recently been supplemented by cardiovascular health data for Alaskan natives (Newman *et al.*, 1993), but one retrospective study (Kromhout *et al.*, 1985) and one prospective study (Burr *et al.*, 1989) have been widely accepted as confirmation of the beneficial effects of eating fish. In the first instance a 20-year survey showed that males frequently eating fish had reduced cardiovascular mortalities, and in the second the introduction of fatty fish to the diet of cardiac patients started to improve their survival after only 100 days. A newly published study of males in Zutphen and their dietary habits has recently been extended from coronary mortalities (Kromhout *et al.*, 1985) to risk of stroke (Keli *et al.*, 1994). Consumption of low levels of fish, perhaps even one portion per week, may be associated with a reduced stroke incidence.

As many as 775 n-3 fatty acid, fish or fish oil biomedical references are officially listed for one year ending in September of 1991 (van de Kamp and Wykes, 1992). Much of this work is repetitive and has failed to impress those regulatory officials who adhere to reducing blood cholesterol and altering the proportions of certain lipoproteins as the ultimate criteria for cardiovascular health benefits. That there are modifications to these complex systems by dietary fatty acids is well established (Welch and Borlakoglu, 1992). However, Nordöy *et al.* (1993) have recently shown that dietary saturated fatty acids and fish oil concentrate (a total of 68% n-3) affect blood lipids as separate effects and by independent mechanisms. Therefore the n-3 fatty acids do not necessarily intervene in the obvious effects of the saturated fatty acids. An indisputable clinical effect of a high intake of the long-chain n-3 polyunsaturated fatty acids such as 20:5n-3 and 22:6n-3 is reduction of serum triglycerides (Harris, 1989; Coniglio, 1992; Schmidt *et al.*, 1992). These are also considered a cardiovascular risk factor. Side-effects of increased fish in the diet, or an intake of fish

Table 4.6. Fat levels in fish commonly found in the marketplace.

Low (<5% fat)		Medium (5-10% fat)	High (>10% fat)
Marine fish			
Anchovy	Mullet[b]	Arctic char	Dogfish (spiny,
Argentine	Ocean perch	Butterfish[b]	US cape shark or
Bass, hybrid striped	Oreo (smooth or black)	Capelin	UK rock salmon)
Black seabass	Plaice	Orange roughy	Greenland turbot[c]
Bluefish	Pollock (Alaska)	Pompano	Halibut (black, European)
Bluefish (Boston)	Pollock (Atlantic)	Salmon (various)[a]	Herring
Cod (Atlantic or Pacific)	Porgy	Salmon (canned)[b]	Mackerel (various)
Croaker	Pout (ocean)	Sea bass[b]	Sablefish
Cusk	Redfish	Sea trout[b]	Shad
Dolfinfish	Red snapper	Shark (mako)	Sprat
Dory (see oreo)	Rockfish (Pacific perch)	Spot	
Drum	Saithe	Sturgeon (white)[b]	
Flounder (various)	Scup	Swordfish[b]	
Grenadier	Shark (most)[b]		
Grouper	Skate		
Haddock	Snapper		
Hake (various, also whiting)	Sole		
Halibut (Atlantic or Pacific)	Striped bass		
Hoki	Surimi products		
Kingclip	Tilefish		
Lemon sole	Tuna (canned in water)		
Ling	Turbot (European)		
Lingcod	Whiting (Pacific)		
Mahimahi	Wolffish[b]		
Monkfish	Yellowtail[b]		

Freshwater fish		
Alewife	Carp[b]	Eel
Barramundi	Drum[b]	Whitefish[a]
Bass (various)	Lake trout[a]	
Burbot	Milkfish	
Catfish (farmed)	Rainbow trout[a]	
Cisco	Trout (various)[b]	
Lake perch (North America)[b]		
Perch (Lake Victoria)		
Pike		
Smelt		
Sucker		
Tilapia		

[a] Medium to high fat fish.
[b] Low to medium fat fish.
Drawn from various sources such as US Department of Agriculture Handbook No. 8-15 (Exler, 1987), US Department of Commerce NOAA Technical report NMFS 55 (Krzynowek and Murphy, 1987), and *The Advanced Seafood Handbook*, Rockland, Maine.
[c] Also known as Greenland halibut or Newfoundland turbot.

oil concentrates in capsules, are occasionally reported but are generally very minor (Nelson *et al.*, 1991; Drevon, 1992).

One of the few things now agreed on by scientists working with polyunsaturated fatty acids is that the two basic families, n-3 and n-6, are both essential. The latter family is typified by linoleic acid (18 : 2n-6), now a major component of the high-fat Western diet, but both it and linolenic acid (18 : 3n-3) are primarily supplied by plant sources (Ackman and Cunnane, 1991) and, as the tables in this review show, are very minor components of marine lipids. The two C_{18} fatty acid families may have opposing biochemical effects once extended to the C_{20} chain length and it is therefore necessary to consider where the n-6 fatty acids fit into the nutritional biochemistry of fish and shellfish lipids.

Arachidonic acid

The developing shortage of high-quality edible fish from some northern latitudes is well known. One result has been an increasing trend to exploit high seas fish resources (Bailey, 1992) and a second has been the appearance of unfamiliar species in some markets and also in prepared food products (Bligh, 1992; Whittle and Wood, 1992). One recent paper on the impact of fish and fish oils on human health (Brown and Roberts, 1991) originated in Australia. The results of including a 'practical' (200 g day^{-1}) amount of lean Australian fish in the diet of healthy men for 6 weeks were improvement in certain haematological parameters implicated in the aetiology of cardiovascular disease. Accordingly Table 4.7 is presented to emphasize a very novel fatty acid feature of Australian fish such as might have been included in the above dietary study.

This feature is the presence of several n-6 fatty acids of longer chain lengths (i.e. 20 : 4n-6, 22 : 4n-6 and 22 : 5n-6) in significant amounts. Many of the fish listed are quite lean and their fatty acids can be compared with those listed for the cold water Atlantic cod in Table 4.2. The saturated and monoethylenic fatty acids are similar, with some possibility that warmer water temperatures in northern Australia could have enhanced the proportions of saturated fatty acids, especially of stearic (18 : 0) acid. This speculation follows from the popular concept of 'membrane fluidity', but it is also probable that the low levels of 20 : 1 and 22 : 1 in lipids of fish from these areas (Evans *et al.*, 1986) allow more saturated acids to be included in the triglycerides and phospholipids in lieu of the exogenous 20 : 1 and 22 : 1 fatty acids (Ackman *et al.*, 1980; Ackman, 1989b). The 22 : 1 is not in any case normally important in fish phospholipids. The real interest lies in the C_{20} and C_{22} n-6 fatty acids. In gas–liquid chromatography of the methyl esters of fatty acids from Atlantic cod muscle fatty acids there is a distinct but very small peak (*c.* 0.1–0.4% of the total) for 20 : 4n-6, but 22 : 4n-6 and 22 : 5n-6 are barely detectable and, if present, are 0.1% or less. In

Table 4.7. Selected data for fillet (flesh and skin) of Australian fish, and edible parts of shellfish, eaten in Australia. Percentages of lipids, of total saturated fatty acids in the fatty acids of those lipids, and similarly percentages of 16:1 and 18:1, and of polyunsaturated acids by family.[a]

	Barramundi	Cod, red (NZ)	Garfish	Gemfish	Grenadier, blue	Groper, blue	Hake, cape (SA)	Leather jacket	Morwong	Mullet, jumper	Mullet, yelloweye	Mulloway	Ruff, tommy	Salmon, Australian	Scallop (FW)	Shark, school	Snapper	Snapper, red	Snook	Whiting, KG	Whiting, sand	Whiting, yellowfin	Scallop, queen	Shrimp (King prawn)	Octopus	Squid
% Lipid	0.8	0.5	1.0	—	0.5	7.8	5.7	0.5	1.0	4.9	1.9	4.4	0.8	1.8	7.7	1.1	2.1	0.8	0.9	1.0	1.0	1.0	—	—	—	—
Total saturated	35.3	27.5	29.4	33.7	33.1	28.8	25.0	27.9	29.5	50.0	33.3	31.5	30.7	28.3	30.6	33.7	34.5	29.6	31.9	28.8	29.6	30.6	33.7	30.5	32.4	35.3
Major monoethylenic																										
16:1	7.0	5.7	3.7	4.4	4.1	2.4	7.7	4.5	6.6	17.1	9.2	15.5	5.1	7.1	12.8	4.4	6.8	4.0	3.8	8.5	17.2	6.6	4.0	8.0	1.4	2.4
18:1	23.9	16.8	11.5	13.9	13.1	9.3	29.9	12.6	17.7	8.6	20.3	22.2	14.1	16.7	18.6	13.9	24.5	14.7	12.7	15.3	21.0	14.1	9.1	14.7	10.0	6.3
Polyunsaturated																										
n-6 family																										
18:2	5.9	2.2	4.4	1.4	2.0	1.2	2.4	2.2	2.7	2.3	3.4	2.3	2.0	2.9	3.4	1.4	1.6	1.9	1.6	1.9	3.8	1.9	1.5	2.2	1.0	0.6
20:4	8.8	2.6	6.1	8.3	2.2	9.3	1.1	15.8	3.1	1.7	4.2	2.7	5.0	5.1	4.8	8.3	5.5	2.5	4.9	10.8	6.2	11.4	4.7	6.4	9.8	2.0
22:4	1.3	0.4	1.2	1.7	0.3	0.9	0.4	2.2	0.8	0.9	1.5	0.8	1.4	1.2	2.1	1.7	2.0	0.3	0.8	2.8	4.3	4.4	2.5	0.8	1.7	0.6
22:5	2.6	0.4	6.1	1.2	0.3	2.2	0.3	2.1	0.8	—	1.3	0.8	1.5	1.6	2.3	1.2	1.2	1.3	1.6	1.5	—	1.5	0.5	0.9	0.6	0.7
n-3 family																										
18:3	2.0	0.7	3.3	0.4	1.0	0.3	0.8	1.1	0.6	2.0	2.2	1.1	0.9	0.8	2.8	0.4	0.5	0.5	0.6	0.4	0.8	0.4	0.2	0.9	0.3	0.2
20:5	2.0	12.7	4.5	5.3	13.3	6.1	4.8	15.9	7.4	7.8	9.3	5.6	7.0	6.6	4.1	5.3	3.4	5.9	4.7	9.2	7.4	13.2	14.8	14.3	13.9	12.5
22:5	1.5	1.0	3.2	3.7	1.1	1.2	1.1	3.3	3.3	3.6	4.5	5.2	2.6	3.2	4.1	3.7	3.8	2.1	1.6	3.7	2.8	4.6	1.1	1.0	2.5	0.9
22:6	8.5	28.7	24.7	25.1	27.5	36.6	9.5	10.8	21.7	4.9	7.7	10.2	26.5	24.6	13.1	25.1	13.0	34.8	34.9	14.8	4.6	9.8	25.8	17.7	20.0	34.6

[a] Adapted from Gibson (1983).

All species are scientifically identified in reference. Samples were of retail origin and marine except for the freshwater (FW) scallop. The barramundi and mulloway are, however, estuarine and may be taken in fresh water. The red cod was from New Zealand and the cape hake from South Africa. KG is King George.

contrast, these three fatty acid can total as much as 20% of the C_{20} + C_{22} acids from some Australian fish of Table 4.7. South African cape hake and New Zealand red cod serve as controls on the accuracy of the fatty acid identifications and quantifications since these fish are from more southerly latitudes and colder waters, and do not readily access these n-6 fatty acids in their diets. Unlike the lean tropical and semitropical fish, the fat contents of South African estuarine fish have been reported to be as high as 16% (Marais, 1990). Although Australia now has fisheries in the comparable Bass Strait – Tasmania latitudes, the data in Table 4.7 were based on samples collected from Sydney and northwards, i.e. from temperate through semitropical to tropical waters. Sinclair *et al.* (1983) have also looked at arachidonic acid (20:4n-6) in fish from the latter area, and Evans *et al.* (1986) report high levels of n-6 fatty acids in marine food organisms from the Northwest Shelf of Australia, definitely a tropical area. The totals for the three C_{20} and C_{22} polyunsaturated fatty acids of the n-3 family in Table 4.7 are reduced in most cases in the lean Australian fish, where the muscle lipid is clearly mostly phospholipids, by amounts corresponding to the higher proportion of C_{20} and C_{22} n-6 fatty acids. If the fish lipid had included very much triglyceride, this effect would not be as clear. Many tropical fish feed continuously on organisms rich in arachidonic acid, whereas cold water fish may be torpid over the winters and food availability is dictated by season (Burt *et al.*, 1992).

The possible physiological consequences in humans of overlooking fish as a source of dietary arachidonic acid have been discussed elsewhere (Ackman 1990; 1991; 1992a; 1992b; 1992c; Sinclair and O'Dea, 1991). Brown and Roberts (1991) also studied a diet group with fish plus fish oil capsules, with fatty acids as described by Brown *et al.* (1990). This provided more eicosapentaenoic acid (EPA) and docosahexaenoic acid (DHA), with the results of Fig. 4.5. The observation that platelet membrane fatty acids of the n-6 and n-3 families can be interchanged is not new (Sinclair *et al.*, 1987), but this finding is of considerable interest since the 200 g of lean mixed Australian fish are reported to contain about 0.6 g of n-3 polyunsaturated fatty acids. The 20:5n-3, 22:5n-3 and 22:6n-3 are of course always quantitatively more important than C_{20} and C_{22} members of the n-6 family (Table 4.7) and therefore their effect on platelet membrane phospholipid exceeds that of the arachidonic acid and its two related fatty acids in the group receiving fish oils in capsule form. These supplements are always low in arachidonic acid and high in EPA and DHA (Sagredos, 1991; Ackman, 1992a). Unfortunately the impact of n-3 fatty acids on platelets is only one factor in the complex aetiology of prevention of cardiovascular disease (Nestel, 1987; Leaf and Weber, 1988) and the n-3 polyunsaturated fatty acids function in other ways as well as in modifying lipids of platelets (Simopoulos *et al.*, 1991; Frölich and von Schacky, 1992).

The commonality of the occurrence of high proportions of arachidonic

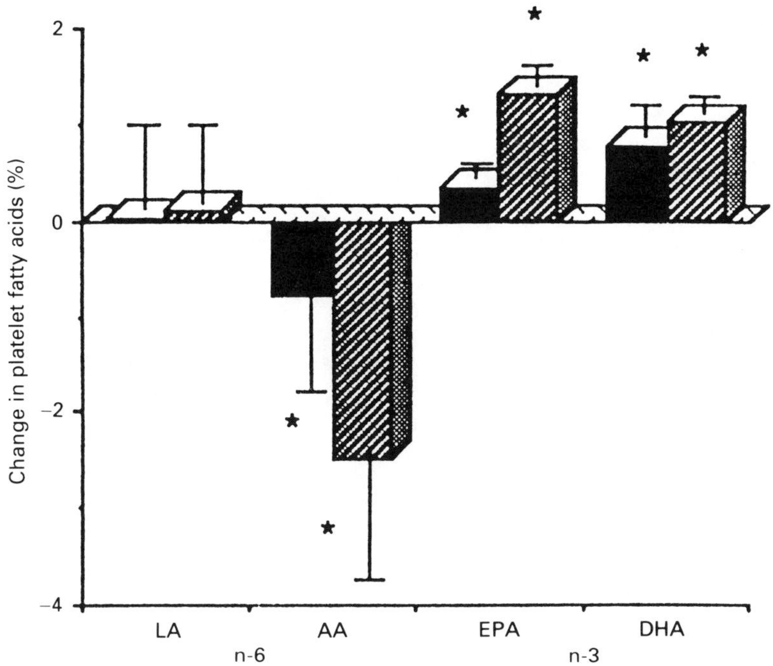

Fig. 4.5. Effect of a 'practical' high-fish diet (200 g day^{-1}) on platelet membrane fatty acids (mean ± SD% proportion) of normal Australians. LA, Linoleic acid; AA, arachidonic acid; EPA, eicosapentaenoic acid; DHA, docosahexaenoic acid; $*P < 0.05$. Reproduced from Brown and Roberts (1991) by permission.

acid in marine warm water fish is shown by fish from the waters of the Indian subcontinent (Nair and Gopakumar, 1978) and from those of the Near East. Thus Ågren *et al.* (1991a) and Eid *et al.* (1992) have published data for fish from the Arabian Gulf. The Ågren group listed arachidonic acid at 2.2–11.7% of the fatty acids in four species, with lipid contents of 0.31–1.89%. The Eid *et al.* fatty acid details are summarized in Table 4.8. The original listing of a polyunsaturated fatty acid as 20 : 3n-3 has been amended to 20 : 4n-6. As recently reviewed elsewhere (Ackman, 1992c), the methyl ester of arachidonic acid is prone to coincide with the methyl esters of other fatty acids such as 22 : 1, depending on the polarity of the column used in the gas–liquid chromatography. This effect includes coincidence of arachidonic acid (20 : 4n-6) with the position for 20 : 3n-3 (Jamieson, 1975; Ackman, 1986). However 20 : 3n-3 is a minor fish fatty acid, whereas 22 : 1 is a major component of fish oils from many marine cold-water species (Ackman, 1982).

Table 4.8. Fatty acid composition (g 100 g^{-1} fat) of fat from fish taken as shrimp by-catch in the Arabian Gulf.[a]

Fatty acid	Therapon	Ponyfish	Goatfish	Catfish	Lizardfish	Trevally	Gizzard shad	Jewfish	Bream	Sole	Anchovy	Mullet	Indian shad
14:0	3.4	6.0	3.2	1.9	1.6	3.1	10.6	2.8	1.9	4.9	4.4	9.5	3.3
16:0	22.5	23.4	26.6	25.2	19.4	26.8	21.3	36.3	33.7	20.8	24.8	20.2	31.2
18:0	8.8	8.4	8.5	10.4	7.5	9.6	3.5	6.7	7.6	6.1	8.0	3.2	7.6
20:0	1.2	2.3	1.3	1.4	0.5	0.7	0.4	0.7	0.7	1.4	1.0	0.5	0.6
24:0	0.2	0.2	0.3	0.2	—	0.5	0.2	0.6	0.3	0.3	0.6	0.2	0.3
Σ Saturated[b]	38.6	43.2	42.2	41.7	30.8	43.0	41.5	49.5	46.3	36.9	41.7	38.8	45.1
16:1n-7	5.9	9.4	6.4	5.9	3.4	4.8	14.6	14.9	18.7	7.0	5.6	15.3	4.7
18:1n-9	15.0	12.5	16.1	16.1	8.5	12.3	8.7	15.0	15.6	12.4	12.6	8.3	15.2
20:1n-9	0.8	1.2	1.4	0.9	0.5	0.5	0.4	0.8	1.0	0.4	0.6	0.3	0.4
22:1n-9	0.9	1.0	0.9	0.5	0.4	0.5	0.7	0.4	0.6	1.4	0.7	0.9	0.5
Σ MUFA[b]	22.8	24.4	24.8	23.4	12.8	18.7	24.6	31.7	36.1	21.6	20.3	25.0	21.1
18:2n-6	0.8	0.7	0.6	0.9	0.8	1.0	0.8	0.8	0.6	0.9	1.0	0.8	0.8
20:4n-6[c]	5.0	2.2	2.4	3.7	3.7	3.0	3.0	2.0	1.8	4.6	2.7	2.2	3.0
18:3n-3	1.0	0.9	0.8	0.8	1.0	1.0	0.6	0.4	—	0.4	1.3	0.4	0.4
20:5n-3	7.9	5.7	7.1	5.0	7.3	6.5	13.6	2.8	3.1	9.4	7.9	15.6	6.4
22:5n-3[d]	1.7	0.7	—	0.9	1.2	0.6	0.6	0.8	0.5	1.1	0.9	0.3	1.1
22:6n-3	14.0	10.0	11.1	13.2	36.6	22.1	7.1	8.4	6.7	13.6	19.6	10.3	18.2
Σ PUFA[b]	32.5	21.9	22.8	25.8	51.6	35.7	27.1	15.6	13.5	31.8	34.8	31.1	30.6
% Fat	2.1	5.2	1.7	2.0	1.1	1.1	4.6	3.0	2.0	0.9	2.4	4.6	0.9

[a] Adapted from Eid *et al.* (1992). Therapon (*Helotes* spp.), ponyfish (*Leiognathus* spp.), goatfish (*Mulloidichthys* spp.), catfish (*Arius* spp.), lizardfish (*Saurida* spp.), trevally (*Caranx* spp.), gizzard shad (*Nematolossa* spp.), jewfish (*Johnius* spp.), bream (*Nemipterus* spp.), sole (*Cynoglossus* spp.), anchovy (*Thryssa* spp.), mullet (*Liza* spp.), Indian shad (*Ilisha* spp.).
[b] Total includes other fatty acids not listed.
[c] Listed as 20:3n-3.
[d] Listed as 22:4n-3.

Freshwater Fish Lipids

It is ironic that many of the wild-caught fish of the cold northern lakes and rivers of Europe and Canada also carry a substantial proportion of arachidonic acid (Ågren *et al.*, 1987; Linko *et al.*, 1992). The important thing to note in Table 4.9 is the higher fat contents of these fish than for most of the marine warm-water fish of Tables 4.7 and 4.8. Obviously the need for overwintering is a controlling factor in muscle depot fat accumulation, although some North American freshwater fish such as the burbot or freshwater cod *Lota lota* and the northern pike *Esox lucius* use the liver for fat storage. Much the same points on the balance between 20:4n-6, 20:5n-3 and 22:6n-3 mentioned above for Australian and similar fish apply. For example, moderate amounts of Finnish freshwater fish in a normal diet had beneficial effects on lipid.metabolism (Hänninen *et al.*, 1989; Ågren *et al.*, 1991b). The effect of the n-3 fatty acids in the range of 0.2–0.9 g $week^{-1}$ was detectable, despite the inclusion of arachidonic acid, in reduced serum triacylglyceride. An unusual aspect of the 1991 report was the inclusion of exercise as well as modified diets. An inverse correlation was found between physical fitness and all of serum triacylglycerides, total cholesterol, low density lipoprotein cholesterol, and platelet aggregation.

In Table 4.9 fatty acids for several species of cold water fish are given, together with bass and wild channel catfish from warmer fresh waters. All wild-caught samples seem to have accumulated arachidonic acid as a relatively high proportion (4–5%) of the total fatty acids. Although the fat content of the fillet is higher in the farmed fish, the arachidonic acid content on a g 100 g^{-1} basis is lower for the farmed channel catfish raised on an artificial diet (see Nettleton *et al.*, 1990). Erickson (1992) has recently shown that within this large industry (154,000 t processed in 1989) there are also genetic effects modifying both polyunsaturated fatty acids and tocopherols.

The question of farmed versus 'natural' food fish is complex. Farmed trout are common, but since the diet is strictly 'artificial' there is no natural source of arachidonic acid and therefore the content of this fatty acid is limited (Ingemansson *et al.*, 1991) compared to wild-caught fish. Carp are widely cultured all over the world. A wild freshwater carp analysis is therefore included in Table 4.9. At 0.035 g 100 g^{-1} fillet it confirms the relatively high USDA Handbook figure for arachidonic acid, although the lipid is much less. The same paper (Aggelousis and Lazos, 1991) includes seven other wild freshwater fish from Greece but only the sheatfish *Silurus glanis* with 3.5% lipid and 3.8% 20:4n-6 in the fatty acids is as rich a source of this acid as is the wild carp.

Table 4.9. Content (g 100 g^{-1}) of lipid and fatty acids in flesh of some freshwater fish species.

	Trout[a] (mixed species)	Burbot[a] (*Lota lota*)	Pike[a] (*Esox lucius*)	Bass[a] (mixed species)	Vendace[b] (*Coregonus albula*)	Catfish, channel[a] (*Ictalurus punctatus*)		Carp (*Cyprinus carpio*)	
						Wild	Farmed	Farmed[a]	Wild[c]
Fatty acid									
14:0	0.18	0.01	0.02	0.09	0.31	0.06	0.10	0.11	0.02
16:0	0.82	0.12	0.08	0.59	0.71	0.44	1.27	0.66	0.22
18:0	0.15	0.04	0.02	0.10	0.14	0.15	0.35	0.18	0.07
Σ Saturated[d]	1.15	0.16	0.12	0.78	1.22	0.72	1.77	1.08	0.35
16:1	0.70	0.03	0.06	0.38	0.31	0.18	0.28	0.66	0.08
18:1	1.44	0.10	0.08	0.99	0.61	0.59	3.17	1.15	0.28
20:1	0.28	0.01	0.01	0.06	0.09	0.02	0.07	0.07	0.02
22:1	0.83	<0.01	0.01	—	—	0.01	0.05	0.40	—
Σ MUFA[d]	3.25	0.13	0.16	1.43	0.95	0.84	3.59	2.33	0.43
18:2n-6	0.18	0.01	0.03	0.09	0.18	0.10	0.88	0.52	0.08
20:4n-6	0.19	0.10	0.03	0.14	0.16	0.15	0.08	0.15	0.04
18:3n-3	0.16	—	0.02	0.11	0.24	0.07	0.10	0.27	0.02
18:4n-3	0.06	—	—	0.04	0.32	0.01	0.04	0.06	<0.01
20:5n-3	0.20	0.07	0.03	0.24	0.39	0.13	0.07	0.24	0.08
22:5n-3	0.18	0.03	0.01	<0.01	0.09	0.10	0.09	0.08	<0.01
22:6n-3	0.53	0.10	0.07	0.36	0.57	0.23	0.21	0.11	0.06
Σ PUFA[d]	1.50	0.30	0.20	1.06	2.27[e]	0.86	1.57	1.43	0.33
Total lipid (g 100 g^{-1})	6.61	0.81	0.69	3.69	4.9	2.82	7.59	5.60	1.4
Cholesterol (mg 100 g^{-1})	58	60	30	68	—	58	47	66	—

[a] Adapted from US Department of Agriculture Handbook No. 8-15 (Exler, 1987).
[b] Adapted from Linko *et al.* (1992), 20 Aug. sample, using the lipid to fatty acid conversation formula of the USDA Handbook 8-15.
[c] Adapted from Aggelousis and Lazos (1991), using the lipid to fatty acid conversion formula and the USDA Handbook 8-15.
[d] Total includes other fatty acids not listed.
[e] Total includes 0.09 g of 22:5n-6 and 0.09 g of 20:4n-3.

Farmed Fish

Studies of farmed fish include industries in many parts of the world (Hillcoat, 1990), in both salt and fresh water (Table 4.10). They are worthy of a long review, but despite a tendency for the muscle fats to reflect dietary fats (Ackman, 1989a; Polvi and Ackman, 1992) the range of rearing conditions is so enormous that only a few examples can be discussed (Table 4.10). The Atlantic salmon is farmed in various parts of the world, especially in Norway, Canada, Scotland and Chile. A characteristic of the fat of this species, farmed or wild, is a high content of 22:6n-3 (Cronin *et al.*, 1991). The popular milkfish *Chanos chanos* of southeast Asia is sometimes raised in ponds naturally flooded by the sea so one stock is from 'brackish' waters. Perhaps this accounts for the 20:5n-3 typical of various marine invertebrates which are part of their diet. The Japanese raise red sea bream *Pagrus major* by several techniques, including sea 'ranching', but all are marine. A comparison of many essential nutrients in wild and farmed fish and shellfish from the USA (Nettleton and Exler, 1992) will suggest how diverse this industry has become.

Shellfish and Other Invertebrates

Shellfish foods broadly include crustacea such as shrimp and lobster, and molluscs. The latter are mostly bivalves such as oysters, scallops and mussels. Technically the octopus and squid are also molluscs. The lipid content of most shellfish food samples is generally in the 1–2% range, although lipid as low as 0.5% is not unusual and as much as 3% has been reported. A basic reason for this low level of lipid is that bivalves store their energy surplus or reserve as glycogen and not as depot fat, while the crustacea, octopuses and squid have digestive glands (sometimes termed the hepatopancreas) where fat can be stored temporarily. In some cultures only selected parts are eaten, for example only the white adductor muscle of the scallop is eaten in North America, whereas the whole animal is eaten if the shellfish are oysters, clams, mussels, etc.

The important fatty acids of typical molluscs are given in Table 4.11. A peculiarity of molluscs is the presence of C_{20} and C_{22} non-methylene-interrupted dienoic (NMID) acids. These are usually the $\Delta5,\Delta11$–20:2; $\Delta5,\Delta13$–20:2; $\Delta7,\Delta13$–22:2; and $\Delta7,\Delta15$–22:2 isomers. In packed column gas–liquid chromatography these have usually been reported as 20:1 and 22:1. The latter fatty acid is present in bivalve lipids in only trace amounts and so 22:1 is not listed in this table. The origin and function of NMID are mysterious (Zhukova, 1986; Zhukova and Svetashev, 1986). Both saturated and monoethylenic fatty acids are in fact rather unimportant in compositions listed for cold-water bivalve molluscs (Takagi *et al.*, 1986;

Table 4.10. Comparison of major or important fatty acids (%w/w) of some farmed and/or wild fish from different parts of the world.

	Salmon, Atlantic[a] (*Salmo salar*)		Milkfish[b] (*Chanos chanos*)		Red sea bream[c] (*Pagrus major*)	American eel[d] (*Anguilla rostrata*)		Ayu[e] (*Plecoglossus altivelis*)		Tilapia[f] (*Tilapia* sp.)
	Wild	Farmed	Brackish	Freshwater	Retail	Wild	Farmed	Wild	Farmed	Wild
Fatty acid										
14:0	2.4	4.9	5.9	4.3	2.0	3.3	2.8	5.0	4.4	5.2
16:0	11.2	13.0	14.0	8.9	17.6	15.5	22.3	26.4	30.0	30.2
18:0	3.8	2.8	5.5	16.5	7.8	4.5	5.3	2.9	3.4	5.6
Σ Saturated	17.4	21.9	23.4	29.7	27.4	23.3	30.4	37.0	39.1	40.9
16:1	4.5	6.7	1.6	15.8	8.6	10.5	13.2	18.6	10.0	15.6
18:1	24.0	10.0	10.2	13.0	20.9	27.3	21.5	11.6	20.6	16.5
20:1	4.0	12.0	5.8	1.9	1.6	2.2	2.7	—	4.9	4.7
22:1	5.0	(12.0)	6.1	1.3	0.9	—	—	—	2.8	—
Σ MUFA	37.4	40.7	23.6	32.0	32.0	40.0	37.4	31.2	38.3	36.2
18:2n-6	3.1	6.7	1.1	2.5	2.1	8.9	0.7	5.6	7.4	2.6
20:4n-6	4.7	(0.8)	—	—	2.6	4.4	2.2	0.9	trace	1.3
18:3n-3	5.2	1.1	3.8	7.9	—	2.6	0.2	10.4	0.9	0.3
18:4n-3	1.5	1.8	4.2	1.8	—	—	—	—	—	2.3
20:5n-3	5.7	6.9	3.0	0.1	4.3	2.9	6.1	4.1	4.8	1.0
22:5n-3	4.5	—	trace	—	3.5	2.7	2.4	1.5	1.7	3.2
22:6n-3	19.8	14.6	0.4	2.2	17.8	5.4	15.0	1.8	7.0	3.4
Σ PUFA	45.2	31.9	12.6	14.4	30.3	30.0	28.6	31.8	22.6	14.7
% Lipid	6.34	10.85	4.5	4.9	2.8	3.4	14.7	10.7	3.1	4.6

[a] Marine. Adapted from Exler (1987). Figures in parentheses adjusted from original report where 22:1 was included in 20:4n-6.
[b] Adapted from Bautista *et al.* (1991).
[c] Marine. Adapted from Ohtsuru *et al.* (1984).
[d] Adapted from Otwell and Rickards (1981).
[e] Muscle triacylglyceride only. Adapted from Nakagawa *et al.* (1991).
[f] Freshwater. Adapted from Steiner-Asiedu *et al.* (1991).

Table 4.11. Major or important fatty acids of some US molluscs (as %w/w of fatty acids).[a]

	Whole body					Adductor muscle		
	Clam, hard (*Mercenaria mercenaria*)	Clam, soft (*Mya arenaria*)	Clam, surf (*Spisula solidissima*)	Oyster (Long Island) (*Crassostrea virginica*)	Oyster, Virginia (*Crassostrea virginica*)	Scallop, bay (*Pecten* sp.)	Scallop, calico (*Argopecten gibbus*)	Scallop, sea (*Placopecten magellanicus*)
Fatty acid								
14:0	2.2	2.2	2.3	8.1	3.4	2.5	2.6	3.2
16:0	8.8	12.5	8.3	22.4	18.4	8.3	10.6	10.0
18:0	4.1	3.0	2.9	3.5	1.9	2.2	2.6	2.0
Σ Saturated	15.1	19.9	13.5	34.0	23.7	13.0	15.8	15.2
16:1	3.1	4.5	3.1	4.3	2.8	3.1	2.1	3.1
18:1	5.3	6.2	5.0	4.6	3.8	3.7	3.0	3.3
20:1	1.2	1.1	3.7	2.4	2.1	2.9	1.2	1.5
Σ MUFA	9.6	11.8	11.8	11.3	8.7	9.7	6.3	7.9
18:2n-6	0.8	0.7	1.9	0.9	0.6	0.9	0.5	0.6
20:4n-6	4.1	4.4	4.2	1.7	2.1	6.8	6.8	3.1
22:4n-6	5.6	3.8	6.0	1.8	1.7	3.1	2.3	3.0
20:5n-3	4.5	11.2	11.4	7.6	12.2	19.3	21.7	20.6
22:5n-3	4.0	3.9	6.9	1.6	1.7	3.9	3.3	2.5
22:6n-3	15.1	13.4	11.5	15.2	15.6	21.6	27.1	27.5
Σ PUFA	34.1	37.4	41.9	30.6	33.9	55.6	61.7	57.3
% Lipid	0.38	2.18	0.88	1.93	1.57	0.84	0.89	1.0

[a] Adapted from Bonnet *et al.* (1974).

Exler, 1987), but possibly higher for those from warmer waters (Johns *et al.*, 1980). The data in Table 4.11 do not include a large number of 'unknowns' registered in the gas–liquid chromatography of that time. Although these animals are primarily plant (algae) feeders, fatty acids of odd-chain and branched-chain structures in such lipids attest to bacteria as being important in their diet (Napolitano and Ackman, 1993). Unusual C_{16} and C_{18} polyunsaturated fatty acids with n-1, n-4 and n-7 structures may be found as well as peaks for some isoprenoid and NMID fatty acids (Ackman, 1989c).

In octopus and squid (Table 4.12) only the outer tube and tentacles are commonly eaten. These seem again to have relatively low monounsaturated fatty acids but their somewhat higher content of saturated fatty acids suggest that the lipid is mostly phospholipids. The polyunsaturated acids are rather interesting since the lipids from the octopuses have some arachidonic acid, while those of the squid have very little. The latter, however, has a much higher level of docosahexaenoic acid (22:6n-3). The tunicate muscle is a food item popular in the Far East, Korean annual landings alone being in excess of 100,000 t. The broad similarity in the fatty acid compositions of the two mussel products of Table 4.12 (one from the Atlantic, the other from the Sea of Japan) is interesting, but these are not radically different from the other mollusc species of Table 4.11. South Pacific (Johns *et al.*, 1980) and North Pacific (Takagi *et al.*, 1986) reports confirm these basic fatty acid patterns for bivalves. The warm-water species may have more 20:4n-6 and less 22:6n-3, but 20:5n-3 of algal origin is almost always an important fatty acid.

An important point in respect to diets and filter-feeding molluscs is that the sterol content has been confused for years by obsolete methods (Gordon, 1982). Total sterols are meaningless and it is now established that in most molluscs over half of the total sterols are a mixture of the presumably beneficial phytosterols (Shimma *et al.*, 1979; Gordon, 1982; Krzynowek, 1985; Tucker, 1989; King *et al.*, 1990). As a result it is possible to say that in most bivalves as little as 20–40 mg 100 g^{-1} fresh meats would be actual cholesterol. The health benefits accruing from the low content of saturated fatty acids and low cholesterol, combined with the high content of n-3 polyunsaturated fatty acids, have been demonstrated (Connor and Lin, 1982).

The crustacea (Table 4.13) are generally good sources of polyunsaturated n-3 fatty acids and are also low in saturated fatty acids, but are not low in cholesterol. Nevertheless they were included in the diet study of Connor and Lin (1982) and can certainly be included in all low-fat diets for greater variety. Shrimp is often served breaded and deep-fried, diluting the content of cholesterol. The Swedish lobster or 'hummer' is recorded as having 0.2 g of cholesterol per 100 g of meat, but the usual figures for shrimp are 0.15–0.165 g 100 g^{-1}. The range 0.135–0.186 g 100 g^{-1} has

Table 4.12. Major or important fatty acids of some edible marine invertebrates.

	Octopus[a] (*Octopus vulgaris*)	Squid[b] (mixed species)	Tunicate muscle[c] (*Halocynthia noretzi*)	Sea-urchin gonad[d] (*Strongylocentrotus nudus*)	Mussel (whole body) (*Mytilus edulis*)		Crab[e]
					Cooked[e]	Fresh[f]	Cooked
Fatty acid							
14:0	4.0	3.6	5.5	8.3	2.5	2.5	1.4
16:0	23.6	26.6	10.5	14.4	12.2	16.6	9.2
18:0	8.0	5.9	5.6	1.9	7.4	4.4	4.3
Σ Saturated	35.6	36.1	21.6	25.5[g]	22.1	23.5	14.9
16:1	10.5	1.0	2.9	5.5	6.6	5.7	5.0
18:1	8.0	4.7	8.3	6.1	7.1	4.5	15.0
20:1	3.8	4.4	1.8	10.7	7.5	6.7	3.5
Σ MUFA	22.3	10.1	13.0	23.3[g]	21.2	16.9	23.5
18:2n-6	1.4	0.2	1.6	2.6	2.0	2.0	3.2
20:4n-6	6.2	0.9	4.3	15.4	4.5	6.1	0.6
22:4n-6	—	0.2	1.1	10.3	1.6	1.4	4.6
20:5n-3	11.5	14.8	22.0	15.6	11.3	13.6	21.5
22:5n-3	1.0	0.4	0.7	0.4	2.0	—	1.4
22:6n-3	12.9	34.6	17.7	1.0	10.1	17.7	10.5
Σ PUFA	33.0	51.1	47.4	51.1[g]	31.5	40.8	41.8
% Lipid	1.04	1.38	—	—	2.5	1.6	1.9
Cholesterol (mg 100 g^{-1})	48	233	—	—	60	—	100

[a] Adapted from Exler (1987).
[b] Adapted from Exler (1987); squid, mixed species, raw, includes Loligoidae and Ommastrephidae.
[c] Adapted from Vysotskii *et al.* (1992).
[d] Adapted from Ohshima *et al.* (1986).
[e] Adapted from Statens Livsmedelsverk (1988). Drained canned meats.
[f] Adapted from Romashina *et al.* (1987); translated (1988) from *Biologiya Morya* by Plenum Publishing Corporation.
[g] Total includes other fatty acids not listed.

Table 4.13 Major or important fatty acids of some US crustacea (as %w/w).

	Crab					Lobster		Shrimp		
	Blue[a] (*Callinectes sapidus*)	King[a] (*Paralithodes amschatica*)		Queen[b] (*Chionoecetes opilio*)	Dungeness[b] (*Cancer magister*)	Northern[b] (*Homarus americanus*) (cooked)	Spiny[a] (*Panulirus argus*)	Maine[a] (*Pandalus borealis*)	Brown[a] (*Penaeus aztecus*)	Freshwater[c] (*Machobrachium rosenbergii*) (farmed)
		Body	Leg							
Fatty acid										
14:0	1.4	1.3	2.3	0.5	0.1	1.7	1.1	1.9	1.8	4.1
16:0	9.6	9.9	9.2	14.5	12.9	22.6	7.6	10.1	10.4	24.8
18:0	4.3	1.5	1.7	2.4	6.6	5.6	6.6	1.0	6.4	11.8
Σ Saturated	15.3	12.7	13.2	17.4	19.6	29.9	15.3	13.0	18.6	40.7
16:1	5.8	3.7	5.6	6.1	6.9	8.4	4.6	6.6	5.1	4.8
18:1	10.9	7.6	9.7	22.4	15.1	26.5	9.9	9.9	7.4	25.7
20:1	3.2	0.7	1.5	2.7	1.8	4.7	1.1	0.7	0.8	1.2
Σ MUFA	19.9	12.0	16.8	31.2	23.8	39.6	15.6	17.2	13.3	31.7
18:2n-6	0.68	0.8	0.3	0.7	—	—	1.0	0.4	0.4	9.7
20:4n-6	5.2	4.7	5.2	4.0	—	—	23.1	2.3	7.5	1.1
22:4n-6	1.4	2.2	1.7	—	—	—	1.8	1.4	3.0	0.1
20:5n-3	19.6	36.0	33.2	31.5	35.6	14.8	16.5	30.8	18.4	6.2
22:5n-3	2.6	2.5	4.4	1.0	0.2	0.6	5.0	2.1	4.2	0.3
22:6n-3	19.9	13.5	13.5	13.8	14.3	8.6	10.2	20.8	17.3	2.1
Σ PUFA	49.3	59.7	58.3	51.0	50.1	24.0	57.6	57.8	50.8	19.5
% Lipid	1.88	1.27	1.73	1.18	0.97	0.59	1.20	1.32	1.19	—
Cholesterol (mg 100 g^{-1})	—	—	—	55	59	72	—	—	—	—

[a] Adapted from Bonnet *et al.* (1974).
[b] Adapted from Exler (1987).
[c] Adapted from Reigh and Stickney (1989).

been reliably reported (Krzynowek and Panunzio, 1989), and for crab and lobster cholesterol is usually in the range 0.06–0.10 g 100 g^{-1}. Very high values are recorded for many shellfish and crustacea by Deslypere (1990) for reasons that are obscure, but such listings illustrate the typical problems of attempts to simplify complex lipid matters for the lay-person. A better approach to such communications will be found in the recent presentation of King *et al.* (1990) combined with the older data of Gordon (1982) and the clinical study of Connor and Lin (1982).

Convenience Foods and the Modern Diet

There are no canned tuna or Pacific salmon products included in this review. The fatty acids of many of these products were included in a recent review (Ackman, 1989a) and others are reviewed by Cronin and O'Sullivan (1990). In addition to canned sardines (Krzynowek *et al.*, 1992), there are several species each of tuna and Pacific salmon and the tuna may be packed in either water or vegetable oil, making it difficult to include tuna in this brief review. Canned salmon has roughly 5% lipid and a fatty acid composition basically similar to that of the Atlantic salmon of Tables 4.5 and 4.10. Older literature (Bonnet *et al.*, 1974) is complemented by newer data (Exler, 1987; Cronin and O'Sullivan, 1990).

The consensus that all saturated fatty acids are bad has lately been eroded (Gurr, 1992), and this is crucial to wider biomedical acceptance of fish and shellfish, since these usually have 20–35% saturated acids, with proportions 16:0 > 14:0 > 18:0. In highly enriched fish oil concentrates of n-3 fatty acids almost all saturated fatty acids are eliminated (Ratnayake *et al.*, 1988; Breivik and Dahl, 1992). In addition to medical reviews (Leaf and Weber, 1988), entire books are devoted in large part to the health benefits of fish oils and their n-3 fatty acids (Stansby, 1990; Simopoulos *et al.*, 1991; Frölich and von Schacky, 1992; Drevon *et al.*, 1993). It is convenient to cite recent reviews on the subject (Rice, 1992; Uauy-Dagach and Valenzuela, 1992). The original findings of reduced cardiovascular mortalities through long-term regular fish consumption (Kromhout *et al.*, 1985) have never been refuted, and the recent studies in the UK (Burr *et al.*, 1989; Burr, 1991), encouraging consumption of fatty fish to reduce cardiovascular mortalities in patients already at risk, indicate that fish should be included as a 'health' food, no matter what the form in which it is eaten. The fact that the likely beneficial components, 20:5n-3 and 22:6n-3, are normal body components with possible different roles (Bønaa *et al.*, 1992), and not drugs with a defined and readily measurable effect, has hindered their acceptance. In particular it should be noted that the modern synthetic drugs often have drastic and objectionable side-effects, but side-effects of high-fish diets are seldom noted. Harris (1992) has summarized in a concise review the most recent research on the effects of fish oils (and hence of fish and shellfish) in humans. Although he points out the need for more research into the basic mechanisms of n-3 fatty acids and their impact on serum lipoproteins (Harris *et al.*, 1990), there seems to be a negative bias against accepting cardiovascular benefits from fish oil based on the fact that n-3 fatty acids do not appreciably reduce serum cholesterol. The reduction in serum lipids is mostly in triglycerides (Coniglio, 1992). Biomedical research is continuing, but few are 'practical' studies such as were recently summarized by Burr (1992). Dietary responses to short-term diets high in fish (Cobiac *et al.*, 1991; Gerhard *et al.*, 1991; Nelson *et al.*, 1991) are not unanimous in their findings, but then fish is becoming scarcer and more expensive, and confusion can be caused by 'farmed' versus 'wild' fish (Ackman, 1989a; Cronin *et al.*, 1991). Fortunately fish oil is plentiful (*c.* 1.4 Mt per annum) and refined concentrates are an alternative permitting more accurate control of intake of n-3 fatty acids for research than with fish alone. Despite a lack of accurate information on health benefits from a long-term intake of fish and shellfish, in terms of fish and shellfish lipids and fatty acids there is no obvious drawback and ample opportunity to enjoy a healthy and varied diet with any of the species listed in this review.

In 1994 the American Heart Association and other bodies convened a meeting (17–19 April) in Houston, Texas. The title 'Scientific Conference on Omega-3 Fatty Acids in Nutrition, Vascular Biology and Medicine' is clear evidence that such authorities have finally outgrown the long-lived view that for the Western diet a reduced intake of both cholesterol and saturated acid were the principal factors for improved cardiovascular health. Almost a decade ago a lecture was published (Ackman, 1988) under the title 'The Year of the Fish Oils'. This reflected the avid enthusiasm of commercial interests of that time for selling fish oils and concentrates. Despite important meetings in the last decade (Simopoulos *et al.*, 1986; 1991), this 1994 meeting may be the breakthrough long awaited for recognition of fish and fish fatty acids as factors for major health improvements.

References and Further Reading

Ackman, R.G. (1980) Fish lipids. Part I. In: Connell, J.J. (ed.) *Advances in Fish Science and Technology*. Fishing News Books, Farnham, pp. 87–103.

Ackman, R.G. (1982) Fatty acid composition of fish oils. In: Barlow, S.M. and Stansby, M.E. (eds) *Nutritional Evaluation of Long-chain Fatty Acids in Fish Oils*. Academic Press, London, pp. 25–88.

Ackman, R.G. (1986) WCOT (capillary) gas–liquid chromatography. In: Hamilton,

R.G. and Rossell, J.B. (eds) *Analysis of Oils and Fats*. Elsevier Applied Science Publishers, London, pp. 137–206.

Ackman, R.G. (1988) The year of the fish oils. *Chemistry and Industry* March 7, 139–145.

Ackman, R.G. (1989a) Nutritional composition of fats in seafoods. *Progress in Food and Nutrition Science* 13, 161–241.

Ackman, R.G. (1989b) Problems in fish oils and concentrates. In: Cambie, R.C. (ed.) *Fats for the Future*. Ellis Horwood, Chichester, pp. 189–204.

Ackman, R.G. (1989c) Fatty acids. In: Ackman, R.G. (ed.) *Marine Biogenic Lipids, Fats, and Oils, Vol. I.* CRC Press, Boca Raton, pp. 103–137.

Ackman, R.G. (1990) Variability of fatty acids and lipids in seafoods. *Omega 3 News* V(4), 1–4.

Ackman, R.G. (1991) Letters to the editor. *Omega 3 News* VI(3), 5–6.

Ackman, R.G. (1992a) Sources of n-3 fatty acids. In: Frölich, J.C. and Schacky, C. von (eds) *Fish, Fish Oil and Human Health, Clinical Pharmacology, Vol. 5.* W. Zuckschwerdt Verlag, München, pp. 14–24.

Ackman, R.G. (1992b) Bioavailability of omega-3 polyunsaturated fatty acids. In: Martin, R.E. and Flick, G.J. (eds) *Advances in Seafood Biochemistry: Composition and Quality*. Technomic Publishing Co., Lancaster, Pennsylvania, pp. 269–289.

Ackman, R.G. (1992c) Gas–liquid chromatography and the arachidonic acid content of Atlantic food fish. *Food Research International* 25, 453–456.

Ackman, R.G. and Cunnane, S.C. (1991) Long chain polyunsaturated fatty acids: Sources, biochemistry and nutritional/clinical applications. In: Padley, F.B. (ed.) *Advances in Applied Lipid Research*. JAI Press, Greenwich, Connecticut (Eurospan Group of Publishers, London), pp. 161–215.

Ackman, R.G. and Eaton, C.A. (1971) Mackerel lipids and fatty acids. *Canadian Institute of Food Technology Journal* 4, 169–174.

Ackman, R.G. and Eaton, C.A. (1976) Variations in fillet lipid content and some percent lipid–iodine value relationships for large winter Atlantic herring (*Clupea harengus harengus*) from southeastern Newfoundland. *Journal of the Fisheries Research Board of Canada* 33, 1634–1638.

Ackman, R.G. and Gunnlaugsdottir, H. (1992) Seafoods and fishery byproducts: Natural and unnatural environments for longer chain omega-3 fatty acids. In: St Angelo, A.J. (ed.) *Lipid Oxidation in Food*. American Chemical Society, Washington, DC, pp. 208–230.

Ackman, R.G. and McLeod, C. (1988) Total lipids and nutritionally important fatty acids of some Nova Scotia fish and shellfish food products. *Canadian Institute of Food Science and Technology Journal* 21, 390–398.

Ackman, R.G., Ke, P.J., MacCallum, W.A. and Adams, D.R. (1969) Newfoundland capelin lipids: Fatty acid composition and alterations during frozen storage. *Journal of the Fisheries Research Board of Canada* 26, 2037–2060.

Ackman, R.G., Sebedio, J.-L. and Kovacs, M.I.P. (1980) Role of eicosenoic and docosenoic fatty acids in freshwater and marine lipids. *Marine Chemistry* 9, 157–164.

Ackman, R.G., Orozco, V.R. and Ratnayake, W.M.N. (1991) Aspects of positional distribution of fatty acids in triacylglycerols of skin, white and dark muscle of mackerel *Scomber scombrus* in relation to hypertension. *Fat Science Technology* 93, 447–450.

Addison, R.F., Ackman, R.G. and Hingley, J. (1968) Distribution of fatty acids in cod flesh lipids. *Journal of the Fisheries Research Board of Canada* 25, 2083–2090.

Aggelousis, G. and Lazos, E.S. (1991) Fatty acid composition of the lipid from eight freshwater fish species from Greece. *Journal of Food Composition and Analysis* 4, 68–76.

Ågren, J.J., Muje, P., Hänninen, O., Herranen, I. and Penttilä, I. (1987) Seasonal variations of lipid fatty acids of boreal freshwater fish species. *Comparative Biochemistry and Physiology* 88B, 905–909.

Ågren, J.J., Al-Amad, H. and Hänninen, O. (1991a) Fatty acid content and composition of five fish species from the Persian Gulf. *Comparative Biochemistry and Physiology* 100B, 339–341.

Ågren, J.J., Pekkarinen, H., Litmanen, H. and Hänninen, O. (1991b) Fish diet and physical fitness in relation to membrane and serum lipids, prostanoid metabolism and platelet aggregation in female students. *European Journal of Applied Physiology and Occupational Physiology* 63, 393–398.

Bailey, R.S. (1992) The global pelagic fish resource and its biological potential. In: Burt, J.R., Hardy, R. and Whittle, K.J. (eds) *Pelagic Fish: The Resource and its Exploitation*. Fishing News Books, Oxford, pp. 1–20.

Bautista, M.N., del Valle, M.J. and Orejana, F.M. (1991) Lipid and fatty acid composition of brackishwater and freshwater-reared milkfish (*Chanos chanos* Froskal). *Aquaculture* 96, 241–248.

Bligh, E.G. (ed.) (1992) *Seafood Science and Technology*. Fishing News Books, Oxford, 406 pp.

Bonnet, J.C., Sidwell, V.D. and Zook, E.G. (1974) Chemical and nutritive values of several fresh and canned finfish, crustaceans, and mollusks. Part II: Fatty acid composition. *Marine Fisheries Review* 36(2), 8–14.

Bønaa, K.H., Bjerve, K.S. and Nordöy, A. (1992) Habitual fish consumption, plasma phospholipid fatty acids, and serum lipids: The Tromsø study. *American Journal of Clinical Nutrition* 55, 1126–1134.

Botta, J.R., Lauder, J.T., Downey, A.P. and Saint, W. (1983) Chemical and sensory assessment of nonspawning capelin (*Mallotus villosus*) subjected to long term frozen storage. *Journal of Food Science* 48, 1512–1515.

Breivik, H. and Dahl, K.H. (1992) Production and quality control of n-3 fatty acids. In: Frölich, J.C. and Schacky, C. von (eds), *Fish, Fish Oil and Human Health. Clinical Pharmacology, Vol. 5*. W. Zuckschwerdt Verlag, Munich, pp. 25–39.

Brown, A.J. and Roberts, D.C.K. (1991) Fish and fish oil intake: Effect on haematological variables related to cardiovascular disease. *Thrombosis Research* 64, 169–178.

Brown, A.J., Roberts, D.C.K., Pritchard, J.E. and Truswell, A.J. (1990) A mixed Australian fish diet and fish-oil supplementation: Impact on the plasma lipid profile of healthy men. *American Journal of Clinical Nutrition* 52, 825–833.

Bruckner, G. (1992) Fatty acids and cardiovascular diseases. In: Chow, C.K. (ed.)

Fatty Acids in Foods and their Health Implications. Marcel Dekker, New York, pp. 735–752.
Burr, M.L. (1991) Is oily fish good for the heart? *Trends in Food Science and Technology* 2(1), 17–20.
Burr, M.L. (1992) Epidemiology and large-scale studies. In: Frölich, J.C. and Schacky, C. von (eds) *Fish, Fish Oil and Human Health. Clinical Pharmacology, Vol. 5*. W. Zuckschwerdt Verlag, Munich, pp. 190–195.
Burr, M.L., Gilbert, J.F., Holliday, R.M., Elwood, P.C., Fehily, A.M., Rogers, S., Sweetnam, P.M. and Deadman, N.M. (1989) Effects of changes in fat, fish, and fibre intakes on death and myocardial reinfarction: Diet and reinfarction trial (DART). *The Lancet*, Sept. 30, 757–761.
Burt, J.R., Hardy, R. and Whittle, K.J. (eds) (1992) *Pelagic Fish: The Resource and its Exploitation*. Fishing News Books, Oxford, 352 pp.
Carlson, S.E., Cooke, R.J., Werkman, S.H. and Tolley, E.A. (1992) First year growth of preterm infants fed standard compared to marine oil n-3 supplemented formula. *Lipids* 27, 901–907.
Charnock, J.S., Abeywardena, M.Y., Tan, D. and McLennan, P.L. (1991) Omega-3 and omega-6 PUFA's have different effects on the phospholipid fatty acid composition of rat myocardial muscle when added to a saturated fatty acid dietary supplement. *Nutrition Research* 11, 1013–1024.
Cobiac, L., Clifton, P.M., Abbey, M., Belling, G.B. and Nestel, P.J. (1991) Lipid, lipoprotein, and hemostatic effects of fish vs fish-oil n-3 fatty acids in mildly hyperlipidemic males. *American Journal of Clinical Nutrition* 53, 1210–1216.
Coniglio, J.G. (1992) How does fish oil lower plasma triglycerides? *Nutrition Reviews* 50, 195–206.
Connor, W.E. and Lin, D.S. (1982) The effect of shellfish in the diet upon the plasma lipid levels in humans. *Metabolism* 31, 1046–1051.
Cronin, D.A. and O'Sullivan, K. (1990) Processed fish products as sources of n-3 polyunsaturated fatty acids. *Irish Journal of Food Science and Technology* 14, 17–24.
Cronin, D.A., Powell, R. and Gormley, R. (1991) An examination of the n-3 and n-6 polyunsaturated fatty acid status of wild and farmed Atlantic salmon (*Salmo salar*). *Irish Journal of Food Science and Technology* 15, 53–62.
Deslypere, J.P. (1990) Effect of fish consumption compared to intake of fish oil. In: Somogyi, J.C. and Hötzel, D. (eds) *Marine Foods*. Bibliotheca Nutritio et Dieta, No. 46. Karger, Basel, pp. 53–69.
Drevon, C.A. (1992) Marine oils and their effects. *Nutrition Reviews* 50, 38–45.
Drevon, C.A., Baksaas, I. and Krokan, H.E. (eds) (1993) *Omega-3 Fatty Acids: Metabolism and Biological Effects*. Birkhäuser Verlag, Basel, 389 pp.
Dyerberg, J. (1992) n-3 Fatty acids: The beginnings. In: Frölich, J.C. and von Schacky, C. (eds) *Fish, Fish Oil and Human Health. Clinical Pharmacology, Vol. 5*. W. Zuckschwerdt Verlag, Munich, pp. 1–13.
Dyerberg, J. and Bang, H.O. (1979) Lipid metabolism, atherogenesis and haemostasis in Eskimos: The role of the prostaglandin-3 family. *Haemostasis* 8, 227–233.
Dyerberg, J. and Jørgensen, K.A. (1992) Marine oils and thrombogenesis. *Progress in Lipid Research* 21, 255–269.

Eid, N., Dashti, B. and Sawaya, W. (1992) Chemical and physical characterization of shrimp by-catch of the Arabian Gulf. *Food Research International* 25, 181–186.

Erickson, M.C. (1992) Variation of lipid and tocopherol composition in three strains of channel catfish (*Ictalurus punctatus*). *Journal of the Science of Food and Agriculture* 59, 529–536.

Evans, A.J., Fogerty, A.C. and Sainsbury, K.J. (1986) The fatty acid composition of fish from the North West Shelf of Australia. *CSIRO Food Research Quarterly* 46, 40–45.

Exler J. (1987) *Composition of Foods: Finfish and Shellfish Products*. United States Department of Agriculture, Human Nutrition Information Service Agriculture Handbook 8-15 (updated 1992). Washington, DC, 192 pp.

Fletcher, A.M. (1989) *Eat Fish, Live Better*. Harper & Row, Publishers, New York, 378 pp.

Food and Drug Administration (1989) *The Fish List, FDA Guide to Acceptable Market Names for Food Fish Sold in Interstate Commerce 1988*. Superintendent of Documents, US Government Printing Office, Washington, DC, 50 pp.

Frölich, J.C. and von Schacky, C. (eds) (1992) *Fish, Fish Oil and Human Health. Clinical Pharmacology, Vol. 5*. W. Zuckschwerdt Verlag, Munich, 198 pp.

Gall, K.L., Otwell, W.S., Koburger, J.A. and Appledorf, H. (1983) Effects of four cooking methods on the proximate, mineral and fatty acid composition of fish fillets. *Journal of Food Science* 48, 1068–1073.

Gerhard, G.T., Patton B.D., Lindquist, S.A. and Wander, R.C. (1991) Comparison of three species of dietary fish: effects on serum concentrations of low-density-lipoprotein cholesterol and apolipoprotein in normotriglyceridemic subjects. *American Journal of Clinical Nutrition* 54, 334–339.

Gibson, R.A. (1983) Australian fish – An excellent source of both arachidonic and ω-3 polyunsaturated fatty acids. *Lipids* 18, 743–752.

Gooch, J.A., Hale, M.B., Brown, T., Jr, Bonnet, J.C., Brand, C.G. and Regier, L.W. (1987) *Proximate and Fatty Acid Composition of 40 Southeastern U.S. Finfish Species*. NOAA Technical Report NMFS 54, US Department of Commerce, Springfield, Virginia, 23 pp.

Gooch, J.A., Hale, M.B. and Galloway, S.B. (1989) *Proximate Chemical Composition and Fatty Acid Contents of 37 Finfish Species of the Southeastern United States*. US Department of Commerce, NOAA/NMFS Southeast Fisheries Center, Charleston Laboratory, Charleston, South Carolina, 11 pp.

Gordon, D.T. (1982) Sterols in mollusks and crustacea of the Pacific Northwest. *Journal of the American Oil Chemists' Society* 59, 536–545.

Gurr, M. (1992) Saturated fatty acids and hypercholesterolaemia: Are all saturated acids equal? *Lipid Technology* 4, 93–96.

Hänninen, O.O., Ågren, J.J., Laitinen, M.V., Jääskeläinen, I.O. and Penttilä, I.M. (1989) Dose–response relationships in blood lipids during moderate freshwater fish diet. *Annals of Medicine* 21, 203–207.

Hardy, R. (1980) Fish lipids. Part 2. In: Connell, J.J. (ed.) *Advances in Fish Science and Technology*. Fishing News Books, Farnham, pp. 103–111.

Hardy, R., McGill, A.S. and Gunstone, F.D. (1979) Lipid and autoxidative changes in cold stored cod (*Gadus morhua*). *Journal of the Science of Food and Agriculture* 30, 999–1006.

Harris, W.S. (1989) Fish oils and plasma lipid and lipoprotein metabolism in humans: A critical review. *Journal of Lipid Research* 30, 785–807.

Harris, W.S. (1992) The absorption of n-3 fatty acids in different chemical forms. In: Frölich, J.C. and Schacky, C. von (eds) *Fish, Fish Oil and Human Health. Clinical Pharmacology, Vol. 5*. W. Zuckschwerdt Verlag, Munich, pp. 40–45.

Harris, W.S., Connor, W.E., Roger-Illingworth, D., Rothrock D.W. and Foster, D.M. (1990) Effects of fish oil on VLDL triglyceride kinetics in humans. *Journal of Lipid Research* 31, 1549–1558.

Hearn, T.L., Sgoutas, S.A., Sgoutas, D.S. and Hearn, J.A. (1987) Stability of polyunsaturated fatty acids after microwave cooking of fish. *Journal of Food Science* 52, 1430–1431.

Hillcoat, J.B. (1990) Fish farming. In: Somogyi, J.C. and Hötzel, D. (eds) *Marine Foods*. Bibliotheca Nutritio et Dieta No. 46. Karger, Basel, pp. 10–21.

Ingemansson, T., Olsson, N.U., Herslöf, B.G. and Ekstrand, B. (1991) Lipids in light and dark muscle of farmed rainbow trout (*Oncorhynchus mykiss*). *Journal of the Science of Food and Agriculture* 57, 443–447.

Innis, S.M. (1992) n-3 Fatty acid requirements of the newborn. *Lipids* 27, 879–885.

Jamieson, G.R. (1975) GLC identification techniques for long-chain unsaturated fatty acids. *Journal of Chromatographic Science* 13, 491–501.

Jangaard, P.M., Brockerhoff, H., Burgher, R.D. and Hoyle, R.J. (1967a) Seasonal changes in general condition and lipid content of cod from inshore waters. *Journal of the Fisheries Research Board of Canada* 24, 607–612.

Jangaard, P.M., Ackman, R.G. and Sipos, J.C. (1967b) Seasonal changes in fatty acid composition of cod liver, flesh, roe, and milt lipids. *Journal of the Fisheries Research Board of Canada* 24, 613–627.

Johns, R.B., Nichols, P.D. and Perry, G.J. (1980) Fatty acid components of nine species of molluscs of the littoral zone from Australian waters. *Comparative Biochemistry and Physiology* 65B, 207–214.

van de Kamp, J. and Wykes, A.A. (1992) *Effects of Fish Oils and Polyunsaturated Omega-3 Fatty Acids in Health and Disease*. Special Bibliography 1992-A, US Department of Health and Human Services, Public Health Service, National Institutes of Health, Bethesda, Maryland, 36 pp.

Ke, P.J. and Ackman, R.G. (1976) Metal-catalyzed oxidation in mackerel skin and meat lipids. *Journal of the American Oil Chemists' Society* 53, 636–640.

Ke, P.J., Ackman, R.G., Linke, B.A. and Nash, D.M. (1977a) Differential lipid oxidation in various parts of frozen mackerel. *Journal of Food Technology* 12, 37–47.

Ke, P.J., Nash, D.M. and Ackman R.G. (1977b) Mackerel skin lipids as an unsaturated fat model system for the determination of antioxidative potency of TBHQ and other antioxidant compounds. *Journal of the American Oil Chemists' Society* 54, 417–420.

Keli, S.O., Feskens, E.J.M. and Kromhout, D. (1994) Fish consumption and risk of stroke: The Zutphen study. *Stroke* 25, 328–332.

King, I., Childs, M.T., Dorsett, C., Ostrander, J.G. and Monsen, E.R. (1990)

Shellfish: Proximate composition, minerals, fatty acid, and sterols. *Journal of the American Dietetic Association* 90, 677–685.

Kinsella, J.E. (1986) Food components with potential therapeutic benefits: The n-3 polyunsaturated fatty acids of fish oils. *Food Technology* 40(2), 89–97, 146.

Kinsella, J.E. (1987) *Seafoods and Fish Oils in Human Health and Disease*. Marcel Dekker, New York, 317 pp.

Kromhout, D., Bosschieter, E.B. and Lezenne Coulander, C. de (1985) The inverse relation between fish consumption and 20-year mortality from coronary heart disease. *New England Journal of Medicine* 312, 1205–1209.

Krzynowek, J. (1985) Sterols and fatty acids in seafood. *Food Technology* 39(2), 61–68.

Krzynowek, J. and Murphy, J. (1987) *Proximate Composition, Energy, Fatty Acid, Sodium and Cholesterol Content of Finfish, Shellfish and their Products*. NOAA Technical Report NMFS 55, US Department of Commerce, Springfield, Virginia, 53 pp.

Krzynowek, J. and Panunzio, L.J. (1989) Cholesterol and fatty acids in several species of shrimp. *Journal of Food Science* 54, 237–239.

Krzynowek, J., Murphy, J., Maney, R.S. and Panunzio, L.J. (1989) *Proximate Composition and Fatty Acid and Cholesterol Content of 22 Species of Northwest Atlantic Finfish*. NOAA Technical Report NMFS 74, US Department of Commerce, Springfield, Virginia, 35 pp.

Krzynowek, J., Uljua, D.S., Panunzio, L.J. and Maney, R.S. (1992) Factors affecting fat, cholesterol, and omega-3 fatty acids in Maine sardines. *Journal of Food Science* 57, 63–65, 111.

Lands, W.E.M. (1986) *Fish and Human Health*. Academic Press, Orlando, Florida, 170 pp.

Leaf, A. and Weber, P.C. (1988) Cardiovascular effects of n-3 fatty acids. *New England Journal of Medicine* 318, 549–557.

Linko, R.R., Rajasilta, M. and Hiltunen, R. (1992) Comparison of lipid and fatty acid composition in vendace (*Coregonus albula* L.) and available plankton feed. *Comparative Biochemistry and Physiology* 103A, 205–212.

Marais, J.F.K. (1990) Body composition of ten marine migrants and one freshwater fish species caught in estuaries of the Eastern Cape, South Africa. *South African Journal of Marine Science* 9, 135–140.

Marshall, D.W. and Currall, J. (1992) Consumer attitudes towards pelagic fish. In: Burt, J.R., Hardy, R. and Whittle, K.J. (eds) *Pelagic Fish: The Resource and its Exploitation*. Fishing News Books, Oxford, pp. 260–266.

Morita, I., Kanayasu-Toyoda, T. and Murota, S. (1993) Beneficial effects of EPA on anti-atherogenesis. In: Drevon, C.A., Baksaas, I. and Krokan, H.E. (eds) *Omega-3 Fatty Acids: Metabolism and Biological Effects*. Birkhäuser Verlag, Basel, pp. 223–229.

Nair, P.G.V. and Gopakumar, K. (1978) Fatty acid compositions of 15 species of fish from tropical waters. *Journal of Food Science* 43, 1162–1164.

Nair, P.G.V, Antony, P.D. and Gopakumar, K. (1979) Oxidative rancidity in the skin and muscle lipids of oil sardine (*Sardinella longiceps*). *Journal of Food Science and Technology* 16, 151–154.

Nakagawa, H., Takahara, Y. and Nematipour, G.R. (1991) Comparison of lipid properties between wild and cultured ayu. *Nippon Suisan Gakkaishi* 57, 1965–1971.

Napolitano, G.E. and Ackman, R.G. (1993) Fatty acid dynamics in sea scallops *Placopecten magellanicus* (Gmelin) from Georges Bank, Nova Scotia. *Journal of Shellfish Research* 12, 267–277.

Nelson, G.J., Schmidt, P.C. and Corash, L. (1991) The effect of a salmon diet on blood clotting platelet aggregation and fatty acids in normal adult men. *Lipids* 26, 87–96.

Nestel, P.J. (1987) Polyunsaturated fatty acids (n-3, n-6). *American Journal of Clinical Nutrition* 45, 1161–1167.

Nettleton, J.A. (1985) *Seafood Nutrition*. Osprey Books, Huntington, New York, 280 pp.

Nettleton, J.A. (1987) *Seafood and Health*. Osprey Books, Huntington, New York, 234 pp.

Nettleton, J.A. (1993) Are n-3 fatty acids essential nutrients for fetal and infant development? *Journal of the American Dietetic Association* 93, 58–64.

Nettleton, J.A. and Exler, J. (1992) Nutrients in wild and farmed fish and shellfish. *Journal of Food Science* 57, 257–260.

Nettleton, J.A., Allen, W.H., Klatt, L.V., Ratnayake, W.M.N. and Ackman, R.G. (1990) Nutrients and chemical residues in one- to two-pound Mississippi farm-raised channel catfish (*Ictalurus punctatus*). *Journal of Food Science* 55, 954–958.

Neuringer, M., Connor, W.E. and Anderson, G.J. (1992) The role of n-3 fatty acids in the development of the retina and brain. In: Frölich, J.C. and von Schacky, C. (eds) *Fish, Fish Oil and Human Health. Clinical Pharmacology Vol. 5*. W. Zuckschwerdt Verlag, Munich, pp. 59–71.

Newman, W.P., Middaugh, J.P., Propst, M.T. and Rogers, D.R. (1993) Atherosclerosis in Alaska natives and non-natives. *The Lancet* 342, 1056–1057.

Nordöy, A., Hatcher, L.F., Ullmann, D.L. and Connor, W.E. (1993) Individual effects of dietary saturated fatty acids and fish oil on plasma lipids and lipoproteins in normal men. *American Journal of Clinical Nutrition* 57, 634–639.

Nunes, M.L., Batista, I. and Morão de Campos, R. (1992) Physical, chemical and sensory analysis of sardine (*Sardina pilchardus*) stored in ice. *Journal of the Science of Food and Agriculture* 59, 37–43.

Obatake, A. and Heya, H. (1985) A rapid method to measure dark muscle content in fish. *Bulletin of the Japanese Society of Scientific Fisheries* 51, 1001–1004.

Ohshima, T., Wada, S. and Koizumi, C. (1986) Lipid deterioration of salted gonads of sea urchin during storage at 5 °C. *Bulletin of the Japanese Society of Scientific Fisheries* 52, 511–517.

Ohta, S., Shinozaki, R., Sasaki, K., Ikuma, T. and Kamimura, S. (1988) Changes in fatty acid composition during the cookings of coho salmon. *Yukagaku* 37, 663–667.

Ohtsuru, M., Fujii, M., Ishinaga, M. and Kito, M. (1984) Fatty acid composition of fish (Fatty acid compositions of fish caught in seas around Yamaguchi Prefecture). *Journal of the Agricultural and Chemical Society of Japan* 58, 35–42.

Orlick, B., Oehlenschläger, J. and Schreiber, W. (1991) Changes in lipids and nitrogenous compounds in cod (*Gadus morhua*) and saithe (*Pollachius virens*) during frozen storage. *Archiv für Fischereiwissenschaft* 41, 89–99.

Otwell, W.S. and Rickards, W.L. (1981) Cultured and wild American eels, *Anguilla rostrata*: Fat content and fatty acid composition. *Aquaculture* 26, 67–76.

Paul, A.A., Southgate, D.A.T. and Russell, J. (1980) *First Supplement to McCance and Widdowson's 'The Composition of Foods'*. Her Majesty's Stationery Office, London, and Elsevier/North-Holland Biomedical Press, Amsterdam, 113 pp.

Polvi, S.M. (1989) Diet and availability of omega-3 fatty acids in salmonids. MSc thesis, Technical University of Nova Scotia, Halifax, 221 pp.

Polvi, S.M. and Ackman, R.G. (1992) Atlantic salmon (*Salmo salar*) muscle lipids and their response to alternative dietary fatty acid sources. *Journal of Agricultural and Food Chemistry* 40, 1001–1007.

Porter, P.J., Kramer, D.E. and Kennish, J.M. (1992) Lipid composition of light and dark flesh from sockeye salmon. *International Journal of Food Science and Technology* 27, 365–369.

Ratnayake, W.M.N., Olsson, B., Matthews, D. and Ackman, R.G. (1988) Preparation of omega-3 PUFA concentrates from fish oils via urea complexation. *Fat Science Technology* 90, 381–386.

Reigh, R.C. and Stickney, R.R. (1989) Effects of purified dietary fatty acids on the fatty acid composition of freshwater shrimp, *Machobrachium rosenbergii. Aquaculture* 77, 157–174.

Rice, R. (1992) Fish oil – How can the balance be redressed? *Lipid Technology* 4, 14–18.

Romashina, N.A., Zhukova, N.V. and Sheina, V.P. (1988) Lipids and fatty acids of common mussels reared in Vostok Bay, Sea of Japan. Comparative Biochemistry, Plenum Publishing Corporation, translated from *Biologiya Morya, Vladivostok*, No. 3, pp. 14–17, May–June 1987. (*Chemical Abstracts* 107 (1987) 216337r.)

Sagredos, A.N. (1991) Fatty acid composition of fish oil capsules. *Fat Science Technology* 93, 184–191 (in German).

Saynor, R. and Ryan, F. (1990) *The Eskimo Diet: How to Avoid a Heart Attack*. Ebury Press, London, 192 pp.

von Schacky, C. and Weber, P.C. (1985) Metabolism and effects on platelet function of the purified eicosapentaenoic and docosahexaenoic acid in humans. *Journal of Clinical Investigation* 76, 2446–2450.

Schmidt, E.B., Varming, K., Svaneborg, N. and Dyerberg, J. (1992) n-3 Polyunsaturated fatty acid supplementation (Pikasol®) in men with moderate and severe hypertriglyceridaemia: A dose–response study. *Annals of Nutrition and Metabolism* 36, 283–287.

Shimma, H., Kinumaki, T. and Ota Y. (1979) Determination of cholesterol in fishery products. *Bulletin of the Tokai Regional Fisheries Research Laboratory* 54(3), 29–44.

Shindo, K., Tsuchiya, T. and Matsumoto, J.J. (1986) Histological study on white and dark muscles of various fishes. *Bulletin of the Japanese Society of Scientific Fisheries* 52, 1377–1399.

Simopoulos, A.P., Kifer, R.R. and Martin, R.E. (eds) (1986) *Health Effects of Polyunsaturated Fatty Acids in Seafoods*. Academic Press, Orlando, Florida, 473 pp.

Simopoulos, A.P., Kifer, R.R., Martin, R.E. and Barlow, S.M. (1991) *Health Effects of ω3 Polyunsaturated Fatty Acids in Seafoods*. World Review of Nutrition and Dietetics, Volume 66. Karger, Basel, 592 pp.

Sinclair, A. and O'Dea, K. (1991) Letters to the editor. *Omega 3 News* VI(3), 4–5.

Sinclair, A.J., O'Dea, K. and Naughton, J.M. (1983) Elevated levels of arachidonic acid in fish from northern Australian coastal waters. *Lipids* 18, 877–881.

Sinclair, A.J., O'Dea, K., Dunstan, G., Ireland, P.O. and Niall, M. (1987) Effects on plasma lipids and fatty acid composition of very low fat diets enriched with fish or kangaroo meat. *Lipids* 22, 523–529.

Somogyi, J.C. and Hötzel, D, (eds) (1990) *Marine Foods*, Bibliotheca Nutritio et Dieta, No. 46. Karger, Basel, 129 pp.

Stansby, M.E. (1990) Nutritional properties of fish oil for human consumption. In: Stansby, M.E. (ed.) *Fish Oils in Nutrition*. Van Nostrand Reinhold, New York, pp. 268–308.

Statens Livsmedelsverk, Sweden (1988) *Livsmedelstabeller* (Food composition tables), 398 pp.

Steiner-Asiedu, M., Julshamn, K. and Lie, Ø. (1991) Effect of local processing methods (cooking, frying and smoking) on three fish species from Ghana: Part I. Proximate composition, fatty acids, minerals, trace elements and vitamins. *Food Chemistry* 40, 309–321.

Takagi, T., Itabashi, Y. and Kaneniwa, M. (1986) Fatty acid composition of bivalves from Japanese waters. *Yukagaku* 35, 517–521.

Toyomizu, M., Hanaoka, K. and Nakamura, T. (1980) Lipid oxidation in the skin during storage of fish in the round at −5°C and the susceptibility predictable for lipid oxidation. *Bulletin of the Japanese Society of Scientific Fisheries* 46, 1011–1017.

Tucker, B.W. (1989) Sterols in seafood: A review. *World Aquaculture* 20(1), 69–72.

Uauy-Dagach, R. and Valenzuela, A. (1992) Marine oils as a source of omega-3 fatty acids in the diet: How to optimize the health benefits. *Progress in Food and Nutrition Science* 16, 199–243.

Vysotskii, M.V., Ota, T. and Takagi, T. (1992) n-3 Polyunsaturated fatty acids in lipids of ascidian *Halocynthia roretzi. Bulletin of the Japanese Society of Scientific Fisheries* 58, 953–958.

Wallace, P.D. (1988) A note on the seasonal change in fat content of the autumn-spawning herring in the northern Irish Sea. *Journal of the Marine Biological Association of the United Kingdom* 66, 71–74.

Welch, V.A. and Borlakoglu, J.T. (1992) Absorption and transport of dietary lipid: Effect on some lipid-related health problems. In: Chow, C.K. (ed.) *Fatty Acids in Foods and their Health Implications*. Marcel Dekker, New York, pp. 559–612.

Whittle, K.J. and Wood, C.D. (1992) Utilization of pelagic fish for human consumption. In: Burt, J.R., Hardy, R. and Whittle, K.J. (eds) *Pelagic Fish: The Resource and Its Exploitation*. Fishing News Books, Oxford, pp. 238–253.

Yamada, J. (1972) Histochemical observation of fish muscle – II. Occurrence and status of lipid and muscle texture of tunas. *Bulletin of the Tokai Regional Fisheries Research Laboratory* 72, 35–42.

Yamada, J. (1979) Lipid oxidation in various tissues of sardine, *Sardinops melanosticta. Bulletin of the Tokai Regional Fisheries Research Laboratory* 99, 23–29.

Yamada, J. (1981) Content and distribution of lipid in sardine flesh. *Bulletin of the Tokai Regional Fisheries Research Laboratory* 104, 103–109.

Yamada, J. and Nakamura, S. (1964) Histochemical studies on fish muscle. I. Distribution of depot fat in fish muscle tissue. *Bulletin of the Tokai Regional Fisheries Research Institute* 39, 21–28.

Yazawa, K. and Kageyama, H. (1991) Physiological activity of docosahexaenoic acid. *Yukagaku* 40, 974–978.

Zhukova, N.V. (1986) Biosynthesis of non-methylene-interrupted dienoic fatty acids from [^{14}C]acetate in molluscs. *Biochimica et Biophysica Acta* 878, 131–133.

Zhukova, N.V. and Svetashev, V.I. (1986) Non-methylene-interrupted dienoic fatty acids in molluscs from the Sea of Japan. *Comparative Biochemistry and Physiology* 83B, 643–646.

Vitamins in Fish and Shellfish

5

SANTOSH P. LALL[1] AND MONINA P. PARAZO[2]
[1]*Biological Sciences Branch, Department of Fisheries and Oceans, Halifax, Nova Scotia, Canada;* [2]*Canadian Institute of Fisheries Technology, Technical University of Nova Scotia, Halifax, Nova Scotia, Canada.*

Introduction

Vitamins are complex organic compounds present in minute amounts as natural components of food that are essential for normal physiological functions. Their absence from the diet causes specific deficiency syndromes. Classically, vitamins have been divided into two groups, fat-soluble (A, D, E and K) and water-soluble (B complex and C), based on their solubilities in fat solvents or in water. Fat-soluble vitamins may be stored in appreciable amounts in the animal body. Except for vitamin B_{12}, water-soluble vitamins are not stored and are readily excreted. A regular dietary supply of the water-soluble vitamins is needed to prevent vitamin deficiencies.

Historically, nutritionists have been concerned with vitamins for their role in preventing vitamin-deficiency diseases such as beriberi, pellagra scurvy and night blindness. The medicinal use of fish liver oil was common long before the discovery of vitamins and their related deficiency syndromes. Early medical records show that fish livers were used in the treatment of rheumatism, bone disorders, tuberculosis and other diseases (Drummond and Hilditch, 1930). In the early 19th century the use of cod liver oil in treating rickets and night blindness was demonstrated (Chabre, 1936). The demand for fish liver oils as vitamin supplements promoted extensive research on vitamins A and D; and created a new industry in the first half of the 20th century. Many fish species, including cod, pollock, haddock, halibut, mackerel, tuna, swordfish and shark, and marine mammals were examined as potential sources of these vitamins. Ultimately, commercial production of cod liver oil became successful and constituted 90% of the concentrated vitamin-containing oil marketed. In the mid-1950s, commercially produced synthetic vitamins gradually replaced the use of fish liver oil as the preferred vitamin supplement.

Data showing the vitamin content of raw fish and shellfish provide useful information when evaluating the adequacy of foods for human consumption. However, estimates of the nutrient intake of individuals based upon these data are seldom accurate because of factors that may alter the vitamin composition of food before it is actually ingested. Sources of error include analytical errors in sampling and measuring the vitamin; variations in the amount of vitamin present; bioavailability of the form of vitamin present; and losses during storage, processing and cooking.

A substantial amount of information on the vitamin composition of fish and shellfish was published during the period from 1930 to the mid-1960s. However, it is well known that the vitamin content of foods is more variable than that of other dietary constituents and for many fish and shellfish the range of published values is very wide. Some of this variability is due to differences in the analytical techniques used to measure these vitamins. Over the last 25 years there have been major advances in many fields of analytical chemistry including vitamin assays. Preliminary isolations have developed from simple extraction and phase separation techniques to swift and powerful chromatographic methods now used.

The first step in any vitamin assay is to extract the raw tissue or processed food matrix to obtain a vitamin-rich fraction. Fat-soluble vitamins are prone to decomposition when exposed to oxygen, heat and light, hence factors affecting vitamin stability require special attention. Purification of vitamin-rich fractions is sometimes necessary before the vitamin(s) can be quantified. Purification procedures rely largely upon some form of chromatography, while quantification procedures range from older, well-established techniques such as colorimetry and spectrophotometry, to sophisticated modern techniques, particularly high-performance liquid chromatography (HPLC). HPLC has been used successfully to measure all fat-soluble vitamins and many water-soluble vitamins in both tissues and foods. HPLC has also improved the accuracy of quantitative measurements by eliminating interfering materials and permitted the measurement of different forms of each vitamin, their isomers and derivatives. A detailed discussion of vitamin assays is beyond the scope of this chapter but the above should provide a description of both the techniques used, and the potential for errors in vitamin analysis.

Although the vitamin content of fish varies with species, age, season, and geographic location, composition of the natural diet and the amount of food consumed may have a greater effect on the deposition of nutrients (Stansby, 1962).

Many species of fish store large amounts of vitamins A and D in the liver. The concentrations of these vitamins vary substantially even within one species (Billis *et al.*, 1935). Some sharks store enormous quantities of vitamin A while others store very little. Many lean fish such as cod store high amounts of fat-soluble vitamins in the liver but not in the flesh. The lipid composition of fish changes with the availability of food; however, an inverse relationship exists between the amount of oil in the liver and its vitamin A potency (Drummond and Hilditch, 1930). Both vitamin A and D continue to accumulate in the liver with increasing age.

Few studies have been conducted to determine the factors affecting the distribution of water-soluble vitamins in fish flesh and liver. Fish nutrition studies show that a plateau is reached when graded levels of water-soluble vitamins are incorporated into fish diets (National Research Council, 1993). During sexual maturity, some mobilization of vitamins from muscle and other tissues may occur, if the diet is deficient in a specific vitamin. Wild fish may also show some differences in water-soluble vitamin composition during migration, maturation, and when food is scarce. However, these differences are not as extreme as those noted for fat-soluble vitamins.

Several reviews (Higashi, 1961; Braekkan, 1962a; 1962b; Cruickshank, 1962; Love, 1970; Sidwell *et al.*, 1978) and survey reports (Dudek *et al.*, 1982; Gordon and Martin, 1982) have been published on vitamin composition of fish and shellfish. Effects of processing on the nutritive value of fish have been dealt with in more detail by Schroeder (1971) and March (1982). Nettleton (1985) summarized the concentration of certain vitamins in selected New Zealand marine fishes. In Tables 5.1 to 5.11, the vitamin composition of common edible fish and shellfish are summarized after a careful review of published data. In some cases, extreme values reported in older publications have been excluded because of problems with analytical methodology and the limited number of samples analysed. Where complete results were not available for individual species, e.g. oysters, data are summarized for the group as a whole. Where only a single value for a vitamin is shown in the tables, this value represents a report from a single publication. In some cases, the range of reported values for a particular vitamin in the flesh of the same species was quite large. It was impossible to cite all the references and it is suggested that reviews of Higashi (1961), Braekkan (1962a; 1962b) and Cruickshank (1962) may be used to locate data published before 1960.

The recommended dietary allowances (RDAs) for various vitamins as established by the Food and Nutrition Board of the National Academy of Sciences, USA, are summarized in Table 5.12. RDAs are used to define a relationship between human populations and the nutrients required at different stages of life. They are intended to serve as a basis for evaluating the adequacy of diets of groups of people rather than individuals. This information, in conjunction with vitamin composition tables, can be used with caution to estimate the proportion of individual vitamins in human diets that can be supplied by fish and shellfish. The limitations of these vitamin composition tables have been already emphasized.

Fat-soluble Vitamins

Vitamin A

Vitamin A has essential roles in vision, growth, bone development, reproduction and normal maintenance of epithelial tissues of humans, mammals, birds, fishes and many lower forms of life. Generally, vitamin A activity refers to β-ionone derivatives, which have the biological activity of all-*trans*-retinol. The most significant retinoids in animal metabolism are the alcohol (all-*trans*-retinol), the aldehyde (11-*cis*-retinal and 11-*cis*-3-dehydroretinal), and the acid (all-*trans*-retinoic acid) forms including retinyl esters (e.g. retinyl palmitate) and retinyl β-glucuronide. All three forms are found in two variants: with the β-ionone nucleus (vitamin A_1) or the dehydrogenated β-ionone nucleus (vitamin A_2). However, the former is both quantitatively and qualitatively more important as a source of vitamin A activity. The term vitamin 'A' usually refers to vitamin A_1.

Vitamin A_1 (retinol) is found in high proportions in marine fishes, whereas vitamin A_2 (3,4-dehydroretinol) is the predominant form in freshwater fish. A small amount ($< 5\%$) of vitamin A_2 is found in most marine fish (Morton and Bro-Rasmussen, 1955). The ratio of vitamin A_1 and A_2 in freshwater fish differs widely. In liver oils of fish caught in Russian rivers, the vitamin A_2 to A_1 ratio was 2:1 (Gillam *et al.*, 1938) whereas the ratio for the liver oils of Indian species was 9:1 (Balasunddaram *et al.*, 1956).

In freshwater fish, oxidative conversion of retinol to 3,4-dehydroretinol occurs (Goswami, 1984), as well as the reversible conversion of retinol to retinal and of 3,4-dehydroretinol to 3,4-dehydroretinal (Wald, 1945–46). Channel catfish convert β-carotene to vitamin A_1 and A_2 in about a 1:1 ratio (Lee, 1987). Lutein is an effective precursor for dehydroretinol in *Heteropneustes fossilis* and anhydrolutein is an intermediate in this transformation (Barua, 1978). In tilapia liver, β-carotene and canthaxanthin are converted into retinol while dihydroxycarotenoids such as astaxanthin, zeaxanthin, lutein and tunaxanthin are converted into 3,4-dehydroretinol (Katsuyama and Matsuno, 1988). In contrast to fish, 3,4-dehydroretinol is found in small quantities relative to retinol in warm-blooded vertebrates. However, this compound supports all vitamin A dependent functions in warm-blooded animals (Sundaresan and Cama, 1961).

Fish oils constitute the chief natural source of vitamin A, but extreme differences in vitamin A levels are observed in various species (Table 5.1). Elasmobranchs (sharks and rays, etc.) accumulate lipid and vitamin A in the liver, but wide variations exist among species. The soupfin shark liver is considered to the richest source of vitamin A; skates and rays accumulate low levels of this vitamin. Unlike elasmobranchs, bony fishes (teleosts) deposit fat in liver and other tissues but the vitamin A content of these

Table 5.1 Vitamin A and D contents of fish liver oils.[a]

Species	Liver oil (%)	Vitamin A (IU g^{-1})[b]	Vitamin D (IU g^{-1})
Marine fish			
Bonito, Pacific		57,000	35,000
Cod	20-70	400-10,000	60-300
Dogfish	13-75	11,000-150,000	3-30
Haddock	40-85	150-3,000	50-75
Halibut	15-25	6,000-330,000	200-5,000
Ling	33-38	2,000-16,000	400
Lingcod	8-20	1,500-30,000	200-6,000
Mackerel	5-20	3,000-163,000	6,300
Pollack, Atlantic	18-30	800-2,500	80-100
Rockfish	7-29	13,000-320,000	
Salmon, Chinook	8-10	25,000-40,000	180-250
Shark, soupfin	45-70	74,000-190,000	13-50
Swordfish	8-35	20,000-400,000	2,000-2,500
Tuna, albacore	7-20	10,000-60,000	20,000-250,000
Tuna, bluefin	9-35	25,000-100,000	16,000-42,000
Tuna, skipjack	4-6	30,000-60,000	25,000-250,000
Turbot	16-20	10,000	
Whiting	50-75	200	1,000
Freshwater fish			
Carp		5,000-7,000	10,000
Eel		5,700-16,500	3,000-4,700
Pike		14,400	1,800
Sturgeon		26,000	10
Whitefish		59,500	

[a] From Butler (1946); Higashi (1948 and 1961); Bailey (1952).
[b] One International Unit (IU) = 0.3 μg retinol.

tissues is not necessarily related to the lipid content. Liver constitutes the major storage organ of both lipid and vitamin A for gadoids (cod, haddock, hake, etc.) but the liver oils of these fish are not as potent as that of sharks. Salmonid fishes and sardines store considerable amounts of fat in the muscle tissues but vitamin A is more concentrated in liver lipid. Halibut accumulate substantial amounts of vitamin A in the viscera as well as in the liver.

Fish and shellfish flesh are poor to moderate sources of vitamin A (Tables 5.2 and 5.3) with some exceptions. The flesh of the deep sea shark (*Heteroscymnus longus*) contains 9000–26,000 IU vitamin A 100 g^{-1} (Higashi *et al.*, 1955). The muscle tissues of oily fish such as mackerel and herring contain higher concentrations of vitamin A than leaner fish.

Table 5.2. Fat-soluble vitamin content of marine fish flesh.

Species	Vitamin A (IU 100 g^{-1})	Vitamin D (IU 100 g^{-1})	Vitamin E (mg 100 g^{-1})
Anchovy	635		
Catfish, sea		96	0.21
Cod, Atlantic	25		0.19 (0.15-0.24)
Cod, Pacific	28	0.20	
Dogfish, spiny	800 (114-1,600)[a]	15	
Dolphin	180 (26-333)		
Drum	200 (31-508)		
Flounder	50 (34-73)	60	0.40
Haddock	50		
Halibut, Atlantic	440	44	0.85 (0.60-1.30)
Herring, Atlantic	130 (100-840)	1,627	1.50 (1.10-2.10)
Herring, Pacific	100		1.10
Lamprey	30,000 (9,060-44,300)	260 (120-400)	
Ling			0.30
Lingcod	7		0.16
Mackerel, Atlantic	105 (32-711)	1,050 (143-2,000)	1.11 (0.70-1.52)
Mackerel, horse or jack	7		0.36
Perch, ocean	125		1.20
Pilchard	155 (50-303)	2,300 (532-5,400)	
Plaice			0.36
Pollack	50 (32-66)		0.26
Redfish			1.25
Rockfish			1.06 (0.79-1.34)
Smelt	96		
Sole (mixed species)	40		0.44 (0.32-0.57)
Swordfish, broadbill	1,585		
Tuna, bluefin	963		1.20
Tuna, skipjack	52		3.00
Turbot	39		0.80
Whiting	55 (12-107)		0.23
Wolf fish	375		
Yellowtail and amberjack	10 (3-22)	27 (0-53)	0.20 (0.12-0.24)

[a] Values in parentheses represent the range of values reported in the literature.
Sources: Braekkan and Boge (1965); Ackman and Cormier (1967); Bunnel *et al.* (1968); Sidwell *et al.* (1978); Statens Livsmedelsverk (Sweden) (1978 and 1988).

Relatively high concentrations (>1000 IU/100 g^{-1}) of vitamin A in the flesh of eels, lampreys, crabs and mussels have also been reported.

Vitamin D

The two major natural forms of vitamin D are cholecalciferol (vitamin D_3, which occurs in animals) and ergocalciferol (vitamin D_2, which occurs predominantly in plants). These forms of vitamin D are known for their antirachitic activity.

Table 5.3. Fat-soluble vitamin content of freshwater fish and shellfish flesh.

Species	Vitamin A (IU 100 g^{-1})	Vitamin D (IU 100 g^{-1})	Vitamin E (mg 100 g^{-1})
Freshwater fish			
Ayu			1.80
Carp	262 (29-1,020)[a]	260	0.56 (0.50-0.63)
Catfish, Channel		500	1.20
Eels	3,295 (1,500-5,700)	200	12.50 (6.30-18.70)
Milkfish	537 (400-673)		
Pike	70		0.70
Salmon, Atlantic	480 (250-970)	45	2.20
Salmon, Chinook	274	40	0.53
Salmon, coho	185	30	
Salmon, sockeye	135	60	
Trout, rainbow	65	35	1.90
Whitefish, Great Lake			2.60
Yellow perch	125		2.20
Shellfish and molluscs			
Clam[b]	370 (11-317)	4.4 (0-8.8)	0.50 (0.20-0.90)
Crab[b]	5,115 (1,360-8,870)		
Crayfish	23		
Lobster	97		1.60 (1.20-1.90)
Mussel	1,226 (33-2,418)		0.54 (0.40-0.60)
Octopus[b]	32		
Oyster[b]	273 (170-366)	40	0.86 (0.60-1.20)
Shrimp and prawn[b]	108 (20-297)		
Snail[b]	83		3.50

[a] Values in parentheses represent the range of values reported in the literature.
[b] Mixed species.
Sources: Bailey (1952); Higashi (1961); Ackman and Cormier (1967); Bunnel *et al.* (1968); Sidwell *et al.* (1978); Statens Livsmedelsverk (Sweden) (1978 and 1988).

In homeotherms, the only clearly established physiological role for vitamin D metabolites, specifically calcitriol (1,25-dihydrocholecalciferol), is in the control of blood plasma calcium and phosphate concentrations. Vitamin D, either ingested or produced in the skin, is carried through the circulatory system to the liver where it is converted to 25-hydroxy vitamin D. This metabolite is the major circulating form under normal conditions and during excessive intake of vitamin D.

Cholecalciferol is widely distributed in fish and other animals, with the highest concentrations reported in some marine teleosts. The liver is a rich source of vitamin D (Table 5.1) but its concentration varies greatly among fishes. Albacore and skipjack tuna show the highest concentrations, which range from 25,000 to 250,000 IU vitamin D per gram of liver oil. In contrast to vitamin A, low concentrations of vitamin D (< 25 IU g^{-1}) are found in liver oils of elasmobranchs. This difference may be due to the lower vitamin D requirement for calcification of structural tissues in elasmobranchs than in teleosts. Cod liver and halibut oils, which were used in early days for medicinal purposes, contain 100 and 3000 IU g^{-1} of vitamin D, respectively. Bailey (1952) reported that the vitamin D content of Pacific salmon ranged from 240 to 540 IU 100 g^{-1}. Cholecalciferol is present in fish both in the free form and as long-chain fatty acid esters. The apparent ability of diverse species of bony fishes to synthesize cholecalciferol from 7-dehydrocholesterol (Takeuchi *et al.*, 1991a; 1991b) exists.

The flesh of many fishes contains small amounts of vitamin D (Tables 5.2 and 5.3). Only low concentrations of vitamin D are found in gadoids, and flatfishes such as flounder. Mackerel, herring and pilchard contain appreciable amounts of this vitamin. Farmed and wild salmon are moderate sources of this vitamin. In general, the magnitudes of vitamin D distribution in fish muscle are similar to those observed for vitamin A.

Oysters and other molluscs in particular are considered to be some of the richest sources of provitamins D, especially 7-dehydrocholesterol (Bergmann, 1952). The average concentration of this provitamin in oysters (*Crassostrea gigas*) was 22 ± 8 mg 100 g^{-1} (Gordon, 1979). This provitamin has not been found in most crustaceans and finfish.

Vitamin E

Vitamin E is a generic term for eight naturally occurring derivatives of dihydrochromanol that are differentiated by the degree of methyl substitution in the ring (α, β, γ, δ) and the presence of unsaturated bonds in the phytyl side chain (tocopherol, tocotrienol). α-Tocopherol has the highest biopotency among the different forms of vitamin E. In biological systems, vitamin E functions as an antioxidant to protect cellular lipid components by donation of the phenolic hydrogen to a free radical, thus inhibiting the chain reaction of free radical propagation. HPLC fluorescence detection for vitamin E is now preferentially used over the once popular Emmerie–Engel colorimetric assay for this vitamin. This new method is highly accurate and precise for quantitative analyses of vitamin E and eliminates the need for exhaustive sample purification which may inadvertently lead to degradation of this readily oxidizable vitamin. In the presence of oxygen, vitamin E oxidizes easily on exposure to ultraviolet (UV) light, heat, alkaline pH conditions and trace minerals like iron (Fe^{3+}) and copper (Cu^{2+}). Ascorbic acid, if present, may inhibit the catalytic effect of ferric and cupric ions on vitamin E by maintaining the metals in their unreactive ground state.

Dietary vitamin E intake values are usually expressed in α-tocopherol equivalents to account for the biopotencies of both α- and non-α-tocopherols and tocotrienols. The concentration of vitamin E in the edible flesh of selected fish species is listed in Tables 5.2 and 5.3. Some of the references on which the tabulation is based reported 'vitamin E' values rather than specifying α-tocopherol content. However, since α-tocopherol is the major form detected in marine animal tissues (Gordon and Martin, 1982), vitamin E values most likely parallel α-tocopherol values.

Fish flesh contains low to modest amounts of vitamin E; a 100 g serving of salmon may provide nearly 14% of the 10 IU recommended dietary allowance for this vitamin. Vitamin E content varies with species and fatty fish tend to have greater vitamin E concentrations than lean fish (Cruickshank, 1962). In the flesh, dark muscle contains more vitamin E than white muscle, reflective of the higher requirement for antioxidants in the metabolically active dark muscle tissues (Ackman and Cormier, 1967). Since vitamin E is fat soluble, fish oils, especially those extracted from the liver, have higher proportions of vitamin E than the flesh. The vitamin E content (mg 100 g^{-1} of oil) of fish liver oils in some species are as follows: Atlantic catfish, 110; Atlantic herring, 15; haddock, 8.8; ling, 9.3 (Cruickshank, 1962); dogfish, 25 (Braekkan *et al.*, 1963); and cod, 30 (O'Keefe and Ackman, 1987).

During processing and frozen storage of fish and fishery products, vitamin E concentrations decrease concomitant with lipid peroxidation. Processed fishery products and fish oils, therefore, have less predictable amounts of vitamin E than the freshly caught fish from which they are taken. Fish oils normally contain significant amounts of polyunsaturated fatty acids and, during storage, the rate of autoxidation is inversely related to its vitamin E content (Cruickshank, 1962). Similarly, rancid and unpleasant flavours develop readily during storage of trout and salmon fillets when tissue vitamin E levels are low (Frigg *et al.*, 1990; Waagbo *et al.*, 1993). Therefore, vitamin E content of fishery products is an important consideration, both from the view of satisfying human dietary requirements as well as preventing lipid peroxidation and thereby improving the keeping quality of food.

Vitamin K

Vitamin K is usually referred to as the 'coagulation vitamin' in reference to its identified role in maintaining normal blood clotting mechanisms in animals. All vitamin K compounds possess the active nucleus, 2-methyl-1,4-napthoquinone (or menadione), and are differentiated from each other only by the nature and length of the side chain. The simplest form of vitamin K is synthetically prepared menadione (K_3), which has no side chain. In nature, vitamin K is synthesized by plants (vitamin K_1 or phylloquinone) or intestinal bacteria in animals (vitamin K_2 or menaquinone). The vitamin K_2, originally isolated from fish meal, is now known to be only one of a series of forms of bacterially synthesized vitamin K with an unsaturated polyisoprenoid side chain at the 3-position. Vitamin K is normally stable to heat and oxidizing conditions but may degrade on exposure to UV light, alkalis, strong acids and reducing agents (Ball, 1988).

Primary vitamin K deficiency in humans is relatively uncommon because the ubiquitous nature of this compound in most foods and adequate microbial synthesis in the normal gut together provide enough dietary vitamin K. Newborn infants may become susceptible to vitamin K deficiency in the period prior to the establishment of their intestinal flora.

To our knowledge, no reliable information exists on the distribution of vitamin K in edible fish products. Animal products contain a mixture of various isoprenalogues of the menaquinone series (Suttie, 1983), which are difficult to separate and accurately quantify. Fish meal produced from mechanically extracted herring contains 0.022 mg vitamin K $100\,g^{-1}$ (National Research Council, 1982).

Water-soluble Vitamins

Thiamin

Thiamin in the form of thiamin pyrophosphate (TPP) functions as a coenzyme in decarboxylation reactions involved in the release of energy from glucose. The enzyme thiamin pyrophosphokinase and adenosine triphosphate (ATP) convert thiamin into its metabolically active enzyme form, TPP. The functions of thiamin are reflected in two symptoms of thiamin deficiency, i.e. increased blood levels of pyruvic acid and decreased red blood cell transketolase activity.

Thiamin is found in most animal tissues mainly in phosphorylated forms, thiamin mono-, di- and triphosphates. Compared to meat products of mammalian origin, seafoods generally contain marginal levels of thiamin and contribute only minor amounts as a human dietary source for this compound. Thiamin concentration varies with species (Tables 5.4 to 5.6). Within the same fish species, individual variation in thiamin content is less pronounced than that observed for fat-soluble vitamins, implying that thiamin uptake by fish tissues is largely governed by the metabolic demand and that no accumulation takes place beyond this limit (Higashi, 1961). Notwithstanding already modest amounts of thiamin in fish flesh, this concentration is further diminished by thiaminase which catalyses the decomposition of thiamin even during frozen storage. Thiaminase is found in the viscera of freshwater fish such as cyprinids, although a few marine forms and shellfish species exhibit this enzymic activity whether it be from endogenous enzymes or those obtained through their diet (Neilands, 1947).

Thiamin, like ascorbic acid, is extremely heat labile and sensitive to oxygen and easily degrades under conventional food processing conditions. Thiamin loss on tuna canning and retorting may reach as much as 70% with the extent of thiamin destruction apparently dependent on sample pH (Gordon *et al.*, 1979; Dudek *et al.*, 1982). Other process conditions, i.e. salting, brining or smoking, also incur thiamin loss relative to the amount of water that is drawn out (Braekkan, 1962b). Some loss of thiamin also occurs during low dose radiation of seafoods.

Riboflavin

Riboflavin functions as a coenzyme in the intracellular conversion of energy from dietary fats and carbohydrates to the form readily used in muscles and tissues. The principal forms occurring in tissues and cells are flavin mononucleotide (FMN, riboflavin 5-phosphate) and flavin adenine dinucleotide (FAD).

Intestinal absorption of riboflavin in humans occurs by a saturable process involving a phosphorylation–dephosphorylation mechanisms (Rivlin, 1984).

Fish products show varying concentrations of riboflavin (Tables 5.4 to 5.6), but as a rule contain less than other meat products. In fish, metabolically active tissues are rich in riboflavin; the dark muscles contain 10–20 times more riboflavin than white muscles and the more active pelagic species may contain 10–20 times more than the sedentary neritic species (Higashi, 1961). Among seafoods, molluscs are particularly rich in riboflavin (Gordon *et al.*, 1979).

Riboflavin is heat stable and losses in seafoods are most often due to leaching of this compound into the cooking juices, wash water or drip during thawing. In shrimp, a 40% reduction in riboflavin content, as well as of thiamin and niacin, occurs during extensive water washing followed by cooking (Gordon *et al.*, 1979). Riboflavin concentrations in the flesh of various fish species remain stable during frozen storage for at least 6 months (Gordon *et al.*, 1979).

Niacin

Niacin is the generic name for nicotinic acid and nicotinamide, both of which may constitute the dietary source for this vitamin. The biologically active forms of niacin, nicotinamide adenine dinucleotide (NAD^+) and nicotinamide adenine dinucleotide phosphate ($NADP^+$), function as coenzymes to many dehydrogenase enzymes found in the cytosol and mitochondria. Thus, they are key components in several pathways of carbohydrate, lipid and amino acid metabolism.

Assessment of niacin value of foods is based not only on niacin but

Table 5.4. Water-soluble vitamin content of marine fish flesh.

Species	Thiamin (mg 100 g^{-1})	Riboflavin (mg 100 g^{-1})	Niacin (mg 100 g^{-1})	Pyridoxine (mg 100 g^{-1})	Biotin (μg 100 g^{-1})	Pantothenic acid (mg 100 g^{-1})	Folates (μg 100 g^{-1})	Vitamin B_{12} (μg 100 g^{-1})
Anchovy	0.01 (0.01-0.04)	0.12 (0.11-0.13)	3.1 (2.3-3.5)	0.18 (0.14-0.26)			12.3	3.3
Bass, rock	0.05	0.05	1.0					
Bream, red sea	0.03	0.25	1.9					
Catfish, sea	0.06 (0.04-0.08)	0.12 (0.08-0.20)	2.4 (0.5-4.5)	0.37		0.57	82.5 (15.0-150.0)	2.4 (2.2-2.5)
Cod, Atlantic	0.07 (0.05-0.18)	0.08 (0.02-0.16)	2 (1.5-2.3)	0.17 (0.12-0.28)	2.6	0.17 (0.08-0.31)	5.0 (1.8-6.7)	1.0 (0.2-1.1)
Cod, Pacific	0.09	0.15 (0.14-0.17)	2	0.22	1.2 (0.2-2.6)	0.15 (0.10-0.18)	5.0 (1.8-6.7)	0.8 (0.7-0.9)
Cusk	0.04 (0.03-0.05)	0.15 (0.10-0.20)	2.8 (2.7-2.8)	0.29	4.5	0.30 (0.30-0.31)		1.2 (1.1-1.2)
Dogfish	0.26	0.27	5.6					0.5
Dogfish, spiny	0.05	0.15 (0.10-0.18)	5.2 (5.1-5.3)	0.75 (0.57-0.80)				1.5 (1.1-2.6)
Dolphin	0.02	0.07	6.1				0.1	
Drum	0.06 (0.02-0.13)	0.13 (0.03-0.53)	3.1 (0.5-8.9)					2.4 (0.2-5.3)
Flounder	0.15 (0.03-0.66)	0.2 (0.11-0.46)	3.5 (2.0-4.3)	0.25 (0.21-0.30)		1.00	5.0	1.0
Grouper	0.11	0.37	1.4					
Haddock	0.07 (0.03-0.10)	0.1 (0.02-0.16)	4 (3.1-4.4)	0.12	2.6 (0.2-4.8)	0.25 (0.04-0.29)	0.8	1.5 (0.5-2.0)
Hake	0.09 (0.04-0.13)	0.08	1.6 (1.0-2.2)	0.88				
Halibut, Atlantic	0.07 (0.03-0.12)	0.08 (0.04-0.18)	6 (3.0-11.0)	0.4 (0.11-0.55)	5 (2.7-8.0)	0.25 (0.15-0.36)	2.6 (2.0-2.9)	0.8 (0.7-0.9)
Herring, Atlantic	0.04 (0.08-0.13)	0.3 (0.09-0.33)	4 (2.0-6.3)	0.45 (0.35-0.52)	10 (9.0-16.0)	0.95 (0.93-0.97)	10.3 (1.7-14.0)	10.0 (8.0-14.0)
Herring, Pacific	0.03 (0.01-0.04)	0.22 (0.18-0.27)	3.8 (0.6-9.6)	0.31 (0.16-0.45)			9.4 (2.4-12.0)	1.6 (0.5-2.4)
Lamprey	0.34 (0.05-0.85)	0.52 (0.43-0.61)	4.7	0.20 (0.17-0.22)		0.43 (0.28-0.57)	25.5 (23.6-29.0)	4.1 (3.8-4.4)
Ling		0.08 (0.06-0.10)	2.5 (2.1-2.5)	0.26	4.2	0.30 (0.24-0.36)		0.6 (0.4-0.8)
Lingcod	0.04 (0.03-0.05)	0.08 (0.04-0.11)	1.9					18.0
Mackerel, Atlantic	0.10 (0.02-0.20)	0.35 (0.16-0.66)	7.5 (4.1-11.4)	0.75 (0.5-0.89)	7 (6.0-8.0)	1.00 (0.97-1.09)		10.0 (2-130)
Mackerel, horse	0.18	0.14	5.2	0.31 (0.27-0.36)		0.35 (0.2-0.5)		12.7
Parrotfish	0.03 (0-0.05)	0.06 (0.05-0.07)	1.4 (1.4-1.5)	0.18				0.6
Pilchard	0.01	0.3	7.4	0.27 (0.24-0.30)	26 (19.0-34.0)	0.66 (0.49-0.83)		17.0
Plaice	0.20 (0.02-0.46)	0.13 (0.09-0.33)	4 (2.1-10.6)	0.25 (0.16-0.30)	90.0	0.80 (0.68-1.12)		
Pollack	0.08 (0.04-0.16)	0.1 (0.09-0.13)	2 (1.6-2.0)	0.12	3.2	0.3 (0.25-0.42)		1.0 (0.8-2.2)
Pomfret	0.12 (0.05-0.19)	0.19 (0.08-0.30)	0.6 (0.4-0.8)					

Pompano	0.11	0.09	3					
Porbeagle	0.08	0.18 (0.09-0.28)	8.0 (7.0-10.0)	0.30 (0.25-0.33)				2.5 (2.5-2.6)
Ray and skate	0.03 (0.02-0.03)	0.40 (0.37-0.41)	2.5 (2.3-2.4)	0.75 (0.46-1.23)				1.0 (0.8-1.0)
Redfish		0.09 (0.07-0.11)	2.0	0.23	1.0	0.36		1.0
Rockfish	0.08	0.13 (0.11-0.15)	0.8			0.15	4.7 (0.4-9.0)	12.0
Saithe	0.10	0.20 (0.17-0.30)	3.5 (3.2-4.4)	0.47	7.2	0.40 (0.35-0.41)		3.5 (2.2-5.0)
Siganid	0.20 (0.16-0.24)	0.13 (0.11-0.15)	4.8 (4.4-5.1)	0.14				0.1
Skipjack	0.06	0.04	7.2					4.2 (2.5-6.0)
Smelt	0.05 (0.01-0.13)	0.14 (0.04-0.36)	1.9 (1.3-3.0)	0.12		0.92 (0.64-1.20)	5.0 (3.7-6.3)	1.8 (0.2-3.4)
Snapper, red	0.10 (0.07-0.13)	0.78	1.5 (0.7-1.8)					
Sole, butter	0.04	0.05	1.2					
Sole, Dover	0.07	0.04	1.6					
Sole, English	0.04	0.04	2.9					
Sole, lemon	0.09	0.08 (0.07-0.09)	3.5 (2.6-4.3)			0.30 (0.25-0.43)	3.0 (2.9-3.1)	0.8 (0.6-1.0)
Sprat, brisling	0.04	0.26	4.8			1.09		11.0
Swordfish, broadbill	0.05	0.50	8.2 (7.4-9.1)		1.2	0.05		5.7
Tuna, albacore	0.05 (0.04-0.07)[a]	0.11	1.2 (1.1-1.4)	0.44	3.0	0.42		1.7
Tuna, bigeye	0.06	0.06	14.3	0.90				5.0
Tuna, bluefin	0.20 (0.14-0.23)	0.12 (0.80-0.22)	8.5 (6.2-10.8)	0.45 (0.37-1.00)	5.0 (2.0-6.0)	0.50 (0.38-0.70)		3.4 (3.2-3.6)
Turbot	0.05 (0.02-0.06)	0.09 (0.04-0.14)	1.1 (0.6-1.5)			0.25		1.3 (1.0-1.5)
Wolf fish	0.07	0.08 (0.07-0.09)	2.0 (1.3-2.5)	0.35	3.7	0.57 (0.50-0.64)		2.0 (1.8-2.8)
Whiting	0.04 (0.02-0.05)	0.08 (0.06-0.10)	6.1 (5.4-6.9)					
Yellowtail and amberjack	0.10	0.04	6.8	0.16 (0.13-0.19)		0.62 (0.46-0.79)		3.5 (3.2-3.7)

[a] Values in parentheses represent the range of values reported in the literature.

Sources: Braekkan (1962a); Sidwell *et al.* (1978); Statens Livsmedelsverk (Sweden) (1978 and 1988).

Table 5.5. Water-soluble vitamin content of freshwater fish flesh.

Species	Thiamin (mg 100 g^{-1})	Riboflavin (mg 100 g^{-1})	Niacin (mg 100 g^{-1})	Pyridoxine (mg 100 g^{-1})	Biotin (μg 100 g^{-1})	Pantothenic acid (mg 100 g^{-1})	Folates (μg 100 g^{-1})	Vitamin B_{12} (μg 100 g^{-1})
Carp	0.11 (0-0.45)[a]	0.08 (0.01-0.18)	3.0 (1.3-11.0)	0.07 (0.01-0.13)		0.15	70.0 (2.8-138.0)	3.2 (0.1-9.0)
Catfish, Channel	0.21	0.07 (0.06-0.07)	1.9 (1.8-1.9)	0.12 (0.11-0.12)		0.76 (0.56-0.97)	10.0	2.2 (2.0-2.4)
Eel	0.15 (0.13-0.20)	0.20 (0.04-0.46)	3.5 (3.3-4.1)	0.30 (0.20-0.36)		0.15 (0.09-0.24)	11.6 (10.0-13.1)	1.0 (0.6-3.0)
Milkfish	0.06 (0.01-0.10)	0.08 (0.05-0.10)	6.1 (5.8-6.4)	0.42			15.9	3.4
Mullet	0.55	0.98	0.7 (0.7-0.8)	0.42 (0.42-0.43)		0.76 (0.69-0.83)		0.2 (0.1-0.3)
Pike	0.09	0.07	1.7	0.14 (0.12-0.15)				
Salmon, Atlantic[b]	0.20 (0.14-0.24)	0.15 (0.06-0.22)	7.0 (5.4-8.8)	0.75 (0.60-0.98)	4.8	2.00		4.5 (4.0-5.0)
Salmon, Chinook[b]	0.10 (0.04-0.11)	0.20 (0.11-0.26)	8.4					
Salmon, coho[b]	0.08 (0.07-0.10)	0.13 (0.12-0.14)	7.3 (8.9-5.8)	0.62 (0.44-0.81)		0.86 (0.79-0.93)	10.0 (10.0-20.0)	4.8 (4.1-5.4)
Salmon, sockeye[b]	0.14 (0.13-0.15)	0.10 (0.04-0.13)		0.19		0.61		2.3
Sucker	0.02 (0-0.04)	0.05	1.2 (1.0-1.5)			0.46		
Trout, rainbow	0.10 (0.60-0.14)	0.08 (0.05-0.27)	5.5 (3.5-5.7)	0.41 (0.40-0.69)		0.97 (0.84-2.33)	10.0	4.8 (4.0-5.6)
Whitefish, Great Lake	0.15 (0.12-0.16)	0.12 (0.11-0.13)	3.0 (2.7-3.3)					
Yellow perch	0.06	0.17	1.8					

[a] Values in parentheses represent the range of data reported in the literature.
[b] Includes brackish water and anadromous fish (juveniles migrate to sea and adults return to fresh water to spawn).
Source: Braekkan (1962a); Sidwell *et al.* (1978); Statens Livsmedelsverk (Sweden) (1978 and 1988).

Table 5.6. Water-soluble vitamin content of shellfish.

Species	Thiamin (mg 100 g^{-1})	Riboflavin (mg 100 g^{-1})	Niacin (mg 100 g^{-1})	Pyridoxine (mg 100 g^{-1})	Biotin (μg 100 g^{-1})	Panthothenic acid (mg 100 g^{-1})	Folates (μg 100 g^{-1})	Vitamin B_{12} (μg 100 g^{-1})
Clam[a]	0.05 (0.04-0.14)[b]	0.18 (0.09-0.29)	1.3 (0.2-2.3)	0.08 (0.04-0.35)	2.3	0.53 (0.44-0.62)	25.1 (2.7-58.0)	3.6 (1.3-6.2)
Crab[a]	0.08 (0.01-0.14)	0.25 (0.06-0.38)	2.5 (1.7-2.8)	0.23	9.8	0.71		13.1
Crayfish	0.10 (0.06-0.15)	0.06	2.4	0.21	3.8	0.41	0.6	2.7
Lobster	0.08 (0.08-0.16)	0.05 (0.04-0.06)	1.6 (1.2-2.1)	0.21	5.0 (4.8-5.2)	0.45	0.8	1.6 (0.5-2.3)
Mussel		0.15 (0.12-0.16)	1.6 (1.2-2.0)	0.10 (0-0.19)			41.8	10.2
Octopus[a]	0.07 (0.02-0.14)	0.07 (0.04-0.11)	3.2 (1.3-5.3)	0.36	5.4			1.6
Oyster[a]	0.12 (0.07-0.29)	0.18 (0.06-0.35)	2.0 (1.4-3.9)	0.17 (0.11-0.32)	8.7	0.36 (0.14-0.53)	84.4 (43.0-32.0)	20.0 (15.0-46.0)
Prawn, deep sea	0.05 (0.04-0.08)	0.05 (0.03-0.07)	2.0 (1.1-4.0)	0.12 (0.10-0.15)	3.1	0.23		4.6
Scallop[a]	0.08 (0.04-0.13)	0.08 (0.06-0.1)	1.3 (1.2-1.4)		0.3	0.14 (0.13-0.14)		8.9 (3.2-24.0)
Sea cucumber[a]	0.05 (0.04-0.07)	0.09	0.5					1.4 (0.5-2.0)
Shrimp and prawn[a]	0.02 (0.01-0.05)	0.02 (0.01-0.07)	2.5 (1.1-5.3)	0.07 (0.06-0.12)	1.0	0.25 (0.12-0.38)	5.2 (3.0-7.4)	3.8 (0.9-8.1)
Snail[a]	0.06 (0-0.21)	0.17 (0.06-0.28)	1.5 (1.3-1.6)	0.12			32.8	26.7 (9.0-77.0)
Squid	0.04 (0-0.15)	0.22 (0.05-0.84)	2.6 (1.2-4.7)	0.62 (0.07-1.30)		0.68	12.5	4.5 (1.3-13.0)

[a] Mixed species.

[b] Values in parentheses represent the range of values reported in the literature.

Sources: Braekkan (1962a); Sidwell *et al.* (1978); Statens Livsmedelsverk (Sweden) (1978 and 1988).

also on tryptophan content since this amino acid is converted to NAD^+ *in vivo* (1 mg niacin equivalent = 1 mg niacin or 60 mg tryptophan). Seafoods contain significant amounts of niacin (Tables 5.4 to 5.6). Fatty fish and those with great locomotive power carry more niacin than lean fish species or those with poor locomotive abilities (Higashi, 1961). For instance, mackerels and tunas are better sources of niacin than gadoids. Fish contain high amounts of the alkali-labile nicotinic acid bound to either protein or carbohydrates (Higashi, 1961). However, bound niacin in fish flesh is clearly distinguished from the bound form occurring in cereals.

Like riboflavin, niacin is a moderately stable compound and minimal losses occur during food processing when care is taken to prevent leaching of this nutrient. For instance, huge losses of niacin are observed during surimi preparation, which involves extensive water washing, while minimal niacin is lost during low dose irradiation or smoking (Nettleton, 1985).

Vitamin B_6

The term vitamin B_6 refers to all 3-hydroxy-2-methylpyridine derivatives exhibiting qualitatively the biological activity of pyridoxine [3-hydroxy-4,5-bis(hydroxymethyl)2-methylpyridine]. The vitamin includes aldehyde (pyridoxal) and amine (pyridoxamine) forms. The metabolically active form of vitamin B_6 is pyridoxal phosphate (PLP), which functions as a coenzyme for reactions (transamination, decarboxylation, desulfhydration and oxidative deamination) involving amino acids. PLP also plays an important role in the biosynthesis of porphyrin, catabolism of glycogen, metabolism of lipid and γ-aminobutyric acid, and synthesis of epinephrine and norepinephrine.

The main form of vitamin B_6 found in plant products is pyridoxine, whereas pyridoxal and pyridoxine constitute the principal form of this vitamin in animal tissues. All three forms are converted in the body to the metabolically active form PLP. Among food sources of animal origin, fish and shellfish are the best sources of pyridoxine (Tables 5.4 to 5.6). Pelagic fishes such as mackerel, herring, tuna and bonito are particularly rich in this vitamin. The flesh of some freshwater fish contains relatively lower concentrations of pyridoxine than marine fish (Yanase, 1956). Canning affects the vitamin B_6 content of fish but most canned fishery products are still considered good sources of this vitamin (Braekkan, 1962b).

Pantothenic acid

Pantothenic acid is a component of coenzyme A, acyl CoA synthetase, and acyl carrier protein (ACP). The coenzyme form of this vitamin is responsible for acyl group transfer reactions. All known derivatives of CoA and related pantothene derivatives are thiol esters, which participate in numerous metabolic reactions. Thus CoA is enzymatically involved in amino acid, carbohydrate and lipid metabolism. Analysis of this vitamin requires enzymatic hydrolysis to liberate pantothenic acid from various coenzyme forms. To date, all reported values for pantothenic acid in fish and shellfish are based on microbiological assays of this vitamin.

Generally fish and shellfish contain less pantothenic acid (Tables 5.4 to 5.6) than most common meats of animal origin with the exceptions of Atlantic salmon and trout, which are rich sources of this vitamin (Braekkan, 1962a). Dietary intake of this vitamin by cultivated fish may also influence the vitamin content of their flesh. The concentration of pantothenic acid in marine fish and shellfish ranges from 0.12 to 0.97 mg 100 g^{-1} of flesh (Braekkan, 1962a). Pantothenic acid in foods is fairly stable to ordinary means of cooking and storage. It is unstable in heat and either alkaline ($pH > 7$) or acid ($pH < 5$) conditions. Appreciable losses in some fishery products may occur due to canning (Higashi, 1961).

Biotin

Biotin is widely distributed in fish and shellfish but most animal products contain low concentrations of this vitamin. Although eight stereoisomers of the biotin molecule are possible, only D-biotin (*cis*) is physiologically active. Biotin is a coenzyme for several enzyme-catalysed carboxylation reactions important in carbohydrate and fat metabolism. Biotin is covalently linked to a carboxyl carrier protein through a peptide bond to the ϵ-amino group of lysine (as biocytin). Thus bioavailability of biotin bound in such forms depends upon the hydrolytic digestion of protein.

The average biotin content of fish and shellfish is highly variable, ranging from 2 to 6 μg biotin 100 g^{-1} of flesh (Tables 5.4 to 5.6). Much higher values (26–90 μg 100 g^{-1}) have been reported for pilchard and plaice. A comparative study of biotin in the muscles of fish showed that red muscle contains 6 to 22% more biotin than white muscle (Braekkan and Boge, 1965). Biotin is destroyed by heat and under conditions that are conducive to lipid peroxidation. Although canning, heat curing and solvent extraction of food products can result in oxidation of this vitamin, use of natural and synthetic antioxidants reduces the loss during processing.

Folate

Folate is the generic descriptor of all compounds that exhibit the biological activity of folic acid (pteroylmonoglutamic acid). In most animal tissues, the predominant forms are polyglutamates. These forms may contain up to eight glutamic acid residues attached to the terminal glutamate of folic acid in an amide linkage as a polyamide. The polyglutamates are reduced active forms in animal tissues. Folic acid is the completely oxidized form of the molecule and it is not found as such in nature. The molecule can be reduced to the dihydro and tetrahydro forms. Folates carry out their metabolic functions as carriers of one-carbon units in tetrahydro (FH_4) forms. The various one-carbon units carried on folate coenzymes are used to synthesize methionine and purine rings and convert deoxyuridine monophosphate to deoxythymidine monophosphate.

Most foods contain a mixture of the mono- and polyglutamate forms of folate. These forms are predominantly in the reduced state. Very little free folate is found in food.

Microbiological assays have provided useful data on total folate contents of fishery by-products, but results are not always satisfactory. For example, analyses of folates may utilize *Lactobacillus casei*, *Streptococcus faecium* or *Pediococcus cerevisiae* but *L. casei* responds to the widest spectrum of folates. Thus some published analyses may have overestimated the biologically useful folate contents of foods. Fish provide moderate amounts of folates (Tables 5.4 to 5.6). The average folate content of fish and shellfish is highly variable, ranging from 0.5 to 10 $\mu g\ 100\,g^{-1}$ of flesh. Much higher values (26–150 $\mu g\ 100\,g^{-1}$) have been found for sea catfish, lamprey and carp. Fish organs including liver, kidney and spleen contain much higher concentrations than flesh (Higashi *et al.*, 1958).

Vitamin B_{12}

Vitamin B_{12} consists of a complex ring structure (corrin ring) in which a cobalt ion is added at the centre, hence its name cobalamin. The commercial preparation includes a cyanide group and is termed cyanocobalamin. *In vivo*, vitamin B_{12} functions as the coenzymes methylcobalamin or 5′-deoxyadenosylcobalamin, and is essential for haemopoiesis and normal function of the nervous system. This vitamin is efficacious in the treatment of pernicious anaemia. Only microorganisms found in animal foods are known to synthesize vitamin B_{12} and deficiency states in humans may arise in individuals who abstain from all animal foods (Herbert, 1984).

Seafoods are excellent sources of vitamin B_{12} (Tables 5.4 to 5.6). A regular serving portion of molluscs, especially oysters, which have remarkably high vitamin B_{12} content, may easily exceed the recommended dietary allowance for adults of 2 μg per day. In fish, dark muscles are particularly abundant in vitamin B_{12} compared to white muscles, and dark-fleshed fish like herring, sardine and lamprey have higher vitamin B_{12} potency than white-fleshed fish such as cod and flatfish species (cited by Higashi, 1961). Unlike other water-soluble vitamins, vitamin B_{12} is stored in substantial amounts in the liver although the extent of accumulation appears inversely related to fat accumulation (Higashi, 1961).

Vitamin C

Most higher animals can synthesize vitamin C via the glucuronic acid pathway from glucose. However, fish, a few species of mammals (humans and other primates, guinea pigs), invertebrates and birds are exceptions. Their inability to synthesize the vitamin appears to result from the congenital absence of the last enzyme in the biosynthetic pathway, L-gulonolactone oxidase. A major function of ascorbic acid is as a cofactor in the biosynthesis of collagen (the main supportive protein of bones, skin, tendon, cartilage and connective tissues). The main biochemical functional property of ascorbic acid relates to its redox potential (+0.08 mV) whereby the compound can reduce transitional metals, thus allowing the vitamin to participate in a number of key hydroxylation reactions.

Vitamin C is widely distributed in fish and aquatic organisms, occurring as both ascorbic acid and dehydroascorbic acid (Ikeda *et al.*, 1963). A significant loss of ascorbic acid in fishery products occurs during storage and processing. Ascorbic acid is susceptible to oxidation to dehydroascorbic acid, which is quickly and irreversibly degraded further by the hydrolytic opening of the lactone ring. These reactions, occurring in the presence of oxygen and traces of metal ions, increase substantially at high temperatures and under neutral to alkaline pH conditions.

Fish and shellfish are not particularly good sources of vitamin C. Ascorbic acid in fish flesh ranges from 1 to 5.1 mg 100 g^{-1} (Hastings and Spencer, 1952; Jacquot, 1961). Recently, Nettleton and Exler (1992) reported values in the same range for channel catfish, coho salmon and rainbow trout. Farmed salmon flesh contains higher concentrations of ascorbic acid (5–10 mg 100 g^{-1}) due to liberal use of this vitamin in fish feeds (S.P. Lall (1993), unpublished data). Liver, kidney and brain contain much higher concentrations of vitamin C than muscle (Fomin *et al.*, 1937). The ranges of ascorbic acid levels reported in crustaceans and molluscs are similar to those in fish; however, mussels may contain higher concentrations (17–29 mg 100 g^{-1}) of vitamin C (Jacquot, 1961). Significant loss of ascorbic acid occurs during ordinary cooking of fish.

Vitamin Content of Farmed and Wild Fish

With the worldwide expansion in aquaculture production, many farmed fish and shellfish are now marketed. The vitamin contents of these cultivated species depend on their diets, which are often liberally supplemented with synthetic vitamins (Woodward, 1994). The limited information available on farmed fishery products suggests that the concentration of most nutrients is relatively uniform as compared with wild fish (Suzuki *et al.*, 1986; Nettleton and Exler, 1992), with the exception of the fat content. Generally, farmed fish contain higher concentrations of fat than their wild counterparts. Vitamin A and E concentrations in the muscle of Atlantic salmon increase with increases in the dietary intake of these vitamins (Lall (1993), unpublished data; Sigurgisladottir *et al.*, 1994). Recently, Nettleton and Exler (1992) reported that differences in the concentrations of nine vitamins in wild and cultivated species in the USA (catfish, coho salmon, eastern oysters and crayfish) were relatively small (Table 5.7).

Effect of Processing on Vitamin Composition

Information on vitamin loss during handling, storage, cooking and other processing conditions such as canning and smoking of fish and shellfish is fragmentary and confined to few species and selected vitamins (Braekkan, 1962b; Schroeder, 1971; March, 1982). The poor handling of fish causes spoilage mainly due to bacterial activity, lipid oxidation and autolysis, which are known to cause destruction of vitamins. The loss of vitamins due to drip loss from the cut surface and washing procedure is relatively small.

Tables 5.8 to 5.10 summarize the vitamin contents of processed fishery products. Generally vitamins are sensitive to oxygen, light, heat and extreme pH or a combination of these. Trace elements, particularly copper and iron, catalyse the oxidative process and accelerate loss of vitamins and other nutrients. During canning of fishery products, the vitamins most affected are thiamin and ascorbic acid (Higashi, 1961). Riboflavin, niacin and vitamin B_{12} also show more significant losses during canning than either freezing or ordinary cooking (Dudek *et al.*, 1982). Higher amounts of vitamin C can be retained when seafood is heated at low temperatures and exposure to oxygen is minimized. In canned seafood, fat- and water-soluble vitamins diffuse into the oil or brine which are often discarded prior to consumption.

Table 5.11 shows the retention of thiamin, riboflavin, niacin and vitamin B_{12} in seafoods as affected by various cooking methods (baking, broiling, microwaving, breading and frying) and the canning process. The retention of these four vitamins under ordinary cooking conditions ranged from 90 to 100% (Dudek *et al.*, 1982). During the cooking of fish products vitamin B_{12} may become bound to protein, thus making extraction difficult. The variability in vitamin B_{12} values is more obvious when microbiological assays are used.

The losses of vitamins as influenced by smoking of fish have been reviewed (Braekkan, 1962b; Tarr, 1962; Nettleton, 1985). Heavy salting of herring prior to smoking results in 50% reduction of the B vitamins. Although the edible portion of fish muscle contains low concentrations of

Table 5.7. Vitamin content of wild and cultivated fish and shellfish.

Species		Moisture (%)	Fat (%)	Vitamin A (RE 100 g^{-1})[a]	Thiamin (mg 100 g^{-1})	Riboflavin (mg 100 g^{-1})	Niacin (mg 100 g^{-1})	Pyridoxine (mg 100 g^{-1})	Pantothenic acid (mg 100 g^{-1})	Folates (μg 100 g^{-1})	Vitamin B_{12} (μg 100 g^{-1})	Ascorbic acid (mg 100 g^{-1})
Fish												
Catfish, Channel[b]	Wild	80.6	2.3	<30	0.21	0.06	1.91	0.12	0.76	10	2.2	<1.0
	Cultivated	72.8	11.3	<30	0.38	0.07	2.24	0.19	0.64	10	2.5	0.6
Salmon, Atlantic[c]	Wild	73.1	4.5	36	0.16	0.10	7.22	0.52	0.92	10	5.3	0.9
	Cultivated	70.8	9.5	78	0.28	0.19	10.1	0.68	1.31	10	5.1	3.5
Salmon, coho[b]	Wild	75.0	3.3	24	0.08	0.13	7.34	0.62	0.86	10	4.8	0.8
	Cultivated	69.4	8.3	55	0.09	0.12	6.65	0.68	1.20	10	2.8	0.9
Trout, rainbow[b]	Wild	74.4	4.6	<30	0.14	0.08	5.48	0.41	0.97	10	4.8	1.8
	Cultivated	72.8	5.4	84	0.20	0.07	8.22	0.62	1.44	10	3.8	2.9
Shellfish												
Crayfish, red swamp[b]	Wild	83.2	0.9	<30	0.06	0.02	2.15	0.10	0.65	30	2.0	<1
	Cultivated	84.4	0.8	<100	0.05	0.04	1.89	0.07	0.65	30	2.4	<1
Oysters, eastern[b]	Wild	87.9	1.2	<30	0.08	0.10	1.41	0.07	0.18	10	21.4	3.1
	Cultivated	88.6	1.2	<30	0.12	0.02	1.26	0.05	0.08	10	14.4	3.8
Prawn[c]	Cultivated	78.8	1.6	189	0.08	0.06	2.51	0.10	0.38	10	3.7	2.1

[a] RE = retinol equivalent.
[b] Nettleton and Exler (1992).
[c] S.P. Lall (1993), unpublished data.

Table 5.8. Vitamin D and E contents of some processed seafoods.

Species	Vitamin D (IU 100 g^{-1})	Vitamin E (mg 100 g^{-1})
Eel		
Poached	40.0	19.3
Baked	35.0	38.0
Smoked	36.0	19.0
Flounder, frozen fillet		0.4
Haddock, broiled		0.6
Herring, Atlantic		
Pickled	5.0	1.0
Salted	5.0	2.3
Canned in tomato sauce	4.1	0.5
Herring, Baltic		
Salted	9.2	0.2
Fried		3.0
Mackerel (Atlantic), smoked	10.0	1.6
Mussel, canned		1.2
Salmon (Atlantic)		
Broiled		1.4
Canned	12.5	1.5
Smoked	5.0	1.3
Scallop		
Fried, frozen, oven-heated	0.6	
Fried, frozen		0.7
Sprat, pickled	0.2	0.6
Sardine		
Canned in oil	15.0	1.7
Canned in tomato sauce	12.0	2.3
Tuna		
Canned in oil	5.8	6.3
Canned in water	4.5	1.2

Sources: Braekkan (1962b); Ackman and Cormier (1967); Bunnel *et al.* (1968); Statens Livsmedelsverk (Sweden) (1978 and 1988).

vitamins A and D, no appreciable loss occurs during the smoking of salmon, herring or mackerel (Lovern, 1944; Junker, 1957; Tarr, 1962; Bhuiyan *et al.*, 1993). The loss of some B vitamins (thiamin, riboflavin, niacin and pantothenic acid) during the smoking process is very small (Hoar and Barberie, 1945; Taarland *et al.*, 1958; Bhuiyan *et al.*, 1993) while vitamin B_{12} concentration is not affected (Taarland *et al.*, 1958).

Preventive measures may control oxidative rancidity and loss of fat-soluble vitamins during processing and storage. Freezing and thawing expels water from seafood and water-soluble vitamins may be inadvertently

Table 5.9. Water-soluble vitamin content of canned seafood products.[a]

Species	Thiamin (mg 100 g^{-1})	Riboflavin (mg 100 g^{-1})	Niacin (mg 100 g^{-1})	Pyridoxine (mg 100 g^{-1})	Biotin (μg 100 g^{-1})	Panthothenic acid (mg 100 g^{-1})	Vitamin B_{12} (μg 100 g^{-1})
Crab paste	0.06	0.19	1.2	0.13	12	1.57	4.4
Mackerel, Atlantic	0.06 (0.04-0.09)[b]	0.24 (0.17-0.31)[b]	5.7 (5-6)[b]	0.21 (0.24-0.4)	3 (3-4)	0.31 (0.24-0.4)	
Mackerel, Pacific	0.02 (0.02-0.03)	0.33 (0.29-0.38)	8.7 (8-9.4)	0.27 (0.25-0.29)	18 (12-23)	0.47 (0.43-0.54)	
Oysters	0.02 (0.01-0.03)	0.20 (0.12-0.33)	1.0 (0.6-1.6)	0.04 (0.01-0.09)			
Salmon, Pacific (in bouillon)	0.03 (0.01-0.05)	0.16 (0.11-0.2)	7.5 (6.2-8.9)	0.45 (0.39-0.5)	15 (11-19)	0.58 (0.47-0.73)	
Sardines, brisling (in oil)	0.03 (0.01-0.03)	0.33 (0.14-0.62)	6.0 (2.5-6.7)	0.25 (0.15-0.26)	15 (14-16)	0.80 (0.6-0.91)	10.0 (8.0-11.0)
Sardines, Pacific (natural)	0.02 (0.01-0.04)	0.30 (0.09-0.37)	7.4 (6.2-8.8)	0.28 (0.24-0.3)	24 (19-34)	0.45 (0.12-0.65)	
Shrimps	0.03 (0.01-0.06)	0.25 (0.04-0.37)	1.5 (0.7-3.4)	0.02 (0.01-0.37)	2	0.25 (0.17-0.35)	3.0
Tuna, Pacific	0.05 (0.04-0.07)	0.11 (0.09-0.15)	13.5 (10.2-15.5)	0.44 (0.37-0.52)	4 (3-6)	0.42 (0.35-0.56)	

[a] Values in parentheses represent the range of values reported in the literature.
[b] Includes values from Gordon and Martin (1982).
Sources: Braekkan (1962b); Statens Livsmedelsverk (Sweden) (1978 and 1988).

Table 5.10. Water-soluble vitamin content of processed seafood products.

Species	Thiamin (mg 100 g^{-1})	Riboflavin (mg 100 g^{-1})	Niacin (mg 100 g^{-1})	Pyridoxine (mg 100 g^{-1})	Biotin (μg 100 g^{-1})	Pantothenic acid (mg 100 g^{-1})	Vitamin B_{12} (μg 100 g^{-1})
Smoked fish							
Cod		0.05 (0.02-0.11)[a]	2.5 (1.3-3.5)	0.22	1.4	0.20 (0.13-0.34)	1.5 (2.0-27.0)
Eel	0.14 (0.07-0.18)	0.30 (0.22-0.48)	4.0 (2.8-5.7)				
Haddock	0.05 (0.04-0.10)	0.02 (0.01-0.04)	4.0 (3.5-5.2)			0.12 (0.07-0.19)	
Halibut	0.06	0.10 (0.04-0.17)	1.5	0.05 (0.03-0.08)	2.5	0.72	0.6
Herring	0.04 (0-0.08)	0.20 (0.16-0.28)	5.0 (3.9-5.8)	0.35 (0.20-0.48)	10.5	0.88	15.0
Herring, salted	0.03 (0.02-0.04)	0.20 (0.16-0.36)	2.5 (2.0-3.4)	0.21 (0.16-0.32)	9.0	0.50	8.0
Mackerel	0.14 (0.10-0.20)	0.35 (0.22-0.47)	9.0 (5.9-13.0)	0.50 (0.34-0.68)	7.7	0.52	12.0
Salmon, Atlantic	0.20 (0.11-0.45)	0.10 (0.07-0.19)	7.5 (5.0-10.8)	0.64 (0.34-0.87)	6.3	0.71	7.0
Miscellaneous							
Cod roe	0.25	0.55	0.8	0.14	17.0	1.97	14.0
Cod roe caviar	0.33	0.46	0.5	0.14	9.0	1.94	9.0
Herring soft roe	0.04	0.75	2.4	0.15	5.0	1.55	10.0
Herring, marinated		0.25	4.7	0.23	9.0	0.82	10.0
Herring, kippered		0.35 (0.20-0.47)	5.0 (2.8-5.8)	0.25	12.0	1.04	1.5

[a] Values in parentheses represent the range of values reported in the literature.
Sources: Braekkan (1962b); Statens Livsmedelsverk (Sweden) (1978 and 1988).

Table 5.11. Effect of different cooking methods and canning on retention of thiamin, riboflavin, niacin and vitamin B_{12} in seafoods.[a,b,c]

Cooking method and canning	Thiamin	Riboflavin	Niacin	Vitamin B_{12}
Bake	96	100[g]	99	100
Broil[d]	93	100	98	90
Microwave	95	96	99	96
Bread and fry[e]	97	100	100	95
Canned[f]	34	66	78	65

[a] Dudek *et al.* (1982).
[b] Average values of six species (flounder, sockeye salmon, Atlantic mackerel, Alaskan pollock, Pacific whiting and shrimp).
[c] Retention (%) = (Average weight of vitamin in processed product/Average weight of vitamin in raw product) × % yield.
[d] All fish species excluding shrimp.
[e] Only pollock, whiting and shrimp.
[f] Only salmon, mackerel and shrimp.
[g] Reported retention value exceeds 100% because of the loss of liquid and fat during processing.

discarded with the fluid. The amount of drip loss in marine fish may range from 5 to 15%. High energy irradiation for the preservation of seafood is not used commercially. Kennedy and Ley (1971) found 47% loss of thiamin upon irradiation of cod fillets but niacin concentration was not affected. Sterilization of scallops with high-voltage cathode rays caused a 10% reduction in thiamin, riboflavin and pyridoxine (Licciardello *et al.*, 1959).

Limited information is available on the vitamin content of convenience foods, particularly those which are coated with heavy batter, composed of flour, starch, dry milk, eggs and seasonings. The vitamin concentrations in these processed seafood products are expected to change with the proportion of fish and batter used as well as the vitamins supplied in batter and fat.

Table 5.12. Recommended dietary allowances (RDAs) (1989).[a]

	Age (years)	Weight (kg)	Height (cm)	Vitamin A (μg RE)[b]	Vitamin D (μg)[c]	Vitamin E (mg α-TE)[d]	Vitamin K (mg)	Vitamin C (mg)	Thiamin (mg)	Riboflavin (mg)	Niacin (mg NE)[e]	Vitamin B_6 (mg)	Folate (μg)	Vitamin B_{12} (μg)
Infants	0-0.5	6	60	375	8	3	5	30	0.3	0.4	5	0.3	25	0.3
	0.5-1	9	71	375	10	4	10	35	4.0	0.5	6	0.6	35	0.5
Children	1-3	13	90	400	10	6	15	40	0.7	0.8	9	1.0	50	0.7
	4-6	20	112	500	10	7	20	45	0.9	1.1	12	1.1	75	1.0
	7-10	28	132	700	10	7	30	45	1.0	1.2	13	1.4	100	1.4
Males	11-14	45	157	1000	10	10	45	50	1.3	1.5	17	1.7	150	2.0
	15-18	66	176	1000	10	10	65	60	1.5	1.8	20	2.0	200	2.0
	19-24	72	177	1000	10	10	70	60	1.5	1.7	19	2.0	200	2.0
	25-50	79	176	1000	5	10	80	60	1.5	1.7	19	2.0	200	2.0
	51+	77	173	1000	5	10	80	60	1.2	1.4	15	2.0	200	2.0
Females	11-14	46	157	800	10	8	45	50	1.1	1.3	15	1.4	150	2.0
	15-18	55	163	800	10	8	55	60	1.1	1.3	15	1.5	180	2.0
	19-24	58	164	800	10	8	60	60	1.1	1.3	15	1.6	180	2.0
	25-50	63	163	800	5	8	65	60	1.1	1.3	15	1.6	180	2.0
	51+	65	160	800	5	8	65	60	1.0	1.2	13	1.6	180	2.0
Pregnant				800	10	10	65	70	1.5	1.6	17	2.2	400	2.2
Lactating	1st 6 months			1300	10	12	65	95	1.6	1.8	20	2.1	280	2.6
	2nd 6 months			1200	10	11	65	90	1.6	1.7	20	2.1	260	2.6

[a] Food and Nutrition Board (1989).
[b] Retinol equivalents (1 retinol equivalent = 1 μg β-carotene).
[c] As cholecalciferol (10 μg cholecalciferol = 400 IU vitamin D).
[d] α-Tocopherol equivalents (1 mg d-α-tocopherol = 1α-TE).
[e] Niacin equivalents (1 NE = 1 mg niacin or 60 mg dietary tryptophan).

References and Further Reading

Ackman, R.G. and Cormier, M.G. (1967) α-Tocopherol in some Atlantic fish and shellfish with particular reference to live-holding without food. *Journal of the Fisheries Research Board of Canada* 24, 357–373.

Bailey, B.E. (1952) Marine oils with special reference to those of Canada. *Fisheries Research Board of Canada Bulletin* 89, 1–413.

Balasunddaram, S., Cama, H.R., Sundaresan, P.R. and Varma, T.N.R. (1956) Vitamin A_2 in Indian fresh-water fish-liver oils. *Biochemical Journal* 64, 150–154.

Ball, G.F.M. (1988) *Fat-soluble Vitamin Assays in Food Analysis: A Comprehensive Review.* Elsevier Applied Science Publishers, Essex, 326 pp.

Barua, A.B. (1978) Biosynthesis of dehydroretinol. *World Review of Nutrition and Dietetics* 31, 89–94.

Bergmann, W. (1952) Sterols: their structure and distribution. In: Florkin, M. and Mason, H.S. (eds) *Comparative Biochemistry*. Academic Press, New York, pp. 103–162.

Bhuiyan, A.K.M., Ratnayake, W.M.N. and Ackman, R.G. (1993) Nutritional composition of raw and smoked Atlantic mackerel (*Scomber scombrus*): Oil- and water-soluble vitamins. *Journal of Food Composition and Analysis* 6, 172–184.

Billis, C.E., McDonald, F.G., Massengale, O.N., Imboden, M., Hall, H., Hergert, W.D. and Wallenmeyer, J.C. (1935) A taxonomical study of the distribution of vitamins A and D in 100 species of fish. *Journal of Biological Chemistry* 109, vii.

Braekkan, O.R. (1962a) B-vitamins in fish and shellfish. In: Heen, E. and Kreuzer, R. (eds) *Fish in Nutrition*. Fishing News Books, London, pp. 132–140.

Braekkan, O.R. (1962b) B-vitamins in some fish products. In: Heen, E. and Kreuzer, R. (eds) *Fish in Nutrition*. Fishing News Books, London, pp. 141–145.

Braekkan, O.R. and Boge, G. (1965) Vitamin B6 og biotin i forskjellige organer fra torskefisker (*Gadidae*) fanget langs Norske kysten. *Fiskeridirektorates Skrifter, Serie Teknologiske Undersøkelser* 12, 4–11.

Braekkan, O.R., Lambertsen, G. and Myklestad, H. (1963) Alpha-tocopherol in some marine organisms and fish oils. *Fiskeridirektorates Skrifter, Serie Teknologiske Undersøkelser* 6, 3–11.

Bunnel, R.H., Keating, J.P. and Quaresimo, A.J. (1968) α-Tocopherol content of feedstuffs. *Journal of Agriculture and Food Chemistry* 16, 659–664.

Butler, C. (1946) Vitamins A and D in fish livers and viscera. *Commercial Fisheries Reviews* 8, 13–19.

Chabre, P. (1936) *Les Huiles de foie de morue, leur teneur en vitamines A et D.* Masson et Cie., Paris, 207 pp.

Cruickshank, E.M. (1962) Fat-soluble vitamins. In: Borgstrom, G. (ed.) *Fish as Food*, Vol. 2. Academic Press, New York, pp. 175–203.

Drummond, J.C. and Hilditch, T.P. (1930) The relative values of cod liver oils from various sources. *Empire Marketing Board Report* No. 35, 129 pp.

Dudek, J.A., Berman, S.C., Behl, B.A., Elkins, E.R., Chin, H.B. and Farrow, R.P. (1982) *Determination of Effects of Processing and Cooking on the*

Nutrient Composition of Selected Seafoods. National Food Processors Association, Washington, DC, 589 pp.

Fomin, S.V., Romanyuk, N.M. and Khvojnitskaya, M. (1937) The ascorbic acid contents of the tissues of various fish. *Ukrainskii Biokhimicheskii Zhurnal* 10, 365–378 (in Russian).

Food and Nutrition Board (1989) *Recommended Dietary Allowances*, 10th edn. National Academy Press, Washington, DC, 66 pp.

Frigg, M., Prabucki, A.L. and Ruhdel, E.U. (1990) Effect of dietary vitamin E levels on oxidative stability of trout fillets. *Aquaculture* 84, 145–158.

Gillam, A.E., Heilbron, I.M., Jones, W.E. and Lederer, E. (1938) On the occurrence and constitution of the 693 μm chromogen (vitamin A_2?) of fish liver oils. *Biochemical Journal* 32, 405–416.

Gordon, D.T. (1979) Cited by Gordon, D.T. and Martin, R.E. (1982) Vitamins and minerals in seafoods of the Pacific Northwest. In: Martin, R.E., Flick, G.J., Hebard, C.E. and Ward, D.R. (eds) *Chemistry and Biochemistry of Marine Food Products*. AVI Publishing Co., Westport, Connecticut, pp. 429–445.

Gordon, D.T. and Martin, R.E. (1982) Vitamins and minerals in seafoods of the Pacific Northwest. In: Martin, R.E., Flick, G.J., Hebard, C.E. and Ward, D.R. (eds) *Chemistry and Biochemistry of Marine Food Products*. AVI Publishing Co. Westport, Connecticut, pp. 429–445.

Gordon, D.T., Roberts, G.L. and Heinz, D.M. (1979) Thiamin, riboflavin and niacin content and stability in Pacific coast seafoods. *Journal of Agricultural and Food Chemistry* 27, 483–490.

Goswami, U.C. (1984) Metabolism of cryptoxanthin in freshwater fish. *British Journal of Nutrition* 25, 575–581.

Hastings, W.H. and Spencer, S.F. (1952) Determination of free and bound ascorbic acid in fishery products. *Journal of Marine Research* 11, 241–244.

Herbert, V. (1984) Vitamin B12. In: *Nutritional Review's Present Knowledge in Nutrition*, 5th edn. The Nutrition Foundation, Washington, DC, pp. 347–364.

Higashi, H. (1948) Studies on vitamins in liver of marine animals – X: Vitamin D content in fish liver. *Bulletin of the Agricultural Chemistry Society of Japan* 22, 15–16.

Higashi, H. (1961) Vitamins contents edible parts. In: Borgstrom, G. (ed.) *Fish as Food*, Vol. 1. Academic Press, New York, pp. 411–486.

Higashi, H. (1962) Relationship between processing techniques and the amount of vitamins and minerals in processed fish. In: Heen, E. and Kreuzer, R. (eds) *Fish in Nutrition*. Fishing News Books, London, pp. 125–131.

Higashi, H., Hirao, S., Yamada, J. and Kikuchi, R. (1955) The vitamin A content of *Heterocymnus longus* Tanaka, a kind of deep sea shark. *Journal of Vitaminology* 1, 200–207.

Higashi, H., Murayama, S., Yanase, M. and Tabei, K. (1958) The folic acid content of fish and shellfish. *Bulletin of the Japanese Society of Scientific Fisheries* 24, 776–780.

Higashi, H., Tarada, K., Nakashira, K. (1972) Studies on roles of tocopherols in fish. *Bitamin* (Vitamins, Japan) 45, 113–120.

Hoar, W.S. and Barberie, M. (1945) Distribution of riboflavin in fresh and processed fish. *Canadian Journal of Research* E23, 8–18.

Ikeda, S., Sato, M. and Kimura, R. (1963). Biochemical studies on L-ascorbic acid in aquatic animals. Distribution in various parts of fish. *Bulletin of the Japanese Society of Scientific Fisheries* 29, 765–770.

Jacquot, R. (1961) Organic constituents of fish and other aquatic animal foods. In: Borgstrom, G. (ed.) *Fish as Food*, Vol. 1. Academic Press, New York, pp. 145–209.

Junker, M. (1957) Vitamin A in Fischen, Krebsen und Muscheln. *Archiv für Fischereiwissenschaft* 7, 248–260.

Katsuyama, M. and Matsuno, T. (1988) Carotenoids and vitamin A and metabolism of carotenoids, beta-carotene, canthaxanthin, astaxanthin, zeaxanthin, lutein and tunaxanthin in tilapia (*Tilapia nilotica*). *Comparative Biochemistry and Physiology* 90B, 134–139.

Kennedy, T.S. and Ley, F.J. (1971) Studies on the combined effect of gamma irradiation and cooking on the nutritional value of fish. *Journal of the Science of Food and Agriculture* 22, 146–148.

Lee, P.H. (1987) Carotenoids in cultured channel catfish. PhD thesis, Auburn University, Auburn, Alabama, USA, 143 pp.

Licciardello, J.J., Nickerson, J.T.R., Proctor, B.E. and Campbell, C.L. (1959) Storage characteristics of some irradiated foods held at various temperatures above freezing. *Food Technology* 13(7), 405–409.

Loughlin, M.E. and Teeri, A.E. (1960) Nutritive value of fish. Biotin, folic acid, pantothenic acid and free amino acids of various salt water species. *Food Research* 25, 479–483.

Love, R.M. (1970) *The Chemical Biology of Fishes*. Academic Press, New York, 547 pp.

Lovern, J.A. (1944) The effect of preservation process on the vitamin content of fish. *Proceedings of the Nutrition Society* 2, 100–106.

Lunde, G., Aschehoug, V. and Kringstad, H. (1937) Fish as a source of vitamins A and D. Herring, brisling and their products. *Industrial Engineering Chemistry* 29, 1171–1175.

March, B.E. (1982) Effect of processing on nutritive value of food: fish. In: Rechcigl, M., Jr (ed.) *Handbook of Nutritive Value of Processed Food*, Vol. 1, CRC Press, Boca Raton, Florida, pp. 363–381.

Morton, R.A. and Bro-Rasmussen, F. (1955) Determination of vitamin A in natural products and especially cod liver oils. *The Analyst* 80, 410–418.

National Research Council (1982) *United States–Canadian Tables of Feed Composition*, 3rd edn. National Academy Press, Washington, DC, 114 pp.

National Research Council (1993) *Nutrient Requirements of Fish*. National Academy Press, Washington, DC.

Neilands, J.B. (1947) Thiaminase in aquatic animals of Nova Scotia. *Journal of the Fisheries Research Board of Canada* 7, 94–99.

Nettleton, J.A. (1985) *Seafood Nutrition: Facts, Issues and Marketing of Nutrition in Fish and Shellfish*. Osprey Books, Huntington, New York, 280 pp.

Nettleton, J.A. and Exler, J. (1992) Nutrients in wild and farmed fish and shellfish. *Journal of Food Science* 57, 257–260.

O'Keefe, S.F. and Ackman, R.G. (1987) Vitamins A_1, D_3 and E in Nova Scotian cod liver oils. *Proceedings of the Nova Scotia Institute of Science* 37, 1–7.

Rivlin, R.S. (1984) Riboflavin. In: *Present Knowledge in Nutrition*, 5th edn. The Nutrition Foundation, Washington, DC, pp. 285–302.

Schroeder, H.A. (1971) Losses of vitamins and trace minerals resulting from process and perservation of foods. *American Journal of Clinical Nutrition* 24, 562–573.

Sidwell, V., Loomis, A.L., Foncannon, P.R. and Buzzell, D.H. (1978) Composition of the edible portion of raw (fresh or frozen) crustaceans, finfish and mollusks. IV: Vitamins. *Marine Fisheries Review* 40, 1–16.

Sigurgisladottir, S., Parrish, C.C., Ackman, R.G. and Lall, S.P. (1994) Tocopherol deposition in the muscle of Atlantic salmon (*Salmo salar*). *Journal of Food Science* 59(2), 256–259.

Stansby, M.E. (1962) Proximate composition of fish. In: Heen, E. and Kreuzer, R. (eds) *Fish in Nutrition*. Fishing News Books, London, pp. 141–145.

Statens Livsmedelsverk, Sweden (1978) *Livsmedelstabeller* (Food composition tables), 584 pp.

Statens Livsmedelsverk, Sweden (1988) *Livsmedelstabeller* (Food composition tables), 398 pp.

Sundaresan, P.R. and Cama, H.R. (1961) Biological activity of crystalline vitamin A_2 aldehyde. *British Journal of Nutrition* 15, 225–230.

Suttie, J.W. (1983) Vitamin K. In: Machlin, L.J. (ed.) *Handbook of Vitamins, Nutritional, Biochemical and Clinical Aspects*. Marcel Dekker, New York, pp. 147–198.

Suzuki H., Okazaki, K., Hayakawa, S. and Tamura, S. (1986) Influence of commercial dietary fatty acids of cultured fresh-water fish and comparison with those of wild fish of the same species. *Journal of Agriculture and Food Chemistry* 34, 58–60.

Taarland, T., Mathieson, E., Ovsthus, O. and Braekkan, O.R. (1958) Nutritional values and vitamins of Norwegian fish and fish products. *Tidsskrift før Hermetikkindustri* 44, 405–412.

Takeuchi, A., Okano, T. and Kobayashi, T. (1991a) The existence of 25-hydroxyvitamin D_3-1-α-hydroxylase in the liver of carp and bastard halibut. *Life Science* 48, 275–282.

Takeuchi, A., Okano, T., Tanda, M. and Kobayashi, T. (1991b) Possible origin of extremely high contents of vitamin D_3 in some kinds of fish liver. *Comparative Biochemistry and Physiology* 100A, 483–487.

Tarr, H.L.A. (1962) Changes in nutritive value through handling and processing procedures. In: Borgstrom, G. (ed.) *Fish as Food*, Vol. II. Academic Press, New York, pp. 235–266.

Tarr, H.L.A., Southcott, B.A. and Ney, P.W. (1952) Vitamin B_{12} active substances in fish products. *Food Technology* 4(9), 354–357.

Waagbo, R., Sandnes, K., Torrissen, O., Sandvin, A. and Lie, O. (1993) Chemical and sensory evaluation of fillets from Atlantic salmon (*Salmo salar*) fed three levels of n-3 polyunsaturated fatty acids at two levels of vitamin E. *Food Chemistry* 46, 361–366.

Wald, G. (1945–46) The chemical evolution of vision. *Harvey Lectures* 41, 117–160.

Woodward, B. (1994) Dietary vitamin requirements of cultured young fish, with emphasis on quantitative estimates for salmonids. *Aquaculture* (in press).

Yanase, M. (1956) The vitamin B_6 content of fish meat. *Bulletin of the Japanese Society of Scientific Fisheries* 22, 51–55.

6

Macro and Trace Elements in Fish and Shellfish

Santosh P. Lall
Biological Sciences Branch, Department of Fisheries and Oceans, Halifax, Nova Scotia, Canada.

Introduction

Fish and shellfish, like other living organisms, contain most of the 90 naturally occurring elements. The greater proportion of their body consists of carbon, hydrogen, nitrogen, oxygen and sulfur. In addition, six macro elements, i.e. calcium, magnesium, phosphorus, sodium, potassium and chlorine, are present in gram per kilogram quantities. The remaining elements occur in the body in much lower concentrations (mg or μg per kg). Initial difficulties in the accurate determination of low levels of many of these elements led inevitably to their description as 'trace elements' and 'ultratrace elements'. Modern analytical techniques such as atomic absorption spectroscopy, neutron activation and inductively coupled plasma atomic emission spectroscopy now permit the accurate analyses of these substances at levels of about 10^{-7} to 10^{-9} g.

Trace elements were at first considered as undesirable contaminants of food, but later it was realized that some of the elements, particularly iron and copper, were essential for normal health. Trace elements were accordingly classified as essential, non-essential and toxic. Fourteen trace elements are now considered essential for human and animal life. The physiological role and deficiency of chromium, cobalt, copper, iodine, iron, manganese, molybdenum, selenium, zinc and fluorine are well recognized. Although deficiencies of nickel, vanadium, silicon and arsenic have been demonstrated in an ultra-clean environment, with the exception of silicon, the physiological functions of these trace elements have not been clearly demonstrated. Other elements including cadmium, lead, bromine and tin have also been claimed essential, but this remains to be confirmed.

Most of these elements are detected in fish and shellfish, which inevitably supply essential macro and trace elements to the human diet. The main functions of essential elements include formation of skeletal structure, maintenance of colloidal systems (osmotic pressure, viscosity, diffusion), and regulation of acid–base equilibrium. They are important components of hormones, enzymes and enzyme activators. Calcium and phosphorus are required in the formation of the skeletal structures of the body. Sodium, potassium and chlorides, along with phosphates and bicarbonates, maintain homeostasis and acid–base balance. Some metals (Fe, Mn, Cu, Co, Zn, Mo, Se, etc.) are firmly associated with specific proteins in metalloenzymes, which produce a unique catalytic function. Certain minerals such as calcium, magnesium and manganese are of particular significance as enzyme activators. A non-metal, i.e. iodine, is necessary for the biosynthesis of thyroid hormones, which in turn greatly affect development and metabolism in all vertebrates.

Inorganic Matter in Fish and Shellfish

Aquatic organisms absorb minerals from their diet and surrounding water and deposit them in skeletal tissue, muscle and different organs (Lall, 1989). Fish muscle contains as much copper as the meat of domestic animals, five to ten times more fluorine, and about 100 times more iodine. Fish and shellfish contain much higher concentrations of selenium than poultry and red meat. Generally fish are at least recommended sources of iron and zinc but dark-coloured flesh of fish contains as much iron as beef. Although nutrient composition data of common foods including fish are widely available, data found in the literature on the composition of essential minerals in fish and fishery products are confined to selected minerals. An extensive report on the elemental composition of marine organisms was published by the Russian scientist Vinogradov (1953). Studies conducted on the mineral composition of fish and some seafoods from 1930 to the mid-1960s were comprehensively reviewed by Causeret (1962). The analytical data on selected North American seafoods are summarized in Agriculture Handbook No. 8-15 (Exler, 1987) and in other publications (Gordon and Roberts, 1977; Sidwell *et al.*, 1977; 1978; Sidwell, 1981; Gordon and Martin, 1982). There are reports from Finland (Nuurtamo *et al.*, 1980), Norway (Julshamn *et al.*, 1978), Sweden (Statens Livsmedelsverk, 1988), the United Kingdom (Lovern, 1964) and the Middle East (El-Faer *et al.*, 1992) on the mineral composition of commercially important fish species in these countries.

The concentration of minerals in fish and fishery products is influenced by a number of factors such as seasonal and biological differences (species, size, age, sex and sexual maturity), food source, environment (water chemistry, salinity, temperature and contaminants) and the method of food processing. Most aquatic organisms accumulate and retain minerals from the environment; however, their incorporation is highly selective. A Russian study on fish from different lakes showed that the quantities of manganese, iron, copper and zinc in the muscle and liver reflect the geochemical characteristics of the earth surrounding the lake (reviewed by Love, 1970). In marine food chains, a unique transport of trace metals has been reported (Bernhard and Andreae, 1984). Most trace metals show the highest concentration increase in the first trophic level (seawater phytoplankton). In zooplankton, only Cd, Cu and Zn concentrations increase. Plankton-feeding fishes have higher Cu and Zn levels than fish preying on invertebrates and fish. The concentration of Cd, Cu and Zn increases along the food chain, reaching a maximum in crustaceans, and then decreasing in fish.

Molluscs are known to accumulate essential and toxic trace metals from the environment. Zinc in oysters, for instance, may range from less than 1 to more than 1000 mg kg^{-1}. The composition of commercial feeds used for farmed fish also influences their elemental composition. The complete inorganic element composition is known for only a few species of fish. The elemental composition of cultured rainbow trout varies with fish size, stage of life cycle, and reproductive status (Shearer, 1984). Wide variations are observed in the reported values of macro and trace element concentrations in the same species of fish. This may be partly due to the above-mentioned factors as well as the variability in sampling procedures and analytical methodology.

The major portion of minerals in the fish body is distributed in skeletal tissues, particularly vertebrae. Morphologically, fish bones consist of dermal bones of the head, the internal skeleton, and scales. The composition of bones in fishes is basically similar to other vertebrates. An extensive review of the composition of fish bones and other skeletal tissues was published by Vinogradov (1953). The main constituents of fish vertebrae include calcium, phosphate and carbonate, with small amounts of magnesium, sodium, strontium, lead, citrate, fluoride, hydroxide and sulfate. Fish bone contains 35 to 58% protein and up to 65% inorganic matter (Roche and Bullinger, 1939). Otoliths also contain minerals. In cod, the otoliths contain mainly $CaCO_3$, but traces of magnesium, phosphorus and other elements may also occur. The elemental concentration of skeletal tissues varies with the age of fish.

Generally, most of the bones and other skeletal tissues of fish are either removed during filleting or prior to consumption. Only smaller bones remain in the fillet, which contains relatively smaller amounts of inorganic matter than whole fish. The average ash content in the edible muscle portion of most fish ranges from 0.5 to 1.8% of wet weight (Sidwell, 1981) unless excessive amounts of fine bones remain in the final product. Raw crab and lobster products contain higher amounts of inorganic matter

(1–2%) than fish and contamination of muscle with shell causes wide variation in the total mineral content of shellfish products.

In Tables 6.1 to 6.7, the mineral composition of common edible fish and shellfish are summarized after a careful review of published data. All data are based on wet weight and macro and trace element contents are expressed as mg $100\,g^{-1}$ and mg kg^{-1}, respectively. It was necessary to summarize trace elements as mg kg^{-1} of flesh because of their very low concentrations found in all fish and shellfish tissues. The extreme values reported in older publications are excluded because of problems with analytical methodology and the limited number of samples analysed. Where complete results were not available for individual species, for example mussels, data are summarized for the group as a whole. Where only a single value for an element is shown in the tables, this value represents a report from a single publication. In many cases, the range of reported values for a particular mineral in fish of the same species was quite large. It was impossible to cite all the references and it is suggested that the reports of Causeret (1962) and Sidwell (1981) may be used to locate data published before 1960.

The recommended dietary allowances (RDAs) for various minerals as established by the Food and Nutrition Board of the National Academy of Sciences, USA, are summarized in Table 6.9. RDAs are used to define a relationship between human populations and the nutrients required at different stages of life. They are intended to serve as a basis for evaluating the adequacy of diets of groups of people rather than individuals. This information, in conjunction with mineral composition tables, can be used with caution to estimate the proportion of individual minerals in human diets that can be supplied by fish and shellfish.

Calcium and phosphorus

Calcium and phosphorus are the most abundant mineral elements in fish, humans and other living organisms. Approximately 99% of the calcium and 80–85% of the phosphorus are present in bones as calcium phosphate ($Ca_3(PO_4)_2$) and hydroxyapatite ($Ca_{10}(PO_4)_6(OH)_2$). The remaining portions of calcium and phosphorus are found in extracellular fluids and intracellular structures and cell membranes. Calcium is responsible for a number of regulatory functions. Phosphorus is directly involved in energy-producing cellular reactions. Thus phosphorus plays an important role in overall metabolism as well as in various metabolic processes involving buffers in body fluids.

Fishery products are a good source of phosphorus and, like other meat products, a poor source of calcium. Fish fillet or other products commonly consumed by humans contain minute proportions of bone and thus supply relatively smaller amounts of calcium and phosphorus than whole fish. In areas where gutted whole fish is consumed, bones and scales provide significant amounts of calcium and phosphorus in the diet. The average calcium concentrations (Tables 6.1 to 6.3) in most finfish and shellfish muscle are generally low and some species show wide differences due to variable amounts of bone and shell. Crustaceans contain more calcium than finfish; the former may require Ca for shell formation and muscle function. Whole sardines, anchovy and other fishery products preserved in oil may contain as much as 580 mg of Ca 100 g^{-1} (Da Costa and Stern, 1956; Sidwell, 1981).

The average phosphorus concentration in fish fillet ranges from 113 to 350 mg 100 g^{-1} (Tables 6.1 to 6.3). Van de Velde (1932) reported higher than average amounts of phosphorus in mackerel (330–370 mg 100 g^{-1}) and plaice (240–410 mg 100 g^{-1}). Phosphates added during the processing of fish increase the phosphorus content of fish. Gibson and Murray (1973) found that phosphate dips prior to freezing increased the phosphate content of fish by 50 to 110 mg 100 g^{-1}.

Magnesium

The distribution of magnesium is similar to the distribution of calcium and phosphorus, the major proportion being located in bone. The remainder is found within the cells of soft tissues. Magnesium is a prosthetic ion in enzymes that hydrolyse and transfer phosphate groups. Hence it is essential for energy-requiring biological functions such as membrane transport, generation and transmission of nerve impulses, contraction of muscles and oxidative phosphorylation. It is also essential for the maintenance of ribosomal structure and thus protein synthesis.

Little is known about the forms in which magnesium is present in animal tissues. Fish red blood cells contain significantly high levels of magnesium and low levels of calcium. In green plants, a high proportion of this element occurs as magnesium porphyrin, chlorophyll. Fishery products, like other animal meat products, are poor sources of magnesium. The average magnesium concentration (Tables 6.1 to 6.3) in the edible portion of most finfish and shellfish muscle is between 21 and 45 mg 100 g^{-1}. Amounts higher than 50 mg magnesium per 100 g of muscle have been found in lobster, shrimps, prawns, snails and squid (Table 6.3). Although magnesium concentrations of fish muscle are similar to beef, lobster, snails and squid supply much higher levels of this element.

Sodium, potassium and chlorine

Sodium and chloride are the principal extracellular cation and anion, respectively, in the human body. Sodium is important in osmoregulation, acid–base balance and the membrane potential of cells, as well as active

Table 6.1. Macro element composition of marine fish.[a]

Species	Calcium (mg 100 g^{-1})	Phosphorus (mg 100 g^{-1})	Magnesium (mg 100 g^{-1})	Sodium (mg 100 g^{-1})	Potassium (mg 100 g^{-1})
Anchovy	195 (20-480)	240 (188-350)	40 (30-54)	115 (70-147)	355 (267-510)
Cod	16 (9-20)	210 (9-240)	23 (20-25)	80 (60-180)	360 (270-460)
Dolphin	15	143	57	170 (98-240)	370
Drum	37 (16-57)	115 (25-200)	27 (24-30)	67 (51-84)	275 (226-300)
Flounder	27 (17-36)	185 (116-250)	24 (10-31)	60 (35-99)	316 (157-394)
Haddock	19 (10-30)	208 (164-320)		57 (49-67)	340 (299-434)
Halibut	23 (13-30)	215 (192-250)	23	65 (36-112)	395 (318-475)
Herring	145 (21-460)	320 (134-742)	38 (30-46)	103 (49-183)	348 (25-512)
Ling	14 (7-20)	200	32	83	300 (259-338)
Mackerel	60 (5-340)	305 (148-778)	30 (25-40)	90 (30-252)	320 (136-471)
Plaice	20	220	25	91 (83-98)	312 (270-333)
Pomfret	130 (23-240)	150 (140-163)	31	114	517
Ray	19 (9-25)	130 (99-170)	38	156 (133-179)	295 (236-352)
Sardine	134 (28-380)	350 (39-580)		90 (60-128)	220 (25-420)
Sole, Dover	12	156	24	110	226
Tuna	30 (6-93)	350 (190-760)	50	70 (31-188)	365 (252-518)
Turbot	49	203	48	68	500
Whiting	45 (16-71)	170 (130-253)	21	85 (50-118)	330 (260-457)

[a] Raw edible part.

Sources: Julshamn *et al.* (1978); Sidwell *et al.* (1978); Nuurtamo *et al.* (1980); Sidwell (1981); Teeny *et al.* (1984); Statens Livsmedelsverk (Sweden) (1988); S.P. Lall (1994), unpublished data.

Table 6.2. Macro element composition of freshwater fish.[a]

Species	Calcium (mg 100 g^{-1})	Phosphorus (mg 100 g^{-1})	Magnesium (mg 100 g^{-1})	Sodium (mg 100 g^{-1})	Potassium (mg 100 g^{-1})
Carp	40 (17-57)	195 (165-225)	37 (30-43)	60 (48-80)	290 (174-397)
Catfish	30 (18-60)	220 (102-327)	45 (26-113)	65 (60-114)	295 (147-440)
Eel	65 (15-188)	295 (196-501)	22 (16-30)	80 (76-81)	360 (214-712)
Mullet	40 (11-99)	260 (140-436)	30 (25-33)	75 (52-100)	300 (259-356)
Perch	110	240	26	60	330
Pike	40 (29-52)	225 (213-240)	30 (15-42)	40 (29-42)	365 (310-416)
Salmon, Atlantic[b]	16 (15-23)	230 (219-241)	28 (25-32)	47 (40-54)	410 (380-440)
Salmon, chinook[b]	17 (14-20)	270 (160-372)	30 (17-95)	42 (38-56)	305 (289-321)
Salmon, coho[b]	35 (28-43)	230 (212-258)	28 (26-29)	38 (26-51)	445 (421-469)
Salmon, sockeye[b]	18	290	30 (21-95)	42 (38-45)	265 (160-372)
Suckers	15	165		55 (50-59)	315 (292-344)
Trout, rainbow	32 (25-51)	260 (196-280)	25 (21-30)	45 (39-52)	455 (420-488)
Whitefish	60	290	30	60	380

[a] Raw edible part.
[b] Anadromous fishes (juveniles migrate to sea and adults return to fresh water to spawn).
Sources: Julshamn *et al.* (1978); Sidwell *et al.* (1978); Nuurtamo *et al.* (1980); Sidwell, (1981); Statens Livsmedelsverk (Sweden) (1988); S.P. Lall (1994), unpublished data.

Table 6.3. Macro element composition of shellfish.[a]

Species	Calcium (mg 100 g^{-1})	Phosphorus (mg 100 g^{-1})	Magnesium (mg 100 g^{-1})	Sodium (mg 100 g^{-1})	Potassium (mg 100 g^{-1})
Clam[b]	100 (59-144)	160 (90-350)	45 (20-50)	190 (180-200)	140 (35-228)
Crab[b]	60 (13-120)	235 (171-350)	50 (42-50)	330 (195-453)	235 (111-322)
Crayfish	50 (41-58)	240 (101-360)	45 (40-80)	120 (60-182)	305 (239-349)
Lobster	40 (16-61)	255 (182-560)	175 (21-442)	300 (194-390)	330 (202-500)
Mussel[b]	105 (71-164)	145 (102-170)	30	215 (140-289)	400 (315-480)
Octopus[b]	30 (12-39)	110 (66-151)		365	235
Oyster[b]	100 (39-210)	155 (76-265)	35 (12-230)	160 (73-418)	250 (90-570)
Scallop[b]	50 (11-115)	210 (176-340)	40 (30-48)	165 (150-176)	280 (162-420)
Shrimp and prawn[b]	145 (16-550)	240 (127-912)	55 (23-111)	135 (45-220)	250 (118-410)
Snail[b]	140 (132-146)	160 (110-191)	65 (61-143)	95 (68-146)	160 (136-179)
Squid	50 (10-109)	220 (153-420)	80 (20-123)	140 (102-176)	275 (246-313)

[a] Raw edible part.
[b] Mixed species.
Sources: Sidwell *et al.* (1978); Sidwell (1981); Statens Livsmedelsverk (Sweden) (1988); S.P. Lall (1994), unpublished data.

transport across cell membranes. Chloride is essential in the maintenance of electrolyte balance and is also the chief anion in gastric juice. Potassium serves as the monovalent cation to balance intracellular anions, and participates in neuromuscular functions. It is the most abundant mineral in muscle tissue.

Wide variations in the sodium content of fishery products have been reported (Thurston and Osterhaug, 1960; Causeret, 1962). For freshwater and marine fish, sodium averages 60 mg 100 g^{-1}. Most shellfish have sodium in the range of 120–400 mg 100 g^{-1}, except squid and scallop where the range is lower, i.e. 102–176 mg 100 g^{-1}. In edible tissue of both marine and freshwater fish, potassium concentration is always higher than that of sodium (Tables 6.1 to 6.3). The potassium content of seafood ranged between 198 and 440 mg 100 g^{-1}, but lower levels occur in some shellfish, e.g. shrimp, crab and lobster (Thurston and Osterhaug, 1960; Dyer *et al.*, 1977; Sidwell *et al.*, 1978; Teeny *et al.*, 1984). Data on distribution of chloride in fish and shellfish are scarce.

Processed fishery products have high sodium contents because of the addition of salt or sodium-containing compounds during processing (Table 6.7). Fish canned in water, bouillon or tomato sauce have more sodium than the same fish canned in oil. Fillets from frozen fish treated with brine and/or sodium polyphosphate to control thaw drip formation contain 165–212 mg Na 100 g^{-1} (Thurston and Osterhaug, 1960). Frozen lobster, queen crab and shrimp packed in brine may contain 500–1000 mg Na 100 g^{-1} (Dyer *et al.*, 1977). Precooked products such as fish sticks, fish cakes and various breaded products contain 300–800 mg Na 100 g^{-1} (Dyer *et al.*, 1977).

Although the early work of Vinogradov (1953) showed that the ratio of sodium to potassium in marine and freshwater fish muscle is 1:3 or 4, more recent reports show a much wider ratio of these elements (Tables 6.1 and 6.2). Gordon and Martin (1982) found that the potassium to sodium ratio in whitefish was 6.8 based on mean sodium and potassium concentrations of 351 ± 45 and 52 ± 15 mg 100 g^{-1}. Most seafoods showed similar ratios with the exception of oyster and processed products.

Iron

Iron is present in all cells of living organisms and plays a vital role in several biochemical reactions. As a component of haem in haemoglobin, myoglobin, cytochromes and other proteins, iron plays an essential role in the transport, storage and utilization of oxygen. Iron is also a cofactor for a number of enzymes. Generally, iron is present in food both in inorganic forms as ferrous and ferric compounds, and in organic forms, the most important of these being haem iron. Haem and non-haem iron have a different mechanism of absorption and often the biological availability of non-haem iron is lower (Carpenter and Mahoney, 1992).

Little is known about the forms of iron present in fish. A considerable proportion will undoubtedly be present in the form of the iron porphyrin, myoglobin, and, depending on the amount of occluded blood, a proportion of the iron in haemoglobin. Total iron content of various fish and shellfish is summarized in Tables 6.4 to 6.6. Extensive studies on the mineral composition of several species of fish caught from the Pacific coast of the USA show values ranging from 2 to 10 mg iron per kg of edible muscle tissues (Gordon and Roberts, 1977; Teeny *et al.*, 1984). A wide range (0.8 to 373) of values has been reported for many marine fish and crustaceans (Tables 6.4 and 6.6) probably due to metal and bone contamination of fishery products and differences in analytical methodology. The removal of skin from mackerel and cod reduces the iron content of the fillet (Gomez-Basauri and Regenstein, 1992).

Dark muscle contains more iron, both haem and non-haem forms, than light muscle (Brown, 1962; Fischer and Deng, 1977; Gomez-Basauri and Regenstein, 1992). Atka mackerel and American shad muscles contain two to three times as much iron and copper as cod, sole, pollock, pomfret and whiting, probably due to a high amount of dark muscle in the former two species (Teeny *et al.*, 1984). The dark flesh of mullet had a total iron content of 5.7 mg kg^{-1}, a much higher concentration than that of beef muscle at 2.6 mg kg^{-1} (Fischer and Deng, 1977). The haem iron constituted 25–44% of the total iron content for the fish compared to 71–89% in beef. The percentage of dark muscle changes with the season, which directly affects the iron concentration of mackerel (Leu *et al.*, 1981).

The iron content of processed fish and fishery products may be influenced by the widespread possibilities for contamination from ferrous metals during cooking and processing. Non-haem iron has been reported to be a major catalyst of lipid oxidation in cooked meat (Love and Pearson, 1974). Direct contact of fillet with a deboning machine or any other metal appears to increase the iron content of minced fillet (Lee and Toledo, 1977). Mechanical de-boning causes an increase in haemoglobin content of minced mullet (Silberstein and Lillard, 1978). The iron content of mechanically separated flesh of several marine fish from the east coast of the USA ranged from 36 to 219 mg total iron per kg of wet tissue (Crawford *et al.*, 1972).

Zinc

Zinc is essential for growth, reproduction, wound healing, and normal functioning of the immune system and other physiological processes. The essential function of zinc is based on its roles as an integral part of a number of metalloenzymes and as a catalyst for regulating the activity of specific zinc-dependent enzymes. Thus zinc can regulate many metabolic

Table 6.4. Trace element composition of marine fish.[a]

Species	Iron (mg kg^{-1})	Zinc (mg kg^{-1})	Manganese (mg kg^{-1})	Copper (mg kg^{-1})	Selenium (mg kg^{-1})	Chromium (mg kg^{-1})	Molybdenum (mg kg^{-1})	Iodine (mg kg^{-1})	Fluorine (mg kg^{-1})
Anchovy	20.0 (8.1-49.0)	15.0	0.4 (0-0.7)	3.0 (0.8-5.0)	0.2	0.1		0.8	
Cod	9.0 (3.4-42.9)	32.0 (9.0-52.5)	0.2 (0-0.5)	3.0 (0.2-5.5)	0.4 (0.2-0.9)	0.1		1.0 (0.2-5.0)	3.0 (0.7-7.0)
Drum	18.0 (0.8-79.2)	4.0 (0.7-7.6)	0.2	1.0 (0.1-2.1)	0.4			0.2 (0-0.6)	
Flounder	8.0 (0.4-22.0)	5.0 (0.8-14.2)	0.4 (0-1.8)	2.0 (0.1-7.0)	0.7 (0.3-1.4)	0.1 (0-0.2)	0.2 (0-0.4)	0.4 (0.1-1.0)	0.4
Haddock	8.0 (3.5-14.0)	8.0 (2.7-17.4)	0.2	2.0 (0.1-2.4)	0.3 (0.2-0.6)	0.1 (0-0.3)	0.2	3.0 (0.3-6.0)	5.0
Halibut	10.0 (6.6-16.0)	2.0	0.5 (0.1-1.3)	1.0 (0.1-2.3)	0.3 (0.2-0.4)	0.1 (0-0.3)	0.1	0.5	
Herring	11.0 (0-31.2)	2.0	0.1 (0-0.2)	2.0 (0.1-2.3)	0.2			0.2 (0-0.5)	
Ling	7.0 (5.0-8.0)	24.0		2.0				2.0 (1.2-3.0)	
Mackerel	30.0 (0.7-84.9)	5.0	0.6 (0.3-1.2)	1.0 (0.3-2.0)	0.3		0.1	0.4 (0.1-1.0)	1.5
Pomfret	16.0	9.0	1.6 (1.4-1.8)	1.0 (0.5-1.2)	0.2	0.1		1.5	
Ray	22.0 (5.0-61.5)	3.0	0.5	2.0 (0.6-3.5)	0.3			1.1	
Sole, Dover	2.0	4.0	0.1	0.2	0.2				
Tuna	28.0 (2.7-80.0)	1.0	0.3 (0.1-0.5)	0.1	0.4	0.1	0.1	0.3 (0.2-0.4)	
Whiting	10.0 (0.9-22.0)	8 (0.6-21.8)	5.0	0.1 (0.1-0.2)	0.3			0.9	

[a] Raw edible part.

Sources: Egaas and Braekkan (1977); Julshamn *et al.* (1978); Sidwell *et al.* (1978); Lo and Sandi (1980); Sidwell (1981); Teeny *et al.* (1984); Statens Livsmedelsverk (Sweden) (1988); S.P. Lall (1994), unpublished data.

Table 6.5. Trace element composition of freshwater fish.[a]

Species	Iron (mg kg^{-1})	Zinc (mg kg^{-1})	Manganese (mg kg^{-1})	Copper (mg kg^{-1})	Selenium (mg kg^{-1})	Chromium (mg kg^{-1})	Molybdenum (mg kg^{-1})	Iodine (mg kg^{-1})	Fluorine (mg kg^{-1})
Carp	11.0	15.0 (8.4-22.0)	0.4 (0.3-0.5)	3.0 (1.2-4.3)	0.2			1.0 (0.1-1.5)	
Catfish	15.0 (3.6-26.9)	4.0 (0.2-9.8)	0.6	1.0 (0.6-1.7)		0.2		1.0	
Eel	20.0 (5.0-54.4)	18.0 (6.2-30.0)	0.3	1.0 (0.2-1.7)				0.8	
Mullet	30.0 (1.0-250.0)	3.0 (0.8-5.3)	0.1	1.0 (0.1-2.0)					2.0 (0.1-4.9)
Perch	3.0	8.0	0.6	0.4	0.3	0.01	<0.1		1.0
Pike	10.0 (0.4-43.1)	13.0 (4.0-19.0)	1.0 (0.2-3.2)	2.0 (0.7-3.4)	0.2	<0.01	<0.1	0.3	0.1
Salmon, Atlantic[b]	4.0	4.0	0.2	0.6	0.3	<0.01	<0.1	0.3	0.3
Salmon, chinook[b]	6.0	7.0	0.2	0.4	0.3			0.4 (0.1-0.6)	
Salmon, coho[b]	13.0	4.0	0.2 (0.1-0.3)	0.6	0.3			0.7 (0.3-1.7)	
Salmon, sockeye[b]	9.0	4.0 (4.0-11.0)	0.2 (0.1-0.4)	1.5 (0.8-1.9)	0.3	<0.01	<0.1		5.0
Trout, rainbow	4.0	5.0	0.2	0.5	0.3	0.01	<0.1		0.3
Whitefish	5.0	12.0	0.5	0.6	0.4	0.01	<0.1		1.0

[a] Raw edible part.
[b] Anadromous fishes (juveniles migrate to sea and adults return to fresh water to spawn).
Sources: Sidwell *et al.* (1978); Nuurtamo *et al.* (1980); Sidwell (1981); Statens Livsmedelsverk (Sweden) (1988); S.P. Lall (1994), unpublished data.

Table 6.6. Trace element composition of shellfish.[a]

Species	Iron (mg kg^{-1})	Zinc (mg kg^{-1})	Manganese (mg kg^{-1})	Copper (mg kg^{-1})	Selenium (mg kg^{-1})	Chromium (mg kg^{-1})	Molybdenum (mg kg^{-1})	Iodine (mg kg^{-1})	Fluorine (mg kg^{-1})
Clam[b]	45.0 (5.0-311.0)	80.0	1.4 (0.9-1.6)	3.0 (2.5-4.9)	0.6	0.3		5.0 (1.4-9.0)	
Crab[b]	25.0 (2.5-61.0)	45.0 (14.0-93.6)	0.3	10.0 (0.1-50.7)	0.6 (0.4-1.1)		0.6	2.0 (0.1-4.6)	2.0
Crayfish	50.0 (4.4-373.0)	30.0 (1.5-66.0)	2.0 (0.1-4.2)	21.0 (0.1-167.0)	0.7	0.1	0.2	1.0 (0.3-1.4)	
Lobster	15.7 (3.0-60.0)	10.0 (1.5-15.0)	1.0 (0.4-2.5)	15 (1.7-33.0)	0.7			1.4	
Mussel[b]	80.0 (10.0-127.0)	34.0 (20.0-44.8)	2.0 (0.1-4.6)	2.0 (0.2-3.6)	0.4 (0.3-0.8)		2.2	1.5 (0.8-1.9)	29.0
Octopus[b]	15.0	16.0	0.1	8.0				0.2	
Oyster[b]	55.0 (8.9-99.6)	850.0 (71.0-2000.0)	3.0 (0.1-6.5)	80.0 (0.3-606.0)	0.7	0.1		0.6 (0.1-1.3)	1.0 (0.7-1.6)
Scallop[b]	20.0 (10.0-30.0)	25.0 (17.0-29.5)	10.0 (1.3-39.0)	1.6 (0.1-3.1)		0.1		0.2	
Shrimp and prawn[b]	20.0 (7.0-40.0)	20.0 (7.4-42.0)	0.7 (0.3-1.9)	5.0 (0.1-13.1)	1.0 (0.6-1.9)	0.1		0.7 (0.2-1.5)	1.0
Snail[b]	70.0 (41.0-86.9)	25.0	17.0 (16.0-17.6)	25.0 (4.0-62.2)				0.1	
Squid	35.0 (5.0-185.6)	25.0 (7.6-84.0)	0.6 (0.2-0.8)	11.0 (2.9-15.0)	0.3			0.3	0.2

[a] Raw edible part.
[b] Mixed species.
Sources: Egaas and Braekkan (1977); Julshamn *et al.* (1978); Sidwell *et al.* (1978); Lo and Sandi (1980); Sidwell (1981); Statens Livsmedelsverk (Sweden) (1988).

processes of carbohydrate, lipid and protein metabolism and synthesis of nucleoproteins.

Among all food sources of animal origin, oyster is the richest source of zinc. The concentrations of zinc in tissues and organs from approximately 650 different species of marine plants, invertebrates, fish and mammals are listed by Eisler (1980). Molluscs contain the greatest concentration of zinc. Recorded values in crassostreid oysters exceed 4000 mg Zn kg^{-1} of dry tissue weight. Several species of algae contain more than 1000 mg Zn kg^{-1} of dry weight. Among marine organisms lowest concentrations (50 to 400 mg kg^{-1} dry weight) are found in fish and mammals. These values represent the concentration in the whole organism and differ from the values found in the edible portion of fish. Liver, viscera, gills, scales and milt contain much higher concentrations of zinc than muscle.

Tables 6.4 to 6.6 summarize the zinc contents of the edible parts of fish and shellfish. In general fish have low zinc contents when compared to molluscs and crustaceans. The average zinc content of marine and freshwater fish is approximately 8 mg kg^{-1}, but salmon and trout contain only 4 mg Zn kg^{-1} of flesh (wet weight). Lobster meat contains a relatively lower amount of zinc than crab and other shellfish. The bioavailability of zinc from most fish and shellfish is relatively high but may be influenced by other dietary components which contain large amounts of calcium, phosphorus and phytates, etc.

Manganese

Manganese functions either as a cofactor activating a large number of enzymes to form metal–enzyme complexes or as an integral part of certain metalloenzymes. Since the chemistry of the manganese ion is similar to that of the magnesium ion, many enzymes can be activated by either manganese or magnesium. Certain enzymes, e.g. glycosyl transferases, are highly specific for manganese activation. The enzymatic function of manganese in lipid and carbohydrate metabolism and brain function is widely recognized.

Manganese is widely distributed in fish and other animal tissues. The highest concentrations are found in bone; however, significant amounts are also present in liver, muscle, kidney, gonadal tissues and skin. Molluscs and crustaceans contain significantly higher levels of manganese than fish (Tables 6.4 to 6.6). Most fish contain 0.1 to 0.4 mg Mn kg^{-1} provided there is no appreciable amount of bones retained in the flesh. Shells in crustacean products also increase their manganese levels. The manganese composition of the edible muscle tissue of seven species of marine fish caught from the Pacific Ocean averaged 0.1 mg kg^{-1} (Teeny *et al.*, 1984). However, some walleye pollock analysed from a different geographic location contained a six times higher manganese concentration.

Copper

Copper is an essential element for humans and a wide range of animal and fish species. It is required for iron utilization, and as a cofactor for enzymes involved in glucose metabolism and the synthesis of haemoglobin, connective tissue and phospholipids. Numerous studies have been directed on copper metabolism in fish and concerned with toxic effects related to heavy metal pollution in the aquatic environment. Molluscs contain more copper in muscle and in their organs than marine plants, crustaceans and fish (Eisler, 1979). Molluscs use copper in a unique way, by producing haemocyanins to carry oxygen to their tissues. The function of haemocyanins is exactly the same as that of haemoglobins. The kidney is one of the major sites where metals accumulate.

The overall concentration of copper in many living organisms including fish is around 2 mg kg^{-1}, with the exception of some shellfish and molluscs (Tables 6.4 to 6.6). Among the molluscs, highest concentrations occur in oysters, particularly in digestive glands and kidneys (Eisler, 1981). In general, copper concentrations among marine fish and mammals are among the lowest recorded for this metal among all groups of marine biota. The reason for this is unknown, but suggests a discrimination against Cu by the highest trophic levels (Eisler, 1981).

An extensive study on the distribution of trace elements in North American finfish revealed that muscle copper concentration ranged from 0.1 to 2 mg Cu per kg fresh weight with most values falling between 0.2 and 0.5 mg kg^{-1} (Hall *et al.*, 1978). However, liver contained significantly higher concentrations of copper than other tissues at values ranging from 1.0 to 110 mg kg^{-1}. Other active sites of copper accumulation in decreasing order include scales, spleen, kidney, gill, swim bladder and muscle. Several factors, including season, temperature, salinity and presence of other metals such as Mn and Fe in the water, affect the accumulation of copper in fish (Eisler, 1981).

Selenium

Selenium is recognized both as an essential trace element and a highly toxic agent. It is an integral component of the enzyme glutathione peroxidase in human and animal tissues, which, together with vitamin E and the enzymes catalase and superoxide dismutase, functions as an antioxidant, thereby protecting cells against oxidative damage. Although many toxic effects of selenium have been described in animals, to date no pathological conditions have been identified in humans resulting from selenium deficiency. Excess selenium promotes dental carries. Selenium reduces the toxicity of methyl mercury; thus selenium deficiency accentuates heavy metal toxicity.

Selenium contents of foods including fish have been reviewed by Lo and Sandi (1980) and Combs (1988). The distribution of selenium in certain fishery products from Canada (Arthur, 1972), Finland (Nuurtamo *et al.*, 1980), Japan (Shinoda *et al.*, 1980) Norway (Egaas and Braekkan, 1977), the UK (Thorn *et al.*, 1978) and the USA (Morris and Levander, 1970; Schroeder *et al.*, 1970) have been reported and typical selenium analyses of fish and shellfish are summarized in Tables 6.4 to 6.6. Fish contain substantially higher levels of selenium than red meats and poultry and are thus considered as major dietary sources of this element. Some fishery products contain relatively high amounts of heavy metals (Chapter 9), which reduce the biological availability of selenium.

A limited amount of information published on the chemical forms of selenium in food suggests that selenium is protein bound, mainly as amino acid analogues (Combs, 1988). Maher (1985) reported that selenium was mainly associated with soluble proteins (72–80%) in muscle tissues of Australian marine organisms (prawn, scallop, squid and fish) and ionic inorganic forms were not detectable. Selenomethionine and selenocysteine appear to be the major forms of selenium found in foods of plant and animal origin, respectively (Combs, 1988). Thus availability of selenium depends on the digestibility of protein from various food sources.

Although the selenium content of most foods of plant and animal origin depends largely on the selenium content of the soil in which the food is grown, the selenium content of fish caught from areas of the midwest USA where excess selenium exists was not high (Olson and Palmer, 1984). Generally, natural waters contain low concentrations of selenium. A survey of fish in natural waters of New York State showed that the selenium levels in freshwater fish were in the range of 0.10 to 1.50 mg kg^{-1} (Pakkala *et al.*, 1972), compared with 0.3 to 3.3 mg kg^{-1} in the Great Lakes of the central regions of Canada (Beal, 1974; Traversy *et al.*, 1975). Unpublished data on selenium content of marine fishes generated by Fisheries and Oceans, Canada, average about 1 mg kg^{-1}. Similar levels have also been reported for marine fish meals and marine organisms in Sweden (Lunde, 1970; 1973). Swordfish caught in the northwest Atlantic Ocean were found to contain 0.3–1.3 mg Se kg^{-1} (Freeman *et al.*, 1978).

Chromium

Chromium is considered an essential nutrient for humans. It may have a role in activating enzymes and in maintaining the structural stability of proteins and nucleic acids, but the primary physiological role of chromium in a biologically active complex is to potentiate the action of insulin. Chromium is present in food in at least two forms, as inorganic Cr^{3+} and as part of a biologically active molecule. The exact structure of the biological molecule is not known but it is postulated to contain nicotinic acid and some amino acids (glycine, glutamic acid, cysteine, glutathione).

Edible tissues of commercially important North American molluscs contain between 0.12 and 0.6 mg Cr kg^{-1} wet weight (Hall *et al.*, 1978). However, levels as high as 3.4 and 5.8 mg Cr kg^{-1} have been reported in oysters and clams, respectively (Shuster and Pringle, 1968). Some molluscs in the coastal areas of Greece have the ability to accumulate substantially higher levels ($>$12 mg Cr kg^{-1}) of chromium in their tissues (Papadopoulou, 1973). In crustacean edible tissues, chromium concentrations seldom exceed 0.3 mg kg^{-1} (Eisler, 1981).

The distribution of chromium has been measured in few fish and shellfish (Tables 6.4 to 6.6). Generally, the edible part of the muscle contains lower levels of chromium than liver and other organs. Muscles of many finfish examined by Hall *et al.* (1978) showed concentrations of chromium ranging from 0.1 to 0.5 mg kg^{-1}. The average chromium levels in individual tissues of many fishes show 0.1 to 0.4 mg kg^{-1} (Eisler, 1981).

Molybdenum

Molybdenum is an essential component of several enzymes, including xanthine oxidase, aldehyde oxidase, and sulfite oxidase, where it occurs in the prosthetic group, molybdopterin. Nutritional deficiency of this metal in humans does not occur under normal circumstances.

In a survey of the trace element composition in marine fishes and shellfish, Hall *et al.* (1978) found that edible muscle of fish did not exceed 0.6 mg Mo kg^{-1}; however, the average concentration was 0.25 mg Mo kg^{-1}. In molluscs, muscle generally contains 0.2 to 0.8 mg Mo kg^{-1}. Some mussel species accumulated as much as 4 mg Mo kg^{-1} in edible tissues. The distribution of this metal in crustacean meat ranged from 0.1 to 0.4 mg Mo kg^{-1}. Reports on molybdenum distribution in aquatic organisms are few and most values fall approximately in the same range as those observed by Hall *et al.* (1978).

Iodine

Iodine, the heaviest essential element in living organisms, is needed for the biosynthesis of the thyroid hormones, thyroxine and triiodothyroxine. Thyroid hormones control cellular oxidation, and influence growth, other endocrine glands, neuromuscular functions, circulatory dynamics, and also the metabolism of major nutrients. A dietary deficiency of iodine causes an enlargement of the thyroid gland resulting in goitre.

Iodine exists naturally covalently bound to tyrosine and thyronine, or in the form of iodate and iodide salts, but it is not found as elemental iodine (I_2). It is present in relative abundance in marine plants, fish and shellfish. Kelp contains 0.1 to 0.3% iodine. Inhabitants of the coastal regions of Hokkaido Island in Japan consume large amounts of seaweeds, which may amount to 50 to 80 mg of iodine intake per day. In these regions goitre induced by excessive intake of iodine has been reported (Suzuki *et al.*, 1965). The edible flesh of marine fish and shellfish may contain 0.3 to 3 mg I kg^{-1} (Tables 6.4 and 6.6). The iodine content of freshwater fish shows wide variations (Table 6.5), but it is generally lower than that of marine fish (Anon., 1952).

Fluorine

Fluorine is an essential trace element that is best known for its beneficial effects in humans, where it has a protecting role against dental caries. Fluorides are widely distributed in nature. Fluorine is a normal component of calcified tissues and its concentration is directly related to fluorine exposure.

Among foodstuffs of animal origin, fish have the highest fluorine content (Toth and Sugar, 1978). Few reports on fluorine content of the edible part of fish and shellfish are available (Tables 6.4 to 6.6). A systematic study of the fluorine composition of some Norwegian fishery products revealed that the fluoride content ranged from 0.16 to 2.06 mg kg^{-1} with an average value of 0.8 mg kg^{-1} (Soevik and Braekkan, 1981). However, canned sardines and 'kippered' herring including skin and bones contain as high as 3.87 mg F kg^{-1}. The concentration of fluoride in the whole fish ranges from 10 to 60 mg F kg^{-1}.

Blue mussels are particularly rich in fluorine (Table 6.6). Fluorine is present in the muscles of fish, crabs, shrimps and prawns from the North Sea in amounts from about 1 to 4 mg kg^{-1}. About five times higher concentrations were found in the muscles of krill-feeding fish from the Antarctic waters. The average fluoride content of the Antarctic fish (*Notothemia gibberifrons* and *N. rossii* Marmorata) in the fillet, skin and backbone were 13, 92 and 1600 mg kg^{-1}, respectively. The Antarctic krill contains exceptionally large amounts of fluoride (mg kg^{-1} dry weight): 1000, whole animal; 70, muscle; 2000, exoskeleton. After capture, a rapid migration of fluoride from the exoskeleton into the muscle takes place.

Cobalt

The only known function of cobalt relates to its role as a component of vitamin B_{12}. Approximately 4.5% of the molecular mass of vitamin B_{12} is contributed by elemental cobalt. Few reports on the cobalt content of fish and shellfish are available. Data summarized on the trace element concentrations in marine organisms by Eisler (1981) show the following range of values for cobalt (mg kg^{-1}) in the edible part: molluscs, 0.05–0.2; crustacea, 0.004–0.009; marine fish, 0.002–0.033. Some marine fish and molluscs accumulate much higher concentrations of this metal. Cobalt accumulates (>15 mg kg^{-1} dry weight) in the kidneys and digestive glands of some molluscs (Bryan, 1973; Bryan and Hammerstone, 1978).

Effects of Processing on Mineral Composition of Fish

A variety of processing methods are used to make fish and shellfish edible, to permit storage, to alter texture and flavour, and to sterilize and pasteurize fishery products. These methods include mincing, cooking, baking, frying, salting, smoking, boiling, freezing, etc. The physical removal of vertebrae and other skeletal tissues causes intentional loss of minerals, particularly calcium, phosphorus, magnesium and some trace elements (Table 6.7). An increase may occur in the calcium content of mechanically prepared fish mince due to the presence of bone fragments and scales (Kryznowek *et al.*, 1984).

Unlike vitamins and lipids, the nutritive values of minerals are not affected during the processing but there may be a net gain or loss of some elements. Losses of inorganic elements during processing are mainly physical, due to liquid discarded from fishery products. Generally the dissolution of mineral salts or complexes depends on the pH. Table 6.8 shows the retention of sodium, potassium, iron, zinc and selenium in seafoods as affected by various cooking methods (baking, broiling (grilling), microwaving, breading and frying) and the canning process. The retention of these five minerals under ordinary cooking conditions ranged from 90 to 100% (Dudek *et al.*, 1982). During the canning process a significant reduction in sodium and potassium levels occurred with net gain in zinc content. A significant loss of iodine (80%) during cooking has been reported (Causeret, 1962).

Salt is traditionally used to preserve fish in many parts of the world. As discussed previously, the sodium content of salted fish is greatly increased. Salt is also known to increase the oxidation of lipid in fatty fishes. Additives such as polyphosphate, sodium glutamate and salt used for surimi production increase the sodium and phosphorus contents of the fish and shellfish product. Washing of fish mince reduces the potassium content of surimi.

Processed fishery products must be free of toxic metals and precautions should be taken to prevent contamination during processing with excessive amounts of essential elements which may also be toxic. Contamination of fish with traces of metals from non-food sources such as food containers and water pipes, etc., may affect the colour, flavour or shelf-life of

Table 6.7. Mineral content of certain processed seafood products.

Species	Calcium (mg 100 g^{-1})	Phosphorus (mg 100 g^{-1})	Magnesium (mg 100 g^{-1})	Sodium (mg 100 g^{-1})	Potassium (mg 100 g^{-1})	Iron (mg kg^{-1})	Zinc (mg kg^{-1})
Smoked fish							
Eel	55	250	25	650	225	10	30
Herring	80	475	35	5700	365	50	35
Mackerel	20	290	30	400	310	15	15
Salmon, Atlantic	15	250	30	1500	330	5	5
Canned products							
Crab	120	140	35	550	110	10	5
Mackerel	20	260	30	390	310	15	30
Mussel	55	140	50	390	140	40	25
Oyster	30	125			70	55	250
Salmon, Pacific	95	240	30	570	300	15	10
Sardines in oil	195	300	40	510	560	30	20
Shrimp	20	145	25	980	170	25	20
Tuna in oil	15	310	45	410	260	15	8
Miscellaneous							
Herring, salted	20	180	25	6000	190	8	10
Herring, marinated	20	155	55	960	155	10	8
Herring, kippered	90	290	35	2300	415	10	20

Sources: Sidwell *et al.* (1978); Sidwell (1981); Statens Livsmedelsverk (Sweden) (1988); S.P. Lall (1994), unpublished data.

Table 6.8. Effect of different cooking methods and canning on retention of sodium, potassium, iron, zinc and selenium in some seafoods.[a,b,c]

Cooking method and canning	Sodium	Potassium	Iron	Zinc	Selenium
Bake	89	93	97	100	100
Broil[d]	88	90	94	97	100
Microwave	94	96	100	100	100
Bread and fry[e]	95	100	98	99	100
Canned[f]	77	63	100	161[g,h]	93

[a] Dudek *et al.* (1982).
[b] Average values of six species (flounder, sockeye salmon, Atlantic mackerel, Alaskan pollock, Pacific whiting and shrimp).
[c] Retention (%) = (Average weight of minerals in processed product/Average weight of minerals in raw product) × % yield.
[d] All fish species excluding shrimp.
[e] Only pollock, whiting and shrimp.
[f] Only salmon, mackerel and shrimp.
[g] Only salmon and mackerel.
[h] Retention value exceeds 100% because of the leaching of zinc from the can.

Table 6.9. Recommended dietary allowances (1989).[a]

	Age (years)	Weight (kg)	Height (cm)	Calcium (mg)	Phosphorus (mg)	Magnesium (mg)	Iron (mg)	Zinc (mg)	Selenium (μg)	Iodine (μg)
Infants	0-0.5	6	60	400	300	40	6	5	40	10
	0.5-1	9	71	600	500	60	10	5	50	15
Children	1-3	13	90	800	800	80	10	10	70	20
	4-6	20	112	800	800	120	10	10	90	20
	7-10	28	132	800	800	170	10	10	120	30
Males	11-14	45	157	1200	1200	270	12	15	150	40
	15-18	66	176	1200	1200	400	12	15	150	50
	19-24	72	177	1200	1200	350	10	15	150	70
	25-50	79	176	800	800	350	10	15	150	70
	51+	77	173	800	800	350	10	15	150	70
Females	11-14	46	157	1200	1200	280	15	12	150	45
	15-18	55	163	1200	1200	300	15	12	150	50
	19-24	58	164	1200	1200	280	15	12	150	55
	25-50	63	163	800	800	280	15	12	150	55
	51+	65	160	800	800	280	10	12	150	55
Pregnant				1200	1200	320	30	15	175	65
Lactating	1st 6 months			1200	1200	355	15	19	200	75
	2nd 6 months			1200	1200	340	15	16	200	75

Food and Nutrition Board (1989).

processed fishery products. Processed seafoods pick up calcium and magnesium from water; iron, copper, zinc, manganese, chromium and nickel from equipment; strontium, iron, aluminium, lead, copper and zinc from cans and food containers; and sodium during brine grading. Aluminium and tin cans cause darkening of foods. Traces of copper, iron and some other metals can act as catalysts in the oxidation of unsaturated fatty acids of fish. The development of stainless steel machinery and limited use of traditional metal cooking ware have contributed to a significant reduction of iron and other mineral contamination of seafoods in industrialized countries.

Some nutrients are unstable during food processing because of their sensitivity to oxygen, light and heat or a combination of these factors. High concentrations of trace elements, particularly copper and iron, and certain enzymes present in fishery products may further enhance the oxidation of lipids and destroy certain vitamins. The stabilities of minerals present in fish are not significantly affected by physical and chemical treatments during food processing, except that some are oxidized to higher valences. To date, there is no convincing evidence that the nutritional value of minerals in fishery products is affected by processing. The processing is known to influence the biological availability of several trace elements, particularly iron, in some foods of animal and plant origin (Fritz, 1976).

References

Anon. (1952) *Iodine Contents of Foods – Annotated Bibliography 1825–1951.* Chilean Iodine Education Bureau, London, 165 pp.

Arthur, D. (1972) Selenium content of Canadian foods. *Canadian Institute of Food Science and Technology Journal* 5, 165–169.

Beal, A.R. (1974) A study of selenium levels in freshwater fish of Canada's central region. *Department of Environment, Technical Report Series* Nr. CEN/T-74-6, 14 pp.

Bernhard, M. and Andreae, M.O. (1984) Transport of metals in marine food chains. *Life Sciences Research Reports* 28, 143–167.

Brown, W.D. (1962) The concentration of myoglobin and hemoglobin in tuna flesh. *Journal of Food Science* 27, 26–28.

Bryan, G.W. (1973) The occurrence and seasonal variation of trace metals in the scallops *Pecten maximus* (L.) and *Chalimys opercularis* (L.). *Journal of Marine Biological Association (UK)* 53, 145–166.

Bryan, G.W. and Hammerstone, L.G. (1978) Heavy metals in the burrowing bivalve *Scrobicularia plana* from contaminated and uncontaminated estuaries. *Journal of Marine Biological Association (UK)* 58, 401–419.

Carpenter, C.E. and Mahoney, A.W. (1992) Contributions of heme and nonheme iron to human nutrition. *Critical Reviews in Food Science and Nutrition* 31, 333–367.

Causeret, J. (1962) Fish as a source of mineral nutrition. In: Borgstrom, G. (ed.) *Fish as Food*, Vol. 2. Academic Press, New York, pp. 205–234.
Combs, G.F., Jr (1988) Selenium in foods. *Advances in Food Research* 32, 85–113.
Crawford, D.L., Law, D.K. and Bobbitt, J.K. (1972) Nutritional characteristics of marine food fish carcass waste and machine-separated flesh. *Journal of Agriculture and Food Chemistry* 20, 1048–1051.
Da Costa, A. and Stern, J.A. (1956) The calcium and phosphorus contents of some foreign and domestic canned sardines. *Food Research* 21, 242–249.
Dudek, J.A., Berman, S.C., Behl, B.A., Elkins, E.R., Chin, H.B. and Farrow, R.P. (1982) *Determination of Effects of Processing and Cooking on the Nutrient Composition of Selected Seafoods*. National Food Processors' Association, Washington, DC, 589 pp.
Dyer, W.J., Fraser Hiltz, D., Hayes, E.R. and Munro, V.G. (1977) Retail frozen fishery products – Proximate and mineral composition of the edible portion. *Canadian Institute of Food Science and Technology Journal* 10, 185–190.
Egaas, E. and Braekkan, O.R. (1977) The selenium content in some Norwegian fish products. *Fiskeridirektoratets Skrifter, Serie Ernaering* 1, 87–91.
Eisler, R. (1979) Copper accumulations in coastal and marine biota. In: Nriagu, J. (ed.) *Copper in the Environment*, Part 1. John Wiley & Sons, New York, pp. 383–449.
Eisler, R. (1980) Accumulation of zinc by marine biota. In: Nriagu, J.O. (ed.) *Zinc in the Environment*, Part II. John Wiley & Sons, New York, pp. 259–351.
Eisler, R. (1981) *Trace Metal Concentrations in Marine Organisms*. Pergamon Press, New York, 687 pp.
El-Faer, M.Z., Rawdah, T.N., Attar, K.M. and Arab, M. (1992) Mineral and proximate composition of some commercially important fish of the Arabian Gulf. *Food Chemistry* 45, 95–98.
Exler, J. (1987) *Composition of Foods: Finfish and Shellfish Products*. United States Department of Agriculture, Human Nutrition Service Agriculture Handbook 8-15 (revised). Washington, DC, 192 pp.
Fischer, J. and Deng, J.C. (1977) Catalysis of lipid oxidation: A study of mullet (*Mugil cephalus*) dark flesh and emulsion model system. *Journal of Food Science* 42, 610–614.
Food and Nutrition Board (1989) *Recommended Dietary Allowances*, 10th edn. National Academy Press, Washington, DC, 66 pp.
Freeman, H.C., Shum, G. and Uthe, J.F. (1978) The selenium content in swordfish in relation to total mercury content. *Journal of Environmental Science and Health* A13, 235–240.
Fritz, J.C. (1976) Bioavailability of mineral nutrients. *Chemtech* 6, 643–648.
Gibson, D.M. and Murray, C.K. (1973) Polyphosphates and fish: Some chemical studies. *Journal of Food Technology* 8, 197–204.
Gomez-Basauri, J.V. and Regenstein, J.M. (1992) Processed and frozen storage effects on the iron content of cod and mackerel. *Journal of Food Science* 57, 1332–1336.
Gordon, D.T. and Martin, R.E. (1982) Vitamins and minerals in seafoods of the Pacific Northwest. In: Martin, R.E., Flick, G.J., Hebard, C.E. and Ward,

D.R. (eds) *Chemistry and Biochemistry of Marine Food Products*. AVI Publishing Co., Westport, Connecticut, pp. 429–445.

Gordon, D.T. and Roberts, G.L. (1977) Mineral and proximate composition of Pacific Coast fish. *Journal of Agriculture and Food Chemistry* 25, 1262–1268.

Hall, R.A., Zook, E.G. and Meaburn, G.M. (1978) National marine fisheries survey of trace elements in the fishery resources. *US Department of Commerce NOAA Technical Report* NMFS SSRF-721, 313 pp.

Hiltner, R.S. and Wichmann, H.J. (1919) Zinc in oysters. *Journal of Biological Chemistry* 38, 205–221.

Julshamn, K., Haugnes, J. and Utne, F. (1978) The contents of 14 major and minor elements (minerals) in Norwegian fish species and fish byproducts, determined by atomic absorption spectrophotometry. *Fiskeridirektoratets Skrifter, Serie Ernaering* 1, 117–135.

Kryznowek, J., Peton, D. and Wiggin, K. (1984) Proximate composition, cholesterol and calcium content in mechanically separated fish flesh from three species of Gadidae family. *Journal of Food Science* 49, 1182–1185.

Lall, S.P. (1989) Minerals. In: Halver, J.E. (ed.) *Fish Nutrition*. Academic Press, San Diego, pp. 220–257.

Lee, C.M. and Toledo, R.T. (1977) Degradation of fish muscle during mechanical deboning and storage with emphasis on lipid oxidation. *Journal of Food Science* 42, 1646–1649.

Leu, S.S., Jhaveri, S.N., Karakoltsidis, P.A. and Constantinides, S.M. (1981) Atlantic mackerel (*Scomber scombrus*, L): seasonal variation in proximate composition and distribution of chemical nutrients. *Journal of Food Science* 46, 1635–1638.

Lo, M.T. and Sandi, E. (1980) Selenium: Occurrence in foods and its toxicological significance – A review. *Journal of Environmental Pathology and Toxicology* 4, 193–218.

Love, J.D. and Pearson, A.M. (1974) Metmyoglobin and nonheme iron as prooxidants in cooked meat. *Journal of Agriculture and Food Chemistry* 22, 1032–1034.

Love, R.M. (1970) *The Chemical Biology of Fishes*. Academic Press, New York, 547 pp.

Lovern, J.A. (1964) The vitamin and mineral contents of fish. In: Bacharach, A.L. and Rendle, T. (eds.) *The Nation's Food*. Society of Chemical Industry, London, pp. 300–308.

Lunde, G. (1970) Analysis of arsenic and selenium in marine raw materials. *Journal of the Science of Food and Agriculture* 21, 242–247.

Lunde, G. (1973) Trace metal contents of fish meal and of the lipid phase extracted from fish meal. *Journal of the Science of Food and Agriculture* 24, 413–419.

Maher, W.A. (1985) Characteristics of selenium in marine animals. *Marine Pollution Bulletin* 16, 33–34.

Morris, V.C. and Levander, O.A. (1970) Selenium content of foods. *Journal of Nutrition* 100, 1383–1388.

Nuurtamo, M., Varo, P., Saari, E. and Koivistoinen, P. (1980) Mineral element

composition of Finnish foods. VI. Fish and fish products. *Acta Agriculturae Scandinavica* 22, 77–87.
Olson, O.E. and Palmer, I.S. (1984) Selenium in foods purchased or produced in South Dakota. *Journal of Food Science* 49, 446–452.
Pakkala, I.S., Gutemann, W.H., Lisk, D.J., Burdick, G.E. and Harris, E.J. (1972) A survey of selenium content of fish from 49 New York state waters. *Pesticide Monitoring Journal* 6, 107–114.
Papadopoulou, C. (1973) The elementary composition of marine invertebrates as a contribution to the sea pollution investigation. *Proceedings of MAMBO Meeting*, Castellabate, Italy, June 18–22, 1973, 18 pp.
Roche, J. and Bullinger, E. (1939) Recherches sur l'ossification. VII. La phosphate du squelette (os, dents, dermatosquelette) chez les poissons osseux ou cartilagineux. *Bulletin de la Société Chimique de Biologique* 21, 166–173.
Schroeder, H.A., Frost, D.V. and Balassa, J.J. (1970) Essential trace metals in man: Selenium. *Journal of Chronic Diseases* 23, 227–243.
Shearer, K.D. (1984) Changes in elemental composition of hatchery-reared rainbow trout (*Salmo gairdneri*) associated with growth and reproduction. *Canadian Journal of Fisheries and Aquatic Sciences* 41, 1592–1600.
Shinoda, T., Aoyagi, Y. and Sugahara (1980) Contents of mineral and proximate compositions in commercial fishes and shellfishes. *Food Chemistry* (Japan) 11, 171–177.
Shuster, C.N., Jr and Pringle, B.H. (1968) Effects of trace metals on estuarine molluscs. *Proceedings of First Mid-Atlantic Industrial Waste Conference*. Department of Civil Engineering, University of Delaware, Newark, USA, pp. 285–304.
Sidwell, V.D. (1981) *Chemical and Nutritional Composition of Finfishes, Whales, Crustaceans, Mollusks and their Products*. Technical Memorandum, National Oceanic and Atmospheric Administration, National Marine Fisheries Service, US Department of Commerce, Washington, DC, 432 pp.
Sidwell, V.D., Buzzell, D.H., Foncannon, P. and Smith, A. (1977) Composition of the edible portion of raw (fresh or frozen) crustaceans, finfish, and mollusks. II. Macroelements: sodium, potassium, chlorine, calcium, phosphorus and magnesium. *Marine Fisheries Review* 39(1), 1–11.
Sidwell, V.D., Loomis, A.L., Loomis, K.J., Foncannon, P.R. and Buxxell, D.H. (1978) Composition of the edible portion of raw (fresh or frozen) crustaceans, finfish, and mollusks. III. Microelements. *Marine Fisheries Review* 40(9), 1–20.
Silberstein, D.A. and Lillard, D.A. (1978) Factors affecting the autoxidation of lipids in mechanically deboned fish. *Journal of Food Science* 43, 764–766.
Soevik, T. and Braekkan, O.R. (1981) The fluoride contents in some Norwegian fish products and other marine products. *Fiskeridirektoratets Skrifter, Serie Ernaering* 2, 1–16.
Statens Livsmedelsverk, Sweden (1978) *Livsmedelstabeller* (Food composition tables). 584 pp.
Statens Livsmedelsverk, Sweden (1988) *Livsmedelstabeller* (Food composition tables). 398 pp.

Suzuki, H., Higashi, T., Sawa, K., Ohtaki, S. and Horiuchi, Y. (1965) Endemic coast goitre in Hokkaido, Japan. *Acta Endocrinologica* (Japan) 50, 161–176.
Teeny, F.M., Gauglitz, E.J., Jr, Hall, A.S. and Houle, C.R. (1984) Mineral composition of the edible muscle tissue of seven species of fish from the Northeast Pacific. *Journal of Agriculture and Food Chemistry* 32, 852–855.
Thorn, J., Robertson, J., Buss, D.H. and Bunton, N.G. (1978) Trace nutrients. Selenium in British food. *British Journal of Nutrition* 39, 391–396.
Thurston, C.E. and Osterhaug, K.L. (1960) Sodium content of fish flesh. *Journal of the American Dietetic Association* 36, 212–215.
Toth, K. and Sugar, E. (1978) Fluorine content of foods and estimated daily intake from foods. *Acta Physiologica Academiae Scientiarum Hungaricae* 51, 361–369.
Traversy, W.J., Goulden, P.D., Sheikh, Y.M. and Leacock, J.R. (1975) Levels of arsenic and selenium in the Great Lakes Region. *Scientific Series* No. 58. Inland Water Directorate, Water Quality Branch, Environment Canada, 18 pp.
Van de Velde, A.J.J. (1932) Investigation on the chemical composition of fish. III. Composition of *Pleuronectes* and *Scomber*. *Natuurwetenschappelijk Tijdschrift* (Ghent) 14, 178–187.
Vinogradov, A.P. (1953) *The Elementary Chemical Composition of Marine Organisms*. Sears Foundation for Marine Research, Yale University, New Haven, Connecticut, 647 pp.

7

Quality and Deterioration of Raw Fish

H.K. Davis
CSL Food Science Laboratory, Torry, Aberdeen, UK.

Perceptions of Quality

Quality is a word much abused by the absence, or disregard for the veracity, of any qualification. Excellence is usually implied when reference is made, simply, to quality products with only vague indications as to the particular attribute being described. The term is sometimes used to promote and exploit how certain species such as Dover sole, brill, turbot and many shellfish species are fashionably perceived. Some of these distinctions have changed over time just as, with declining catches of wild fish, they are changing now. *Nephrops norvegicus* was frequently discarded until someone exploited the term 'scampi', and monkfish increased in value when a television presenter commented on its value as a scampi substitute. Mackerel, long held in poor esteem, became an important and more highly valued resource when herring stocks in UK waters declined so badly as to become subject to a temporary ban on fishing. Like most other freshwater fish, pike are little valued in Britain as food, but, before modern roads and rapid transport systems were developed, any fresh fish was a rare commodity away from coastal areas. The produce of rivers, streams and 'stew-ponds' was an important source of fresh protein food, and pike is said to have been more valuable than salmon. Nowadays, such distinctions are made between salmon from different sources, with implicit assumptions and, sometimes, explicit claims that wild fish are intrinsically superior to the farmed product. Other such distinctions which are intended to enhance the appeal of one type of product by denigration of the alternative, and which may be strengthened through oft-repetition more than informed understanding, include the terms: 'fresh' (i.e. meaning chilled) as opposed to frozen; 'white' fleshed species as opposed to others which have a darker flesh; and lean as opposed to fatty (though, perhaps, recent interest in the n-3 unsaturated fatty acid content of fish lipids might reverse this view).

Such perceptions are probably vital in determining the place of any product on the market, and have their place within definitions of consumer quality which refer to fitness for purpose. Though purveyors and advertisers need to know how to form and exploit these views, the fish technologist is concerned more with specific attributes which may suffer as a result of poor control as fish are grown, harvested, processed, transported and stored. The hygiene aspects of quality, the sources and control of parasites and pathogenic microorganisms, and the presence or absence of contaminants and toxins are dealt with elsewhere in this book. Here, the primary concern is with the remaining major quality attribute, degree of freshness: the changes which take place once the fish have been caught, and the ways in which major processing variables influence these changes.

Deterioration of Fish Stored in Ice

The raw material

The number and diversity of fish species commonly exploited for food is much greater than that of mammalian meats. For many people, the words 'edible fish' bring to mind the gadoids: marine, demersal, lean species from temperate waters, such as cod and haddock. But many others are utilized and they may be very different in form and structure. In almost all cases the main edible tissue, and the part which has been most studied as a food, is skeletal muscle, which can be cut away from the spine as fillets. Fillet meat of lean species consists primarily of white muscle with only a small amount of dark muscle immediately under the skin. Pelagic species live a more active, continuously swimming life and have a much thicker band of dark muscle, which is richer in lipids and haem pigments. The flesh of these fish (e.g. herring and mackerel) has a much higher, seasonally variable lipid (fat) content than white fish. Other commonly utilized marine fish include: various flatfishes such as plaice, which, lying on the seabed, are inactive for much of their time and contain even less dark muscle than the gadoids; the elasmobranchs or cartilaginous fish (e.g. skates, dogfish and sharks), which contain a high concentration (can be over 2%) of urea; and shellfish, which contain relatively high concentrations of glycogen (in molluscs) and free amino acids. Salmonids migrate between marine and freshwater environments although some (e.g. rainbow trout) are adapted to living entirely in fresh waters. Many of the salmonids are now reared in intensive farm systems. A similarly large range of fish and shellfish species, caught from the wild or intensively reared, is taken from warmer waters in tropical and subtropical regions.

Differences in composition of fish tissues can have considerable effects on the patterns and rates of spoilage. Within a species, size, condition and methods of catching and handling can also have significant influence. There have been systematic studies of the changes which take place during storage of representative species of all the groups which have been mentioned but, almost certainly, gadoids have been studied the most. They have provided a relevant, readily obtainable raw material for many researchers, and present a basic pattern of change which is common to all species. To this basic pattern is added a variety of additional spoilage characteristics dependent upon the compositional differences of other types of fish.

Storage in ice

Stability studies need some constant set of conditions which can provide readily reproducible reference material. For whole fish, the appropriate conditions are to surround them with melting ice but, even for such an apparently simple procedure, further effort is needed to make sure that optimum conditions prevail. The aim is to ensure that fish are constantly bathed with a film of water at 0°C. Thus, after packing carefully in ice, the fish are held in a chill room which can maintain temperatures just above freezing point so that the ice will melt slowly, providing a gentle trickle of melt-water.

Experimental work commonly employs much higher ratios of ice to fish than the 1:3 minimum recommended for practical commercial use. Prolonged storage studies require the ice to be replenished, and care has to be taken to ensure, for example, that containers are not at risk of freezing from being too close to refrigeration units, and that they drain freely, and are agitated occasionally to disrupt the tendency for ice bridges to form as melting creates cavities between the fish and remaining ice.

General patterns of deterioration

Stored in this way, spoilage is slowed almost to the minimum possible without actually freezing. Note, the temperature of fish in ice is slightly below 0°C and some further reduction is obtainable using the less readily controlled techniques of storage in refrigerated sea water or 'superchilled' fishrooms. Although small amounts of ice are formed in tissues below 0°C, freezing is generally regarded as beginning between −1 and −2°C, when the amount of frozen water present in flesh increases to the point where the fish begins to stiffen.

Deteriorative changes begin as soon as the fish dies. To some extent, they may be regarded as starting even before death as fish which have had time to struggle have used some of their energy reserves with immediate post-mortem consequences. The principal agents of decay are endogenous enzymes and bacteria, and their effects are most apparent at different times. They produce two overlapping phases: the first, autolysis, can be regarded in a more passive sense as a loss of freshness, a decline consequent upon disruption of the balance of the fish's metabolic functions; later, bacterial activity produces more pronounced effects of overt spoilage and the fish eventually becomes inedible.

Autolysis

Cellular activity continues after death, driven by the remaining energy reserves. These reserves, mainly glycogen, may already have been depleted in muscle cells during the capture struggle and, with blood no longer circulating, cannot be replenished. The breakdown of glycogen to glucose and lactic acid is similar to that which occurs in mammalian meats except for the initial stage, which produces glucose via, predominantly, hydrolytic rather than phosphorolytic reactions (Burt, 1966). Accumulation of lactic acid causes muscle pH to fall. Being close to neutral in live cod (McCallum *et al.*, 1967), the flesh becomes slightly acidic. Although the fall in pH may be faster in fish than in mammals, there is less glycogen in the muscles of most live fish and much is lost during capture, so the ultimate pH is relatively high. There are significant differences between species, and muscle pH can vary considerably within a species depending upon season and biological condition. Values below pH 6.0 have been observed in some species, including halibut, mackerel and tuna (Buttkus and Tomlinson, 1966) and also in sturgeon (Izquierdo-Pulido *et al.*, 1992), but for cod and most other fish they tend to be between 6 and 7. As pH decreases, muscle proteins approach their isoelectric point. At the lower ultimate pH values, denaturation begins; water-holding capacity decreases and small amounts of exudate ('drip') are released. In the most extreme cases the flesh, normally translucent in freshly killed fish, becomes opaque to the extent of being referred to as 'chalky' (Tomlinson *et al.*, 1965) and is accompanied by a soft, 'flabby' texture.

For a while, glycolysis can support some resynthesis of adenosine triphosphate (ATP), the main carrier of energy in cell metabolism, but not enough to prevent the concentration declining. When ATP concentration falls below a critical, low level, rigor mortis occurs. The time between death and rigor mortis varies due to the variations in glycogen and ATP content, and because the critical concentration of ATP varies with temperature. When this level is reached, enzymes which keep muscle cells in a relaxed state can no longer function and the muscles begin to contract. If fillets are removed before the onset of rigor, the contractions can cause them to shrink by as much as 40% and even more when cooked. Attached to the spine, muscles cannot contract as they would in live fish because they are pulling with about equal force on opposite sides, causing the whole structure to stiffen.

Eventually, the loss of cellular energy reserves results in a chemical imbalance within muscle cells, which appears to activate certain endogenous, proteolytic enzymes, causing peptide bonds to rupture and, hence, the muscle structure to loosen (Lawrie, 1992). Rigor is said to be resolved, and it is at this stage that fish is usually regarded as being at its most palatable. The time taken for fish to go into and through rigor varies greatly and a compilation by Stroud (1969) gives examples of total elapsed times of 20–65 hours for trawled cod, 72–96 hours for rested cod, and 110 hours for saithe.

There is little difference in the appearance of iced fish before and after rigor mortis. Externally, the skin still has a covering of transparent slime, which glistens. The flesh is firm and the pattern of underlying blocks of muscle is easily seen. Eyes are bright and have a convex shape; gills are uniformly pink to deep red and have characteristic, subtle 'marine' odours. These odours are thought to be products of rapid lipoxygenase activity (Josephson *et al.*, 1984a; 1984b) in enzymic hydroperoxidation of fatty acids via pathways similar to those which have been established as causing the characteristic odours we detect from mushrooms, cucumbers and melons. A 'green-plant-like' note appears almost immediately in all white fish due to the production of C_6 aldehydes and alcohols (e.g. hexanal and hexanol). Within just a few minutes, production of C_8 and C_9 aldehydes, alcohols and ketones modifies the initial odour by addition of characteristic mushroom-like aromas (e.g. 1-octen-3-one and 2,6-nonadienal) and cucumber or melon-like notes (e.g. 2-nonenal; 2,6-nonadienal; 6-nonen-1-ol and 3,6-nonadienal) (Josephson *et al.*, 1987). Some of these volatile aroma substances appear to be associated with the water-temperature dependencies of certain species of fish, and some appear only in freshwater species (Josephson *et al.*, 1984a). They might also vary with other physiological differences (Josephson *et al.*, 1984b). A small, rythmical, annual cycle in panel scores for smell (and taste) which was found in data obtained by Shewan, and reported by Jason and Lees (1971), might be linked to variations in these constituents.

Provided fish have been bled thoroughly before storage and subsequent filleting, the flesh is usually translucent with a slight bluish hue. However, the taste of fish cooked after rigor has been resolved differs from that of pre- (or in-) rigor fish. Pre-rigor, there is less sweetness but, often, a metallic note, and the texture tends to be tougher and somewhat 'stringy'. Immediately post-rigor, the texture is firm but easily chewed, and the flavour has changed to become distinctly sweeter with characteristic notes which give different species their own recognizable flavours (though not so reliably discernible as is often thought). Note that, whilst there has been an improvement in acceptability, in that post-rigor fish may be

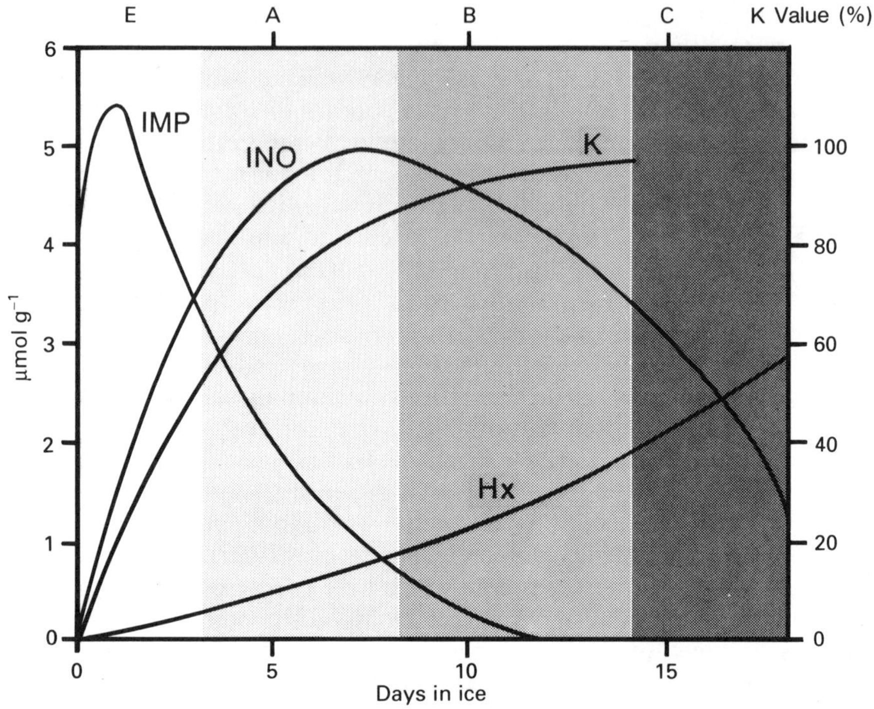

Fig. 7.1. Patterns of change in the concentrations of nucleotide degradation products in muscle of fish stored in melting ice. E, A, B and C refer to the EC Freshness Grades for fishery products.

more palatable, there has also been a loss of freshness quality because the fish is older, less fresh than before. At this time, continuing activity of endogenous enzymes has effected a near-total degradation of ATP via adenosine diphosphate (ADP) to the monophosphate (AMP) followed by rapid deamination to inosine monophosphate (IMP). Further degradation to inosine (INO) is slower than the preceding reactions, and IMP is close to its maximum concentration (Kassemsarn *et al.*, 1963; Dingle and Hines, 1971; Fig. 7.1).

The sweetness of cooked, fresh fish is mostly due to the presence of sugars, including glucose and fructose, and their phosphates (Jones, 1961). Additional sweetness, and much of the species-characteristic flavour, arises from synergism between IMP and free amino acids (Jones, 1969), which produce effects known by the Japanese term 'umami' (variously translated as savouriness or deliciousness). IMP also makes its own contribution to flavour, with a threshold in fish flesh reported as less than 1 $\mu g\ g^{-1}$ (Spinelli and Miyauchi, 1968). At the same concentrations as occur naturally in fish, up to 7 $\mu g\ g^{-1}$ (Kassemsarn *et al.*, 1963), most of which persists through normal cooking regimes, IMP imparts strong meaty flavours to aqueous solutions. More evidence of the important contribution of this nucleotide to the flavour of fresh fish has been presented in the following: direct studies of the consequences of interfering, chemically, with the further degradation of IMP (Groninger and Spinelli, 1968); comparisons of the taste of fish which had been stored until all naturally occurring IMP had been degraded, with similarly treated samples 'spiked' with fresh IMP (Spinelli and Miyauchi, 1968); and in the relationships derived by Fletcher *et al.* (1990) between flavour rating of finfish and their IMP and hypoxanthine (Hx) contents:

$$\text{Flavour score (1 to 6)} = 3.834 + 0.195\,\text{IMP} - 0.294\,\text{Hx}$$
(very stale to very fresh)

Removal of data for very fresh and very stale fish, when IMP might not yet have reached its maximum or would have long since been entirely lost, reduced the variance and increased the slope for IMP, giving:

$$\text{Flavour score} = 3.53 + 0.231\,\text{IMP} - 0.212\,\text{Hx}$$

From this attractive, succulent peak, there is a gradual decline in appearance, smell, texture and taste of fish caused by continuing enzyme activity. Eyes flatten and skin loses the visible impression of underlying muscle blocks though the flesh may still feel quite firm. When live fish are stressed there is a massively increased rate of production of transparent mucus (or slime), which, after death, becomes more apparent as it loses transparency and develops a diluted milk-like opacity. Gill colour fades, and changing concentrations of C_6, C_8 and C_9 alcohols and carbonyl compounds cause the aroma to change.

A suppressed, neutral or 'flat-sweet' odour develops (Josephson *et al.*, 1987). Similarly, as IMP concentration declines, flavour intensity fades (Jones, 1961) to a state referred to as neutral, probably the result of remnants of some original flavour components without the enhancing effects of IMP. In a delicately flavoured species such as cod, further loss of flavorous constituents can result in a tasteless or 'cotton-wool'-like stage.

Although the changes so far described may be considerable in chemical terms, due almost solely to the activities of endogenous enzymes, they have been said to be relatively unimportant organoleptically (Herbert *et al.*, 1971). However, they make the difference between a highly attractive product and one which, at best, may be said to be insipid. A more dramatic consequence of autolytic activity of enzymes which can occur in ungutted fish, particularly in small pelagic species which have been feeding heavily before being caught and stored with little or no ice, is known as 'belly burst'. High concentrations of digestive enzymes begin to digest the gut tissues which previously contained them. Free to diffuse outward, disruption of the belly wall can occur, releasing digestive enzymes and gut bacteria over the external surfaces of adjacent fish.

Microbial spoilage

Muscles of healthy animals are sterile and invasion into fish tissues by microorganisms during ice spoilage is very limited. Hence, it has been possible to use aseptically prepared samples of fish muscle to distinguish between effects due to spoilage bacteria and those due solely to chemical and enzymically controlled changes. In this way, Herbert *et al.* (1971) found that during storage at 2°C the odours and flavours characteristic of fresh fish were gradually lost, leaving a neutral or bland state which changed little during a total storage period of 24 weeks. This is in distinct contrast with the changes which occur in normally spoiling fish.

Large numbers of microorganisms inhabit the alimentary canal and, inevitably, many are spread about the gut cavity during evisceration. Smaller numbers are also found on the skin, and ice used for storage also contributes to the bacterial load. As fish are poikilothermal animals, they exert less of a modifying effect on the pattern of microbial populations than do warm-blooded animals. Hence, the predominant flora largely reflects that of the environment of the live fish (Shewan, 1977). Removal of fish from the water, followed by slaughter and stowage in ice, presents the contaminant microbial flora with a changed environment to which they must adapt. Also, organisms on the skin may be inhibited by protective mechanisms such as lysozymes present in the slime of some fish such as plaice (Murray and Fletcher, 1976).

The bacterial flora of fish from cool waters of arctic and temperate regions consist mainly of aerobic Gram-negative psychrophiles and psychrotrophs. Large seasonal and other variations occur between batches as well as between species, but typical observations have found that the greater part (60 to 80%) of the external flora consists of organisms in two main generic groups: *Pseudomonas/Alteromonas/Shewanella* and *Moraxella/Acinetobacter* (Hobbs, 1991). They are described in this way because of developments in bacterial classification which have occurred since much of the original work when the predominant groups were recorded as *Pseudomonas* and *Achromobacter* (Shewan, 1962). Other aerobes found on freshly caught fish include members of the Gram-negative genera *Vibrio, Flavobacterium* and *Cytophaga*, and also some Gram-positive *Micrococcus* and coryneforms. Low oxygen tension of intestines supports a different distribution of the same bacterial groups; *Vibrio* strains form a much higher proportion, and small numbers of strict anaerobes, including *Clostridium* spp., are usually present.

At some stage during storage, there are increased numbers of most of these bacterial groups. However, under changed and changing conditions,

Table 7.1. Changes in the flora of the skin of cod stored in melting ice (from Hobbs, 1991). Figures given are counts after storage for variable numbers of days in melting ice at 0°C.

	Days							
	0		5		10		15	
	P	*N*	*P*	*N*	*P*	*N*	*P*	*N*
Pseudomonas/Alteromonas/ Shewanella	26	24	33	163	84	11,088	82	45,805
Moraxella/Acinetobacter	33	31	26	128	7	924	13	7,262
Flavobacterium/Cytophaga	0	0	8	39	0	0	0	0
Coryneform bacteria	25	24	12	59	8	1,056	3	1,676
Micrococcus	14	13	21	104	0	0	0	0
Unidentified	2	2	0	0	1	132	2	1,117
Total count		94		493		13,200		55,860

P, % of total flora; *N*, no. of organisms × 10^3 per cm^2 of skin.

their different requirements usually result in further increases in the proportions of those groups which dominate the flora of newly caught fish. Eventually, the *Pseudomonas/Alteromonas/Shewanella* strains account for over 80% of the total flora (Table 7.1). By this stage, total numbers have increased from 10^3–10^5 cm^{-2} to 10^7–10^8 cm^{-2}, but not all bacteria which flourish produce the metabolites which we find so offensive. Attempts to distinguish between bacterial groups on this basis led to the concept of 'active spoilers'. In experiments with cod spoiling in ice, Shaw and Shewan (1968) found that they remained a fairly constant proportion (below 25%) of the total viable population. Further studies with addition of bacterial cultures to sterilized muscle preparations and, later, aseptically excised samples have demonstrated many of the sensory consequences of the activity of individual bacterial groups and species (e.g. Castell and Greenhough, 1957; Herbert *et al.*, 1971). Most are psychrophiles, able to grow at temperatures as low as −5 or −6°C. Thus, although present initially in fairly small numbers, they can grow sufficiently fast during chilled storage to reach 10^5 to 10^7 g^{-1} (of gill tissue) within 7 days (Shewan and Murray, 1979). It is at this stage that bacterial spoilage starts to become sensorily detectable. The greater numbers of active spoilers are, by this time, generating greater amounts of waste products, and the concentrations start to exceed comparative odour and flavour thresholds. The term 'comparative' is used because the threshold is not the usual detection threshold whereby the diluent medium, clean air or water, provides a neutral or minimally interfering background. Where the stimulus is some substance or mixture within fish tissues, then the background is complex in its composition and in the sensory effects of many of the constituents. As well as masking effects due to direct competition from other odorous substances, there may be some synergistic effects. Here, we return to possible influences of IMP, which has been shown to suppress sulfide odours and flavours and suppress sourness (Wagner *et al.*, 1963) and bitterness (Woskow, 1969) in some foods. Further, our perception of odour and flavour of many substances changes with concentration. Sulfides, which add to the objectionable effects of the later stages of spoilage, are probably detected much earlier, at lower concentrations, with a very different response.

The more distinct odours and flavours of spoilage are, primarily, the result of bacterial utilization of 'extractives', relatively small, water-soluble molecules in the fish tissues, particularly the non-protein nitrogen constituents in the fish tissues. These are, mainly, amino acids, small peptides and, in marine fish, trimethylamine oxide (TMAO). Protein breakdown does not add significantly to odour and flavour of fish until spoilage is well advanced, though denaturation by other organisms may cause the texture to soften (Herbert *et al.*, 1971). There is a small amount of ammonia in the flesh of fresh white fish, a product, in part, of the enzymic degradation of ATP. However, at normal tissue pH, i.e. below 7, most NH_3 will exist in the protonated form with little free ammonia which has the characteristic smell. In bony fish, the amount of NH_3 remains low until microbial spoilage is quite advanced. As spoilage progresses, muscle pH rises above neutral, allowing increasing amounts of ammonia to be released and contribute to the odour of staling fish. However, the predominant cause of ammoniacal odour is trimethylamine (TMA), which is produced as a degradation product of TMAO. Thought to have some osmoregulatory function, TMAO is lost from the muscle of salmonids as they move from the sea and swim up river to spawn. Most freshwater fish contain little or no TMAO but there are exceptions such as pike (Love *et al.*, 1959). The amounts in marine fish vary widely from around 100 to over 1000 mg per 100 g (Love *et al.*, 1959). TMAO is thought to have particular significance because it can be used by some aerobic spoilage organisms as a terminal electron acceptor, enabling them to grow under anaerobic conditions which would, otherwise, be inhibitory. Other products of bacterial metabolism which add significantly to the sensory perception of spoilage include: lower fatty acids such as lactic and butyric acids from glucose and ribose sugars; aldehydes and ketones from tissue lipids; volatile sulfides such as hydrogen sulfide, dimethyl sulfide and methyl mercaptan from sulfur-containing amino acids (cysteine, methionine); amines (e.g. indole and skatole) and the so-called 'biogenic amines' (such as histamine, putrescine, cadaverine) as well as ammonia from amino acids and, later, proteins. Bacterial production of proteinase enzymes appears to be repressed at first but is derepressed in the later stages of spoilage when increased hydrolysis of proteins rapidly replenishes the amino acid pool (Liston, 1980).

Variations from the general spoilage pattern

Examination of a typical sensory scoring scheme (Tables 7.2a and 7.2b) shows how the chemical products of spoilage affect odour and flavour. The changes may be summarized as a progressive post-rigor decline in the sweet, characteristic odours and flavours of the fresh fish, through a neutral phase, followed by increasing sourness, then bitterness and eventual inedibility. In many fish, the same underlying processes still occur but other factors, mostly compositional, cause subtle additions to, or may overwhelm, that common pattern.

Species

As already mentioned, some fish produce antimicrobial substances in their slime which may help to prolong shelf-life; and bivalve molluscs may benefit from a spoilage flora modified by lower ultimate pH values arising from their relatively high glycogen content. Conversely, the greater amounts of free amino acids which contribute to the highly prized characteristic flavours of many shellfish species also make them more vulnerable to rapid bacterial spoilage and earlier release of ammoniacal odours. More substantial differences in the perceived pattern of spoilage occur with fatty fish and elasmobranchs. The appreciably greater amounts of lipid in species such as salmon, herring and mackerel make them vulnerable to deteriorative processes which give rise to rancid flavours. Endogenous enzymes hydrolyse both structural phospholipids and depot fat triglycerides, releasing fatty acids and glycerol (and various amino and phosphoric acid derivatives). The consequences are often slight but are thought to account for certain 'rotten fruit' odours and soapy, non-oxidative rancid flavours. The high degree of unsaturation which confers great nutritional value on fish lipids also makes them particularly vulnerable to autoxidation. Free-radical chain reactions, mainly affecting the triglycerides, give rise to a large number of products which contribute to the characteristic sensations of oxidative rancidity: a sharp odour and taste of oxidized linseed oil or paint, and a prickly, choking sensation at the back of the throat. Oxidative rancidity is a particular problem with frozen-stored fish; whilst it can be detected in chill-stored fatty fish it is not, normally, a dominant feature.

High concentrations of urea and TMAO exert considerable influence on the smell and taste of elasmobranchs. Another osmoregulatory agent,

Table 7.2. Score sheet for white fish (cod, haddock, whiting, saithe) (from Shewan *et al.*, 1953).

	Score
(a) Raw fish	
General appearance	
Eyes perfectly fresh, convex black pupil, translucent cornea; bright red gills (colour depending on species) - no bacterial slime; outer slime water white or transparent; bright opalescent sheen, no bleaching	5
Eyes flat, very slight greyness in pupil; slight loss of colour of gills	4
Eyes slightly sunken, grey pupil, slight opalescence of cornea; some discoloration of gills and some mucus; outer slime opaque and somewhat milky; loss of bright opalescence and some bleaching	3
Eyes sunken, milky white pupil, opaque cornea; thick knotted outer slime with some bacterial discoloration	2
Eyes completely sunken; shrunken head covered with thick yellow bacterial slime; gills showing bleaching or dark brown discoloration and covered with thick bacterial mucus; outer slime thick yellow-brown; bloom completely gone; marked bleaching and shrinkage	0
Odour of gills	
Fresh seaweedy odours	10
Loss of fresh seaweediness, shellfish odours	9
No odours, neutral odours	8
Slightly musty, mousy, acetamide-like, milky or caprylic acid-like odours, garlicky, peppery	7
Bready, malty, beery, yeasty odours	6
Lactic acid, sour milk or oily odours	5
Some lower fatty acid odours (e.g. acetic or butyric acid), grassy, 'old boots', slightly sweet, fruity or chloroform-like odours	4
Stale cabbage water, turnipy, 'sour sink', wet matches, phosphene-like odours	3
Ammoniacal (trimethylamine and other lower amines) with strong 'byre-like' ('*o*-toluidine') odours	2
H_2S and other other sulfide odours, strong ammoniacal	1
Indole, ammonia,faecal, nauseating, putrid odours	0
Flavour	
Fresh, sweet flavours characteristic of the species	10
Some loss of sweetness	9

Table 7.2. (*Contd*).

	Score
Slight sweetness and loss of the flavours characteristic of the species	8
Neutral flavour, definite loss of flavour but no 'off' flavours	7
Absolutely no flavour, as if chewing cotton-wool	6
Trace of 'off' flavours, some sourness but no bitterness	5
Some 'off' flavours, and some bitterness	4
Strong bitter flavours, some rubber-like flavour, slight sulfide-like flavours	3
Strong bitterness, but not nauseating	1
Strong putrid flavours (e.g. sulfides), tasted with difficulty	0
(b) Cooked fish	
Odour	
Strong fresh seaweedy odours	10
Some loss of fresh seaweediness	9
Lack of odour or neutral odours	8
Slight strengthening of the odour but no sour or stale odour, wood shavings, woodsap, vanillin or terpene-like odours; slight salt-fish or cold-storage odours	7
Condensed milk, caramel or toffee-like odours	6
Milk jug odours, or boiled potato or boiled clothes-like odours	5
Lactic acid and sour milk, or 'byre-like' odours	4
Lower fatty acids (e.g. acetic or butyric acids), some grassiness or soapiness, turnipy or tallowy odours	3
Ammoniacal (trimethylamine and lower amines) odours	2
Strong ammoniacal (trimethylamine) and some sulfide odours	1

urea, is produced as an alternative means of detoxifying ammonia, and much of that ammonia is released when urea is acted upon by bacterial urease. If properly handled, the fish do not become ammoniacal until bacterial spoilage develops. However, fresh dogfish has a strong, astringent-bitter taste which decreases in intensity but is a common feature throughout spoilage. Ammoniacal odours dominate as a result of release of both ammonia and TMA, and the taste loses the clean, metallic-bitterness of the fresh fish as sourness and 'off' flavours of the general spoilage pattern develop.

It is a commonly held view that iced fish from tropical waters spoil at a slower rate than cold water fish. Longer shelf-lives are thought to be a consequence of greater temperature differences on transfer of fish from warm waters to storage in ice (Liston, 1992). As Shewan (1977) suggested, the cause appears to be absence from the initial (predominantly mesophilic, Gram-positive) bacterial flora of many of the 'active spoilers' found on fish from cooler waters. However, a review by Lima dos Santos (1981) concluded that it was unwise to state that tropical fish species have a longer shelf-life in ice because, although some keep relatively well, some cold water species have the same property. The picture is more complicated and other factors must exert significant influences on rates of spoilage. An important contributor may be the ultimate pH which is known to affect not only bacterial growth generally but, more specifically to fish spoilage, can affect bacterial reduction of TMAO and autolytic degradation of IMP (Davis, 1992). A phenomenon known as cold shock which affects some tropical fish species adds to these complexities. Unlike 'cold shortening', which can occur when some mammalian meats are chilled, cold shock is a rigor-like stiffening which does not involve contraction of muscle fibres. However, such fish suffer lower processing yields, higher pH and spoilage rates, and it has been found that delaying icing for a few hours gives significant improvements (Curran *et al.*, 1986).

Within-species differences

Despite being stored under identical or closely comparable conditions, there is variation within, as well as between, different species. The differences within a species are generally small but on occasions may be measurable, above the normal scatter inherent in any biological system.

Size

Permitted sizes which may be caught and used for human consumption can vary greatly, e.g. the minimum size for cod is 350 mm (~ 400 g) whilst the biggest size grade denoted in European Community Regulations is for fish which exceed 7 kg. Halibut can be well over 100 kg (though few such specimens are landed nowadays). Such differences in mass affect the rates of cooling and warming when transferred to and from ice storage, with consequences to autolytic and bacterial stages of spoilage. Of greater consequence, perhaps, are differences in the ratio of surface area (exposed to bacterial contamination and available to heating and cooling) to volume (of fish meat). Small fish have a much greater area of skin and gut-cavity wall in relation to the amount of skeletal muscle; and their deepest regions of muscle are more quickly affected by inward diffusion of metabolic products of bacterial activity.

Condition

Before capture, fish may be in poor condition through disease, injury, starvation or, as is more commonly meant, through diversion of resources to production of gonads. Such fish are intrinsically of poorer food quality than fit, healthy, well-nourished specimens. The flesh of spent fish tends to be softer, has a reduced water-holding capacity and, when cooked, may remain soft and quite watery. In extreme cases, mostly affecting some flatfish species, the cooked flesh has an unpalatable, sloppy-gelatinous texture. However, apart from such extreme examples, the flesh of spent fish remains acceptable but keeps less well. The skin, together with a reduced amount of subcutaneous fatty tissue, presents a less effective defence against mechanical damage and bacterial penetration; and the flesh is more vulnerable to bacterial activity because of the higher muscle pH arising from depleted levels of glycogen.

Catching, processing and storage conditions

For at least part of the catch, most capture techniques for wild fish involve long delay between death and landing on the fishing vessel. As a result, some fish may be slaughtered and processed almost immediately after becoming trapped, whilst others may have a prolonged struggle and die some hours before being hauled from the water. Further long delays may then occur on the processing deck. The main consequence to spoilage is relatively high temperatures which accelerate spoilage reactions. Excepting where large numbers of small fish are involved, evisceration (gutting) and washing before packing in ice is generally recommended. Washing is often thought to be detrimental because of the loss of mucus, a natural, protective barrier. The balance tips in favour of washing in order to reduce the high bacterial load which otherwise occurs from contamination by detritus, release of fish faeces during catching and, more especially, from smearing by gut remnants. There have been many studies designed to measure the benefits of gutting in terms of reduced spoilage rates. As might be expected from the patterns of autolytic change and microbial growth described earlier, differences in eating quality of flesh from gutted and ungutted fish are small. Where differences have been detected, fairly long storage periods have been required to give bacterial effects sufficient time to become apparent. The main benefit of gutting soon after catching is to avoid the problems which arise during subsequent processing of ungutted fish. As visceral organs will soften or even liquefy under the influence of continuing enzyme activity, they can become unpleasant to work with, and a far greater source of contamination than when the fish are gutted earlier.

Occasionally, situations may arise when fish are unintentionally stored anaerobically, conditions which favour hydrogen sulfide-producing bacteria (Jensen *et al.*, 1980). With shorter voyages to less distant waters, the classic manifestation of anaerobic spoilage, known as 'stinkers' or 'bilgy' fish, is less likely to be found. Long voyages, with fish badly packed in pounds contained by wooden boards, sometimes led to prolonged storage with fish pressed directly against the pound boards. Depletion of oxygen within the contact area, together with inadequate cooling, could result in stale cabbage water, cess-pool or turnip-like odours and flavours. Ironically, the greatest danger of such effects probably arises nowadays with frozen fish. Similar reactions appear to take place in the relatively short time that frozen blocks of fish remain in the warm, moist air of thawing plants if shelves are not cleaned thoroughly between batches.

After further processing, products such as fillets and steaks also need to be stored. Processing inevitably causes some warming and additional contamination, whilst removal of the protection afforded by skin makes products more vulnerable to spoilage.

Effects of Storage Temperatures above 0°C

Although fish spoilage bacteria may be quite active at 0°C, optimum temperatures for many are closer to normal room temperatures; and spoilage, both autolytic and bacterial, accelerates as fish warm up. Also, Shewan and Murray (1979) found evidence of bacterial penetration of flesh occurring at higher temperatures. Of the various attempts to derive mathematical relationships which properly describe the effects of storage temperature on rates of spoilage, the most successful is that of Ratkowsky *et al.* (1982). Working originally from data on the growth of spoilage bacteria in culture media, they found that growth rates fitted a square law. Storey (1985) solved their equation for spoilage rates of (temperate water) fish stored between 0 and 15°C, and obtained the relative rate function:

$$R = (0.1T + 1)^2 \text{ where } T = \text{temperature in } ^\circ\text{C}$$

and used sensory data to confirm its relevance. Thus, one can readily see that fish spoils more than twice as fast (2.25×) at 5°C than at 0°C, and four times as fast if stored at 10°C.

Storage below 0°C

Fish begins to freeze at about −1.5°C and most bacterial activity ceases. Rapid freezing and storage at low temperatures (−30°C) can maintain the fish in good condition for long periods but, at relatively high cold store temperatures, various deteriorative changes take place. Some significant amounts of enzyme activity can continue and may be aggravated because of increased substrate concentrations in decreased amounts of unfrozen water. Of particular significance are the effects on TMAO and on lipids.

TMAO is degraded by the action of endogenous enzymes to dimethylamine (DMA) and formaldehyde (Hultin, 1992a). *In vitro* studies have shown them to be produced in stoichiometric amounts, but the reactivity of formaldehyde with proteins and other tissue constituents reduces the amounts which are recoverable from muscle. Partly as a result of reaction between formaldehyde and proteins, the texture of fish stored at high temperatures becomes tough and fibrous.

A characteristic cardboard-like smell and taste which develops in frozen lean species, and is known, simply, as cold store odour and flavour, appears as a result of lipid oxidation. McGill *et al.* (1977) found several oxidation products which contributed to the flavour, the main component being hept-*cis*-4-enal. In fatty fish, the greater concentrations of lipid with various pro-oxidants in close proximity make them more vulnerable to oxidative rancidity (Hultin, 1992b). When frozen, this becomes the primary degradative process. Simple exclusion of oxygen, by wrapping in impermeable films and/or coating with an ice glaze, provides a very effective protection. Products which are not so protected inevitably suffer desiccation due to sublimation of water under the influence of fluctuating temperatures. This can quickly lead to 'freezer-burn', seen as patches of white, sponge-like tissue, which cannot be reversed. In addition to being unsightly, freezer-burn allows access of air to a greater surface of fish tissue, thus aggravating the development of oxidative rancidity.

Uncharacteristic Flavours – Taints

Any commercially utilized species will have a range of characteristic odours and flavours regarded as typical of that species, but all are vulnerable to change by the uptake of substances which cause the edible parts to be tainted. Taint is defined as 'a taste or odour foreign to the product' (ISO, 1983) and, therefore, includes any odour or flavour outside the typical range. Unpleasant characteristics such as oil or other chemical pollution are often taken as synonymous with taint, but the definition includes any foreign odour or flavour, whether or not they are derived from 'natural' sources, or might be regarded as pleasant by some consumers. Difficulties occur where the intensity of an effect is close to an assessor's threshold of detection but not strong enough to be identified with certainty. There can also be confusion at high levels of odour or flavour intensity when a similar characteristic can occur from different possible sources.

One of the major naturally occurring taints is dimethyl sulfide (DMS), which, at low concentrations, contributes to the normal flavour of a few fish species such as clams. At higher concentrations, it contributes to the unpleasant characteristics of bacteriologically spoiled fish. However, DMS can also occur at these higher, usually unpleasant, concentrations after the live fish have ingested certain planktonic organisms which contain DMS and a precursor, dimethylsulfoniopropionate (or dimethyl-β-propiothetin). Sensory responses vary; such fish are frequently described as 'turnipy' or strongly 'sulfide', and may be thought to have been subjected to bacterial spoilage. Other, quite different, descriptions have been applied, probably due to the different sensory perceptions which can be produced by different concentrations of a chemical stimulus. As well as 'gunpowder', 'blackberry', 'weedy' and 'iodine', fish affected in this way have been described as having 'petrol' and 'diesel' odours, with resultant suspicion of contamination by these mineral oil products.

Another important, naturally occurring taint, and one which is more readily recognizable as a taint rather than a loss of 'freshness', is the earthy taste which can severely affect freshwater fisheries (and domestic water supplies). Earthiness is mainly the result of ingestion by fish of two substances, geosmin and/or methyl iso-borneol, metabolites of some species of blue-green algae (cyanobacteria) and actinomycetes, which thrive in warm weather. The main route of ingestion by fish is uptake via the gills after these odoriferous compounds have been released into the water.

Measurement of Deterioration

Sensory tests

Chill-stored fish

The sensory scales shown in Table 7.2 were developed by Shewan *et al.* (1953) to enable rates of spoilage to be measured in sensory terms, and to make more precise comparisons of different treatments. A multilingual guide to freshness grades (Howgate *et al.*, 1992) uses a modified version of Shewan's scales for grading of fish at first sale under EC grading regulations. Construction of the scales was governed by the characteristics observed without constraint towards linear rates of change. Nevertheless, for the species on which the work was based (cod stored in ice), a linear rate is observed over the main region of interest, i.e. between scores of about 8 (some loss of characteristic fresh notes) and 4 (distinct off odours, sourness and slight bitterness). A crude approximation gives a decrease of one score point per 3 days of storage in melting ice at 0°C (Fig. 7.2).

Like any other instrument operating near to its limit of sensitivity, there is some variation. Even with extensive training, individuals have systematic biases as well as day-to-day differences. Both are usually small and their significance is further reduced by averaging the results from panels of several people. Ideally, panels of six or more assessors should be used but the costs involved often lead to smaller panels for industrial work.

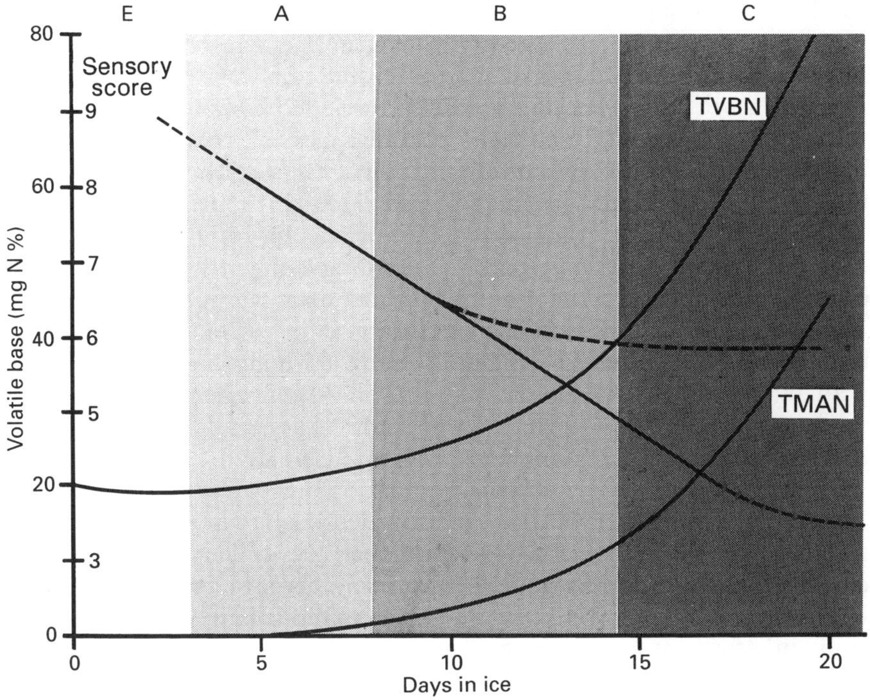

Fig. 7.2. Patterns of change in sensory scores and concentrations of volatile nitrogenous bases in muscle of fish stored in melting ice.

Baines and Shewan (1965) determined error variances (for fish × tasters, i.e. the scoring variation arising from differences between both samples and assessors) in data from an unspecified number of experiments conducted by coworkers. With the most frequently assessed species, cod and haddock, error variances for raw (gill) odour scores were between 0.1 and 0.2, and between 0.2 and 0.5 for cooked flavour scores. Position in the range depended mainly on experience and regularity of tasting. Variances within batches of fish from the same catch were about 0.1 and 0.15 (raw and cooked, respectively), and for species which were examined less often the variances were higher. Raw odour assessment is often used as a predictor of cooked flavour score and the relationship was found to have remained consistent over a long period. In experiments conducted to compare the performance of two panels which normally worked independently on different tasks in separate locations, mean scores usually differed by no more than 0.3 units.

The results mentioned earlier, of studies on rates of spoilage at different temperatures, might suggest that, although rates may change, spoilage patterns and sensory characteristics do not. This is not entirely true and it is safer to say that the differences have proved to be within the scope of experienced panellists to interpret particular attributes within the scoring scheme. Nevertheless, users of expert sensory panels need to remember that such scoring schemes are necessarily developed by monitoring spoilage under a particular set of conditions, and that radically changing the conditions may change the proportions of chemical stimuli in a significant way. Left to their own devices, panellists will struggle to fit their perception of the product to what they consider to be the appropriate point on the scale. Where this can be particularly misleading is that relatively high scores can be given to experimental samples which many consumers would find unpleasant. With panellists trained to set aside personal preferences, a poorly constructed experiment could reach a wholly unwarranted (and, perhaps, expensive) conclusion.

Frozen-stored fish

To some extent, assessment of changes which take place during storage is simpler, but less precise, for frozen fish than for chilled fish. Assessment of either form of deterioration has to be made against a background which is dependent upon the extent of the other. Often, there is a need for both chill and frozen storage deterioration to be assessed in the same sample. Whole fish which have been frozen and stored under good conditions not only look different from unfrozen fish but gill odours change appreciably and can be very unreliable. Flesh tends to lose its opacity and there are textural changes with associated increases in drip loss. Baines *et al.* (1969) examined the various changes which occurred during frozen storage of fish of differing pre-freezing iced storage history, and included a study of the frequency with which different classes of flavour description occurred. Apart from a tendency for 'acidic' flavours to appear more frequently as high temperature frozen storage continued, no particular class of descriptors was associated with any particular icing or frozen storage treatment. The nature of the odours and flavours produced during frozen storage was found to be independent of initial freshness. The single descriptive term, 'cold store', for odour and flavour, was used in an intensity scale along with similar quantitative scales for firmness and dryness. Subsequently, assessment of degree of fibrousness has been included as a useful index of deterioration in frozen fish. Although intensity scales appear to be relatively simple, panellists still need training to achieve some uniformity of interpretation of quantitative terms such as slight, moderate and strong. Human senses are much less discriminating between different intensities rather than types of stimuli, and it is not surprising, therefore, that the error variances are larger than those obtained when scoring for freshness of unfrozen fish.

Non-sensory tests

Chill-stored fish

Despite training and experience, problems can arise with panel performance particularly under commercial conditions. Consequently, non-sensory tests are needed for assessment and comparison. However, it must be remembered that chemical and physical analyses will also have errors and be equally dependent upon the integrity of the analyst. Nevertheless, such tests will always be regarded as more objective and reliable in cases of dispute. Also, in the preparation of purchase specifications, non-sensory criteria are more easily described for application at different locations (which may be in different countries with very different cultures). This of course leads to other problems. An analyte can usually be measured by a number of different methods, and standardization of techniques and analytical performance needs to be established for comparison of analytical results between laboratories. For comparative purposes, relationships with sensory properties need to be known and the potential for variability to be understood. Sensory stimuli from any given sample arise from a large number of constituents within the tissue, and sensory tests enable their total impact to be assessed. Non-sensory tests select individual, or small groups of, components which may have no direct bearing on the sensory response. Similar cautions to those given earlier for sensory assessments apply here also. As well as the variations between individual fish, batches, grounds, season, etc. which may contribute additional errors, radical differences in product storage conditions can disrupt the relationships between sensory and non-sensory tests (Davis, 1992).

Physical methods

The two main physical techniques for assessment of deterioration have measured systematic changes in electrical conductive properties of chilled fish. They offer the benefit of ease and speed of use in industrial premises without the expense of training groups of assessors. Hennings (1965) developed the Intelectron Fish Tester V, which measured the ratio of impedances at two different frequencies between graphite electrodes applied across a specified region of the fish. A more convenient device, less dependent upon the geometry of fish samples or precise location of the region of measurement, was developed by Jason and Lees (1971). Changes in the dielectric properties, capacitance and resistance, of fish muscle are made via a small, hand-held unit which can give readings on individual fish or, more sensibly, compute the average of 16 successive measurements across a batch of fish. In an extensive examination of methods of freshness assessment under experimental conditions, Burt *et al.* (1976a) found both the Torry Fish Freshness Meter and the Fish Tester V correlated best with scores for general appearance, though re-calibration to particular kinds of fish was recommended for use with market fish (Burt *et al.*, 1976b). Freezing, irradiation or skinning radically alters the dielectric properties of fish muscle. In all cases, much lower 'Torrymeter' readings are given regardless of pre-treatment history.

CHEMICAL TESTS

Of the various attempts to use changing amounts of chemical constituents as freshness indicators of spoiling fish muscle, the main ones to have been adopted are based on degradation either of TMAO or ATP. Burt *et al.* (1976a) found the highest association of non-sensory and sensory tests to be between concentration of hypoxanthine (Hx) and cooked assessment. Hx concentrations in deproteinized extracts can be determined by: photometric measurements after reaction with an oxidation–reduction indicator, i.e 2,6-dichlorophenol indophenol (Burt *et al.*, 1969); or after conversion to uric acid by treatment with an enzyme, i.e. xanthine oxidase (Jones *et al.*, 1964); and, more recently, by high performance liquid chromatography, which separates all the components of ATP catabolism and enables them to be quantified (Murray and Thomson, 1983; Kennish and Kramer, 1987). The last-mentioned has the advantage of avoiding interference from additional components in some shellfish and allows all the catabolites to be summed, thus permitting results to be presented as proportions of the total. Such an index, the 'K value', was first proposed by Saito *et al.* (1959), where

$$K = \frac{\text{INO} + \text{Hx}}{\text{ATP} + \text{ADP} + \text{AMP} + \text{IMP} + \text{INO} + \text{Hx}}$$

Several more convenient and rapid devices have been developed for making measurements of nucleotide catabolism using enzyme sensors based on oxygen electrodes, and immobilized enzymes coupled with a redox indicator on paper test strips (Gill, 1990).

Use of Hx as a freshness index benefits from the fact that, from a near-zero baseline, it begins to increase immediately after death. The patterns of change in Fig. 7.1 show how K values increase sensitivity during the early stages of spoilage, improving the index as a measure of 'freshness' as apposed to 'staleness'. Conversely, measures of volatile base nitrogen are dependent upon microbial degradation of TMAO to (mainly) TMA. These measures cannot discriminate between categories of fresh fish but have had the advantage over many years of being relatively simple and cheap to perform. Again, deproteinized extracts can be used and the bases are released by steam distillation after addition of a strong alkaline solution (AMC, 1979). The amount of base in an acidic receiving solution is then titrated. Ammonia may be compensated for by complexing with formaldehyde, but these procedures cannot distinguish between TMA and DMA. A small amount of the latter is produced as fish spoil and, in frozen fish, TMA production ceases as TMAO degrades enzymically to DMA and formaldehyde. There have been many alternative approaches to analyses of total bases and individual amines (Gill, 1990) but gas–liquid chromatography techniques such as that developed by Keay and Hardy (1972) are the main means of distinguishing between TMA and DMA. Ruiter and Weseman (1976) adapted an autoanalytical procedure to obtain measurements of TMA, DMA and ammonia, but the non-sensory technique most often used in industrial quality assurance work is to measure the total amount of all volatile bases (TVB), by distillation and titration. As with all these measures of volatile base, the determination is calculated and quoted as the nitrogen component and, hence, is more correctly known as TVBN. Because total bases include ammonia which is present in the fresh fish, initial values are well above zero. Further, not only do these initial values vary between batches and, particularly, between species, but they are also very method-dependent. Additional ammonia is generated through breakdown of nitrogenous extractives in hot alkaline solution. For lean white fish such as cod, analysed by preparation of an acid extract rendered alkaline with strong basic solution, values for very fresh fish are 20 mg N% (i.e. 20 mg N per 100 g of sample, the unit commonly used). Alternatively, a muscle homogenate is sometimes used instead of an acid extract, and magnesium oxide powder, a weak base, is added to release the volatile bases, giving results which are generally lower by about 5 mg N%. As fish spoil, increases in TVBN then follow an exponential curve similar to the patterns seen for TMA nitrogen (TMAN) (Fig. 7.2) and bacterial counts.

Frozen-stored fish

Although measures of DMA produced during frozen storage of fish have proved of value to experimenters, they are little used commercially because of cost, variability between batches, and the relatively small amounts produced. Similarly, none of the other techniques which have been examined have a direct relationship with flavour or texture. Some, such as measures of protein solubility and cell fragility, are correlated with sensory attributes because both are dependent on time and temperature (Love, 1969), but they require very closely controlled conditions to obtain consistent results. Another physical indication of cold-store deterioration in lean fish can be obtained by examination of the relative fluorescence of muscle. Fluorescence due to reduced nicotinamide adenine nucleotides present in fresh muscle fades slowly during chilled storage, and rapidly during thawing of frozen products. Conversely, two different, stable fluorochomes may appear during frozen storage. As well as varying between species, the final result depends upon time, temperature and pre-freezing history, and proved too variable to use as an instrumental technique. However, direct visual observation of fluorescent emission can be a useful adjunct in the sensory evaluation of lean fish (Davis, 1982) and in differentiating between frozen and non-frozen samples. Chemical estimations of fat oxidation, such as peroxide and thiobarbituric acid values, as measures of rancidity in fatty fish are similarly limited (Hardy, 1980). Consequently, the main practical form of assessment of frozen fish has to be sensory testing.

Consequences to Human Health from Consumption of Spoiled Fish

Risks to health from consumption of fishery products are discussed in Chapters 8 and 9. The hazards from contamination by pathogens are similar to other foods, but there is only one type of illness, scombrotoxic poisoning, in which spoilage seems to be an important factor. Some pelagic fish (mostly scombroid species such as mackerel and tuna) contain relatively large amounts of histidine, which some spoilage organisms can decarboxylate to histamine. This has often been regarded as the cause of scombrotoxic poisoning because many of the symptoms are the same as those produced by high blood levels of histamine. However, the work of Murray *et al.* (1982) and Ijomah *et al.* (1992) has shown that histamine alone is not the cause. Although fish lipids are important nutritionally, oxidation products present some potential for harm. There may be some nutritional loss, particularly through oxidation of these unsaturated lipids and lipid-soluble vitamins, and small losses of protein and water-soluble B vitamins via drip loss. Lysine availability decreases in frozen fish which have been frozen and stored at relatively high temperatures but, otherwise, the tendency is in the opposite direction – probably due to some unfolding of proteins. However, consumer rejection of spoilt fish, leading to a general rejection of a valuable part of a varied diet, is, arguably, the major consequence of consumption of fish which has been allowed to lose the attractive, succulent properties of the fresh product.

References

AMC (1979) Recommended general methods for the examination of fish and fish products. A report prepared by the fish products sub-committee of the Analytical Methods Committee. *The Analyst* 104, 434–450.

Baines, C.R. and Shewan, J.M. (1965) Sensory methods for evaluating the quality of white fish. *Laboratory Practice* 14, 160–163.

Baines, C.R., Connell, J.J., Gibson, D.M., Howgate, P.F., Livingstone, E.I. and

Shewan, J.M. (1969) A taste panel technique for evaluating the eating quality of frozen cod. In: Kreuzer, R. (ed.) *Freezing and Irradiation of Fish*. Fishing News Books, London, pp. 361–366.

Burt, J.R. (1966) Glycogenolytic enzymes of cod (*Gadus callarias*) muscle. *Journal of the Fisheries Research Board of Canada* 23, 527–538.

Burt, J.R., Stroud, G.D. and Jones, N.R. (1969) Estimation of hypoxanthine concentrations in fish muscle by a rapid, visual modification of the enzymatic assay procedure. In: Kreuzer, R. (ed.) *Freezing and Irradiation of Fish*. Fishing News Books, London, pp. 367–370.

Burt, J.R., Gibson, D.M., Jason, A.C. and Sanders, H.R. (1976a) Comparison of methods of freshness assessment of wet fish. II. Instrumental and chemical assessments of boxed experimental fish. *Journal of Food Technology* 11, 73–89.

Burt, J.R., Gibson, D.M., Jason, A.C. and Sanders, H.R. (1976b) Comparison of methods of freshness assessment of wet fish. III. Laboratory assessments of commercial fish. *Journal of Food Technology* 11, 117–128.

Buttkus, H. and Tomlinson, N. (1966) Some aspects of post-mortem changes in fish muscle. In: Bruskey, E.J., Cassens, R.G. and Trautman J.C. (eds) *The Physiology and Biochemistry of Muscle as a Food*. University of Wisconsin Press, pp. 197–203.

Castell, C.H. and Greenhough, M.F. (1957) The action of *Pseudomonas* on fish muscle. 1. Organisms responsible for odours produced during incipient spoilage of chilled fish muscle. *Journal of the Fisheries Research Board of Canada* 14, 617–625.

Curran, C.A., Poulter, R.G., Brueton, A., Jones, N.R. and Jones, N.S.D. (1986) Effect of handling treatment on fillet yields and quality of tropical fish. *Journal of Food Technology* 21, 301–310.

Davis, H.K. (1982) Fluorescence of fish muscle: description and measurement of changes occurring during frozen storage. *Journal of the Science of Food and Agriculture* 33, 1135–1142.

Davis, H.K. (1992) Fish. In: Parry, R.T. (ed.) *Principles and Applications of Modified Atmosphere Packaging of Foods*. Blackie Academic and Professional, Glasgow, pp. 189–228.

Dingle, J.R. and Hines, J.A. (1971) Degradation of inosine 5′-monophosphate in the skeletal muscle of several North Atlantic fishes. *Journal of the Fisheries Research Board of Canada* 28, 1125–1131.

Fletcher, G.C., Bremner, H.A., Olley, J. and Statham, J.A. (1990) Umami revisited: the relationship between inosine monophosphate, hypoxanthine and Smiley scales for fish flavour. *Food Reviews International* 6, 489–503.

Gill, T.A. (1990) Objective analysis of seafood quality. *Food Reviews International* 6, 681–714.

Groninger, H.S. and Spinelli, J. (1968) EDTA inhibition of inosine monophosphate dephosphorylation in refrigerated fishery products. *Journal of Agricultural and Food Chemistry* 16, 97–99.

Hardy, R. (1980) Fish lipids. Part 2. In: Connell, J.J. (ed.) *Advances in Fish Science and Technology*. Proceedings of the Jubilee Conference of the Torry Research Station, Aberdeen, 1979. Fishing News (Books), Farnham, pp. 103–111.

Hennings, C. (1965) The 'Intelectron Fish Tester V'. A new electronic method and device for the rapid measurement of the degree of freshness of 'wet' fish. In: Kreuzer, R. (ed.) *The Technology of Fish Utilisation*. Fishing News Books, London, pp. 154–158.

Herbert, R.A., Hendrie, M.S., Gibson, D.M. and Shewan, J.M. (1971) Bacteria active in the spoilage of certain sea foods. *Journal of Applied Bacteriology* 34, 41–50.

Hobbs, G. (1991) Fish: Microbiological spoilage and safety. *Food Science and Technology Today* 5, 166–173.

Howgate, P., Johnston, A. and Whittle, K.J. (1992) *Multilingual guide to EC Freshness Grades for Fishery Products*. Torry Research Station, Aberdeen, 32 pp.

Hultin, H.O. (1992a) Trimethylamine-*N*-oxide (TMAO) demethylation and protein denaturation in fish muscle. In: Flick, G.J. and Martin, R.E. (eds) *Advances in Seafood Biochemistry: Composition and Quality*. Papers from the American Chemical Society Meeting, New Orleans. Technomic Publishing Co., Lancaster, Pennsylvania, pp. 25–42.

Hultin, H.O. (1992b) Lipid oxidation in fish muscle. In: Flick, G.J. and Martin, R.E. (eds) *Advances in Seafood Biochemistry: Composition and Quality*. Papers from the American Chemical Society Meeting, New Orleans. Technomic Publishing Co., Lancaster, Pennsylvania, pp. 99–122.

Ijomah, P., Clifford, M.N., Walker, R., Wright, J., Hardy, R. and Murray, C.K. (1992) Further volunteer studies on scombrotoxicosis. In: Burt, J.R., Hardy, R. and Whittle, K.J. (eds) *Pelagic Fish*. Fishing News Books, Oxford, pp. 194–199.

ISO (1983) Sensory analysis: vocabulary. *International Standards Organization* 5492-5.

Izquierdo-Pulido, M.L., Hatae, K. and Haard, N.F. (1992) Nucleotide catabolism and changes in texture indices during ice storage of cultured sturgeon, *Acipenser transmontanus*. *Journal of Food Biochemistry* 16, 173–192.

Jason, A.C. and Lees, A. (1971) Estimation of fish freshness by dielectric measurement. *Department of Trade and Industry, Special Report* T71/7. Torry Research Station, Aberdeen, 32 + 17 pp.

Jensen, M.H., Petersen, A., Røge, E.H. and Jepsen, A. (1980) Storage of chilled cod under vacuum and at various concentrations of carbon dioxide. In: Connell, J.J. (ed.) *Advances in Fish Science and Technology*. Proceedings of the Jubilee Conference of the Torry Research Station, Aberdeen, 1979. Fishing News Books, Farnham, pp. 294–297.

Jones, N.R. (1961) Fish flavours. In: *Proccedings of Flavour Chemistry Symposium 1961*. Campbell Soup Co., Camden, New Jersey, pp. 61–81.

Jones, N.R. (1969) Meat and fish flavours: Significance of ribomononucleotides and their metabolites. *Journal of Agricultural and Food Chemistry* 17, 712–716.

Jones, N.R., Murray, J., Livingstone, E.I. and Murray, C.K. (1964) Rapid estimations of hypoxanthine concentrations as indices of the freshness of chill-stored fish. *Journal of the Science of Food and Agriculture* 15, 763–774.

Josephson, D.B., Lindsay, R.C. and Stuiber, D.A. (1984a) Variations in the occurrences of enzymically derived volatile aroma compounds in salt- and freshwater fish. *Journal of Agricultural and Food Chemistry* 32, 1344–1347.

Josephson, D.B., Lindsay, R.C. and Stuiber, D.A. (1984b) Biogenesis of lipid-derived volatile aroma compounds in the emerald shiner (*Notropsis atherinoides*). *Journal of Agricultural and Food Chemistry* 32, 1347–1352.

Josephson, D.B., Lindsay, R.C. and Olafsdottir, G. (1987) Measurement of volatile aroma constituents as a means for following sensory deterioration of fresh fish and fishery products. In: Kramer, D.E. and Liston, J. (eds) *Developments in Food Science, Vol. 15: Seafood Quality Determination*. Proceedings of an International Symposium, Anchorage, Alaska, 1986. Elsevier Science Publishing, Amsterdam, pp. 27–47.

Kassemsarn, B.-O., Perez, B.S., Murray, J. and Jones, N.R. (1963) Nucleotide degradation in the muscle of iced haddock (*Gadus aeglefinus*), lemon sole (*Pleuronectes microcephalus*) and plaice (*Pleuronectes platessa*). *Journal of Food Science* 28, 28–37.

Keay, J.N. and Hardy, R. (1972) The separation of aliphatic amines in dilute aqueous solution by gas chromatography and application of this technique to the quantitative analysis of tri- and dimethylamine in fish. *Journal of the Science of Food and Agriculture* 23, 9–19.

Kennish, J.M. and Kramer, D.E. (1987) A review of high-pressure liquid chromatographic methods for measuring nucleotide degradation in fish muscle. In: Kramer, D.E. and Liston, J. (eds) *Developments in Food Science, Vol. 15: Seafood Quality Determination*. Proceedings of an International Symposium, Anchorage, Alaska, 1986. Elsevier Science Publishing, Amsterdam, pp. 209–220.

Lawrie, R.A. (1992) Conversion of muscle into meat: biochemistry. In: Ledward, S.A., Johnston, D.E. and Knight, M.K. (eds) *The Chemistry of Muscle Based Foods*. Royal Society of Chemistry, Cambridge, pp. 42–61.

Lima dos Santos, C.A.M. (1981) The storage of tropical fish in ice – a review. *Tropical Science* 23, 97–127.

Liston, J. (1980) Microbiology in fishery science. In: Connell, J.J. (ed.) *Advances in Fish Science and Technology*. Proceedings of the Jubilee Conference of the Torry Research Station, 1979. Fishing News Books, Farnham, pp. 138–157.

Liston, J. (1992) Bacterial spoilage of seafood. In: Huss, H.H., Jakobsen, M. and Liston, J. (eds) *Developments in Food Science. Vol. 30: Quality Assurance in the Fish Industry*. Proceedings of an International Conference, Copenhagen, Denmark, 1991. Elsevier Science Publishing, Amsterdam, pp. 93–105.

Love, R.M. (1969) Time–temperature tolerance. *Institute of Food Science and Technology (UK) Proceedings* 2(3), 62–67.

Love, R.M., Lovern, J.A. and Jones, N.R. (1959) The composition of fish tissues. *Food Investigation Special Report No. 69*. HMSO, London, p. 37.

McCallum, W.A., Jaffray, J.I. and Churchill, D.N. (1967) Post mortem physicochemical changes in unfrozen Newfoundland trap-caught cod. *Journal of the Fisheries Research Board of Canada* 24, 651–678.

McGill, A.S., Hardy, R. and Gunstone, F.D. (1977) Further analysis of the volatile components of frozen cold stored cod and the influence of these on flavour. *Journal of the Science of Food and Agriculture* 28, 200–205.

Murray, C.K. and Fletcher, T.C. (1976) The immunohistochemical localization of lysozyme in plaice (*Pleuronectes platessa* L.) tissues. *Journal of Fish Biology* 9, 329–334.

Murray, C.K., Hobbs, G. and Gilbert, R.J. (1982) Scombrotoxin and scombrotoxin-like poisoning from canned fish. *Journal of Hygiene, Cambridge* 88, 215–220.

Murray, J. and Thomson, A.B. (1983) Reverse phase ion-pair separation of nucleotides and related products in fish muscle. *Journal of High Resolution Chromatography and Chromatographic Communication* 6, 209–210.

Ratkowsky, D.A., Olley, J., McMeekin, T.A. and Ball, A. (1982) Relationship between temperature and growth rate of bacterial cultures. *Journal of Bacteriology* 149, 1–5.

Ruiter, A. and Weseman, J.M. (1976) The automated determination of volatile bases (trimethylamine, dimethylamine and ammonia) in fish and shrimp. *Journal of Food Technology* 11, 59–68.

Saito, T., Arai, K. and Matsuyoshi, M. (1959) A new method for estimating the freshness of fish. *Bulletin of the Japanese Society of Scientific Fisheries* 24, 749–757.

Shaw, B.G. and Shewan, J.M. (1968) Psychrophilic spoilage bacteria of fish. *Journal of Applied Bacteriology* 31, 89–96.

Shewan, J.M. (1962) The bacteriology of fresh and spoiling fish and some related chemical changes. In: Hawthorn, J. and Leitch, J.M. (eds) *Recent Advances in Food Science Vol. 1: Commodities*. Butterworths, London, pp. 167–193.

Shewan, J.M. (1977) The bacteriology of fresh and spoiling fish and the biochemical changes induced by bacterial action. In: *Conference Proceedings; Handling, Processing and Marketing of Tropical Fish*. Tropical Products Institute, London, 1976, pp. 51–66.

Shewan, J.M. and Murray, C.K. (1979) The microbial spoilage of fish with special reference to the role of psychrophiles. In: Russell, A.D. and Fuller, R. (eds) *Cold Tolerant Microbes in Spoilage and the Environment*. Society for Applied Bacteriology, Technical Series No. 13, pp. 117–136.

Shewan, J.M., MacIntosh, R.G., Tucker, C.G. and Ehrenberg, A.S.C. (1953) The development of a numerical scoring system for the sensory assessment of the spoilage of wet white fish stored in ice. *Journal of the Science of Food and Agriculture* 4, 283–298.

Spinelli, J. and Miyauchi, D. (1968) Irradiation of Pacific coast fish and shellfish. 5. The effect of 5′-inosine monophosphate on the flavour of irradiated fish fillets. *Food Technology* 22(6), 123–125.

Storey, R.M. (1985) Time temperature function integration, its realization and application to chilled fish. In: *Storage Lives of Chilled and Frozen Fish and Fish Products*. International Institute of Refrigeration, Commissions C2 and D3 1985-4, Aberdeen, pp. 293–297.

Stroud, G.D. (1969) Rigor in fish: the effect on quality. *Torry Advisory Note* No. 36. Ministry of Agriculture, Fisheries and Food, Torry Research Station, Aberdeen.

Tomlinson, N., Geiger, S.E. and Dollinger, E. (1965) Chalkiness in halibut in relation to muscle pH and protein denaturation. *Journal of the Fisheries Research Board of Canada* 22, 653–663.

Wagner, J.R., Titus, D.S. and Schade, J.E. (1963) New opportunities for flavour modification. *Food Technology* 17(6), 52–55, 57.

Woskow, M.H. (1969) Selectivity in flavour modification by 5′-ribonucleotides. *Food Technology* 23(11), 32, 37.

Hygiene and Safety of Seafood 8

D.M. Gibson
Formerly of MAFF, Torry Research Station, Aberdeen, UK.

Introduction

The microbiological safety of fish, shellfish and their products depends on the interactions of the normal spoilage flora with pathogenic agents. The spoilage flora arises from the environment the fish have been taken from and exposed to post-mortem. The spoilage or natural flora of such fish consists of rather unspecialized bacteria with few outstanding features (Cahill, 1990). Fish taken from temperate and cold waters have a Gram-negative flora whose metabolism consists of reactions in and close to the tricarboxylic acid cycle and of amino acids. Because fish are low in carbohydrates, fermentations without additions are not usual. Fish taken from warmer waters have a high proportion of Gram-positive bacteria, which are severely inhibited by chilling. The flora that develops during storage is dependent on the storage temperature, Gram-negative types predominating at chill temperatures, Gram-positives at higher temperatures. Liston (1992) discusses this in some detail.

The spoilage of fish leading to rejection is mainly due to the activities of bacteria, except in the case of fatty fish, where lipid oxidation occurs. The object of most techniques in fish processing is to inhibit the growth of the spoilage flora. Thus, chilling and, more especially, freezing, salting, smoking, marinading, and pasteurization by heat or irradiation all reduce the growth rate of the flora. However, because they are better adapted to the storage environment, the normal flora of fish and shellfish acts as a barrier to the growth and survival of any contaminating bacteria.

The microbiological safety record of fish and shellfish and their products is generally good, except where raw or undercooked products are consumed, for example shellfish. Fish taken from unpolluted waters are

Table 8.1. Incidence of reported food poisoning cases attributed to fish and shellfish in Europe, 1985-1989 (Anon., 1992a).

Country	Incidence (%)
Albania	2
Belgium	10
Bulgaria	3
Denmark	17
Finland	16
France	7
Germany (East)	3
Germany (West)	6
Hungary	0
Israel	2
The Netherlands	6
Poland	1
Portugal	10
Romania	2
Spain	5
Sweden	7

Table 8.2. Foodborne disease outbreaks implicating seafood products (after Bryan, 1987).

	Fish	Molluscs	Crustaceans
Botulism	12	0	0
Clostridium perfringens	1	0	0
Salmonellosis	3	1	3
Scombrotoxicity	76	0	0
Shigellosis	5	1	2
Staphylococcus spp.	1	0	2
Hepatitis non-B	2	4	0
Bacillus cereus (emetic)	0	1	2
Cholera	0	2	1
Streptococcal infections	0	2	0
Staphylococcal toxins	0	0	2
Vibrio parahaemolyticus	0	7	6
Algal toxins	56	6	0
Diseases of unknown aetiology	12	76	5

usually free of pathogens but shellfish are often taken from polluted areas where, due to the increased availability of nutrients, they grow well.

The incidence of food poisoning attributed to the consumption of seafood in Europe from 1985 to 1989 varies from 2 to 17%, depending on the country (Table 8.1). Since then, there has been a large increase in food poisoning caused by *Salmonella* spp. and *Campylobacter* spp.

associated with poultry (Anon., 1990; 1991) and so these figures may be an overestimate. However, two of the largest ever outbreaks of food poisoning have fish connections. These are the cholera epidemic in South America and the hepatitis incident in China. Bryan (1987) has summarized data from the USA for the period 1979–1984 as to the aetiological agents involved (Table 8.2). The number of cases per outbreak is not stated, nor the seriousness of the illness. Besides the category of 'diseases of unknown aetiology' in which seafood consumption was thought to have played a role, it must be stressed that, in most cases of food poisoning, no food or agent is identified and, in the case of illness of short duration affecting only one or a few people at a time, the incidents are not reported. New agents of bacterial food poisoning have since become evident. These include *Campylobacter* spp., *Listeria monocytogenes*, *Aeromonas* spp. and *Yersinia enterocolitica*. More algal toxins have also been identified or suspected but, due to surveillance and control measures, cases of intoxications have not increased.

Fish can carry parasites which can affect consumers (Olson, 1987), but again control measures, such as freezing susceptible fish, have been instigated. Viral infections have been associated with the consumption of raw or lightly cooked shellfish and viruses are probably the cause of many of the 'unknowns' in Table 8.2.

The safety of seafood (Gibson, 1992) ultimately depends on what agents of disease are present on newly caught or harvested fish, that is the agents that are part of the aquatic ecosystem, those resulting from coastal pollution such as improperly treated sewage, and those introduced during processing from, for example, the workers and the environment. How well they fare and affect the consumer depends ultimately on their survival during food preparation operations such as cooking, or marinading, etc., and the susceptibility of the consumer. Certain people are more likely to succumb to illness because they are immunocompromised. They include the young, whose immune system is not fully developed, the old, whose system is becoming less efficient, the pregnant, whose system is in abeyance, and those with recognized diseases affecting their immune systems such as those with AIDS, cancers, etc. (Mossel, 1991). Seafood is often a preferred food of those already ill, being part of a 'light, easily digested' diet.

Pathogenic Microorganisms Naturally Present in the Aquatic Environment

Clostridium botulinum

Botulism, the disease caused by the toxins of *C. botulinum*, is of rare

Table 8.3. Some properties and characteristics of *C. botulinum*.

	Proteolytic strains	Non-proteolytic strains
Toxin types	A, B, F	B, E, F
Minimum pH for growth	4.6	5.0
Minimum NaCl concentration for growth at:		
(a) ambient temperature	10%	5%
(b) minimum growth temperature	10%	3.5%
Temperature range for growth	10-48°C	3.3-45°C

occurrence in most countries. It is a paralytic disease as the toxins, neurotoxins, affect the nervous system, eventually causing death. Since the syndrome was recognized, food technologists have tried to ensure that the processing and storage of foods known to contain the organism or its spores do not increase the risk of disease by permitting its growth and toxicity. The toxins are heat sensitive and destroyed by adequate cooking.

Botulinum organisms are classified into two groups, those which can hydrolyse proteins and are therefore referred to as proteolytic, and those which cannot, the non-proteolytics (Hauschild, 1989). Their properties are shown in Table 8.3. The types relevant to fish consumers are non-proteolytic types B and E. The latter are found only in marine/aquatic environments. Spores have been found worldwide (Huss, 1981), especially in muds, and can and do contaminate fish. Proteolytic types grow only at temperatures above 10°C. Their metabolism or digestion of the protein fragments results in compounds with very offensive odours such that foods containing them are regarded as spoiled and would be rejected by potential consumers. At temperatures allowing growth of *C. botulinum*, fish spoil quickly. The non-proteolytic types do not produce malodours during growth. They can grow at chill temperatures and produce toxins. With conventionally handled fish, they are not a problem as the normal spoilage flora can outgrow them, by competing better for nutrients, and the fish products are spoiled before they are toxic. But there are instances where some traditional products have become toxic and caused the death of their consumers. Some outbreaks are listed in Table 8.4.

Generally, it has been found that the processing was inadequate, and many outbreaks concern home rather than commercial processing. In such cases, the raw material has, of course, contained spores or vegetative cells of the organism, and the processing has been adequate to suppress the normal flora but not *C. botulinum*, because, for example in fermentation, the pH has been too high. In fish there is very little carbohydrate to ferment to acid. In other cases, the salt level has been too low. Such products have then been eaten without any or adequate cooking. Such

Table 8.4. Outbreaks of botulism incriminating fishery products worldwide.

Region	Outbreaks	Percentage (number) involving fish	Product, type
North America (Canada and USA)	279	94 (262)	Fermented fish, E
South America (Argentina)	30	10 (3)	
Japan	94	99 (93)	Fermented fish, E
China	958	0 (0)	
Europe	142	13 (10)	Various, E
East Europe	72	1 (1)	
Former USSR	83	67 (55)	Cured fish, E
Iran	63	97 (62)	Fish, eggs, E

incidents are, fortunately, rare but changes to existing fish processing technologies and the introduction of new ones require careful investigation to ensure that the risk of botulism to the consumers is not increased. The organism, an anaerobe, does not need the complete absence of oxygen and can grow well a few millimetres below the surface of a product where conditions are anoxic due to the oxygen demand from the rest of the flora. In addition, in comparison with other similar proteinaceous foods, marine fish contain trimethylamine oxide (TMAO), which the active spoilage flora can use as electron acceptor under conditions of low oxygen tension (Strøm *et al.*, 1979; Barrett and Kwan, 1985). TMAO is reduced to trimethylamine, which has a stale-fish-like odour. The growth of these organisms is not inhibited to any significant extent by the absence of oxygen as occurs in meat for example, and the extension in shelf-life of fish by packaging is small. Some strains of *C. botulinum* can reduce TMAO (Castell, 1946).

While *C. botulinum* has to be regarded as ubiquitous in the environment in which fish live, control of the risk of botulism by the application of known information is generally successful, but the introduction of new technologies, especially those resulting in products eaten without cooking, must be carefully investigated (Conner *et al.*, 1989; Notermans *et al.*, 1990). Similarly, with the trend towards the sale of chilled prepared meals in which fish is but one component, contamination of the fish with microbes from the other components, for example vegetables contaminated with soil, in which they cannot grow, should be carefully assessed because fish is a very good medium for *C. botulinum* growth and toxigenesis.

As a follow-up to some incidents, botulism was found in some northern European trout farms (Huss *et al.*, 1974). The fish themselves were susceptible and the disease spread due to cannibalism. It became apparent that there were two sources of infection, the feedstuffs used, and muds on the bottom of the ponds. Ponds are now constructed of concrete or suitably lined, so there are no niches to harbour the organism. The clostridia also require for growth some carrion; improved farm management ensures that dead fish are promptly removed and this has helped to control the problem.

Vibrio spp.

In countries such as Japan where there is a tradition of eating raw fish, *V. parahaemolyticus* tops the food poisoning statistics table (Twedt, 1989). The organism is found in the marine environment especially in waters where the temperature is always above 10°C. It requires salt for growth and is therefore called a halophile. It is very heat sensitive, being killed even with the lightest of cooking. It can grow very rapidly at ambient and body temperatures but not below 10°C. Therefore, chilling of fish is sufficient to stop it growing, and is lethal to it. In temperate waters, it is detected in summer but not in winter because, then, it attaches to chitin particles and sediments. It is found on freshly caught fish at levels of $10^2\,g^{-1}$. For infection, a dose of 10^5–10^7 organisms is required. As the organism does not grow under chill conditions, it is evident that the fish must be stored at relatively high temperatures (temperature abuse) before they become toxic. The generation time for the organism can be as short as 10 min.

In recent years, another vibrio, *V. vulnificus*, has been recognized as a pathogen from shellfish (Oliver, 1989). It has been associated with oysters and crabs taken from coastal waters off the southeast of the USA and has caused illness in consumers and also through wounds to fish process workers. It secretes many degradative enzymes and has been called one of the most aggressive organisms known. Infections of wounds have been so severe that the victims have required amputations. One predisposing condition of victims is iron overload due, for example, to liver damage caused by alcoholism.

Cholera has a long association with the aquatic environment and the organism responsible, *V. cholerae*, in its various types, has been found in fish and fishery products. At present, there is a cholera epidemic raging through South America (Rice and Johnson, 1991). It started in Peru and was linked originally to the consumption of a speciality fish product, civenche. This is a white fish product lightly marinaded in lemon or lime juice for 30 min. Evidence has emerged that the source of the cholera was probably bilge water from a freighter from Bangladesh (Anderson, 1991). Fish became contaminated and were used as the raw material for the production of civenche. The vibrio survived processing, perhaps due to insufficient contact time with the acid. The civenche was distributed countrywide and caused the primary infection. It appears that the Peruvian authorities had discontinued chlorination of domestic water supplies, which were not totally separate from sewage and waste waters, apparently because of a USA agency report which linked chlorination with the formation of some carcinogens in the waters. The result of this decision has been thousands of deaths and probably a million cases of cholera, not all, it must be pointed out, as a consequence of fish consumption, but in which a fish product was an early vehicle of transmission. The cholera vibrio does not survive for long in processed fish, and is easily destroyed during cooking. In addition, the infectious dose is large.

Toxins

Toxicity following the consumption of seafoods, especially shellfish and certain fish from tropical waters, has long been recognized. There is an increase in reports of the presence of toxins worldwide, perhaps due to more efficient surveillance. When toxins are suspected present, fisheries are closed or regulated so that public health is protected. The main groups of toxins are described in Motohiro (1992) and Wright (1992). Paralytic shellfish poisons (PSP) are a group of 18 toxins based on saxitoxin. They are water soluble. They are produced by dinoflagellate algae of the genera *Alexandrium*, *Gonyaulax* and *Protogonyaulax* but there are reports that they can be produced by marine bacteria. Symptoms in victims include tingling of the extremities and around the mouth, progressing to more extensive paralysis and death. Outbreaks in the northern hemisphere tend to occur in early summer, confirming the folklore that shellfish should be eaten only when there is the letter 'r' in the month. Control is by regular monitoring of suspect areas, of mussels in coastal waters and scallops in deeper waters with closure of the fisheries as appropiate. Toxicity is determined by mouse bioassay; chemical assays are presently inadequate (Hungerford, 1993).

Diarrhoeic shellfish poisons (DSP) are produced by dinoflagellates of the genera *Dinophysis* and *Protocentrum*. They are soluble in non-aqueous solvents and comprise a group of seven polyethers, the best characterised being okadaic acid and its methyl derivative. They cause gastrointestinal disease, often within a short time of consumption. Large outbreaks have occurred affecting hundreds of people. Symptoms abate quickly and no deaths have been reported. A bioassay is used for control.

Ciguatera toxins are responsible for most poisoning attributed to fish consumption in the Pacific and Caribbean areas; they also occur in the Mediterranean and the Arabian Gulf. They are also produced by dinoflagellates, and can pass into the food chain and become concentrated in fish such as barracudas and snappers. Control is based on local knowledge, and on discarding large specimens of the fish with a reputation for toxicity. Unfortunately, while imports of large fish can be rejected, some exporters have taken to portioning such species and it is not easy to tell if a portion has come from a large, potentially toxic fish or a small, safe one. The toxin is a small lipid-soluble molecule and acts as a sodium channel blocking agent with a host of neurological effects.

Two other shellfish toxins have recently been described. One is amnesia shellfish poisoning (ASP) and the other brevetoxin, a neurotoxin (NSP). ASP was first reported in Canada in 1987, where there were cases of toxicity following the consumption of cultured blue mussels. The victims lost their short-term memory, in some cases permanently. The toxin is domoic acid, a water-soluble amino acid produced by a diatom, *Nitzschia pungens*. NSP occurs in bivalve shellfish in the Gulf of Mexico following red tides of the dinoflagellate *Ptychodiscus brevis*. The toxin is often called brevetoxin, and affects the peripheral neural system. It is fatal to fish.

Pufferfish poisoning is due to tetrodotoxins, bacterial metabolites. It is not restricted to these fish but it occurs only around Japan.

Parasites

Fish can harbour many parasites (Olson, 1987), of which a few can affect consumers of raw or semi-raw products such as some marinades. The most important are the nematodes or round worms of *Anisakis* species. They cause gastrointestinal infection and severe pain. They were first described in persons who had eaten raw herring. The worms are killed by cooking or by freezing the fish; salting and marinading (acid treatment) are effective but only after long exposure times (Anon., 1986). Worms can sometimes be seen with the unaided eye on the surfaces of fillets. Dead worms are harmless. The nematodes use fish muscle in part of their complex life cycle, in which marine mammals are the final hosts.

Pathogens Contaminating Fish during Processing

The division between zoonotic and other infections is blurred in that, for example, bacteria of human origin can contaminate fish in sewage-polluted waters or during processing in factories. The data in Table 8.1 would indicate that such contamination rarely leads to illness in consumers, but food poisoning is under-reported and, even when reported, often no attributable source is found. Illness caused by some microbes occurs soon after their consumption and in these cases the victims are likely to remember what they had eaten and there may well be some of the incriminated food remaining for testing. When the incubation time is long, as with listeriosis, it is difficult to be definitive as to the cause. Some of the microbes of current interest are discussed, including those which have been called emerging pathogens because of their recent description or involvement in human disease. Some of them can grow well at chill temperatures.

Listeria monocytogenes

There are two forms of listeriosis, the disease caused by *L. monocytogenes*, one a mild gastrointestinal illness of short duration occurring soon after the ingestion of large numbers of the organism, and the other a serious infection of immunocompromised patients with high mortality rate, of 30%, perhaps months after the initial infection, which shows as a mild flu-like illness. In the USA, there are about 1600 cases per year with 400 deaths. Fish and fishery products can carry the microbe, usually in low numbers, $< 1\,g^{-1}$, but *Listeria* can grow even at 0°C. With long storage times or shelf-lives, products can become hazardous. The incidence of *L. monocytogenes* is variable. Kvenberg (1988) reported that 5% of all fishery product types examined in the USA were positive, while Weagant *et al.* (1988) found 26% of frozen seafood positive. Dillon and Patel (1992) have summarized recent data. The source is the environment. *L. monocytogenes* can be isolated from most waters, both fresh and marine, as well as from sewage. Seabirds can carry and excrete it. Soils and plants are also involved, and a small percentage of people are symptomless carriers, usually transiently rather than permanently. It can be carried in the air and so control of the source of air in processing establishments and of its movement has become important. The organism is heat sensitive, being killed by exposure to 70°C for 2 min or the equivalent. Hazardous products are, therefore, those eaten without further cooking such as some cold smoked products. The organism can grow in up to 9% NaCl and survive in brine for over 100 days. Until 1993, there were no well-authenticated reports of cases of listeriosis implicating fish or shellfish products as the source of the infection. However, a link has been established between two cases of perinatal listeriosis and the consumption of smoked mussels in New Zealand (Baker *et al.*, 1993). Listeria cultured from the victims and from the products have been shown to be identical using advanced molecular biological techniques.

Many microbiological specifications for foods require the complete absence of listeria, an impossibility in view of its ubiquitous distribution in nature (Lund, 1990; Dillon and Patel, 1992).

Salmonella spp.

There is a worldwide epidemic of gastrointestinal salmonellosis mainly due to *S. enteritidis*, phage types 4 and 13. Fish and fishery products have not been involved. Salmonella can be found in fish and shellfish taken from sewage-polluted waters. However, about 5% of fish meal samples have been found to contain salmonellas. A major source of infection is avian, due to the exposure of the raw material to birds and cross-contamination or recontamination of finished products.

The heat treatments used to produce the meal and to dry it should be sufficient to eliminate salmonellas. As fish meal is used in animal and poultry feed, it has an important role in animal and hence human salmonellosis.

Scombrotoxicity

The most commonly reported illness following consumption of fish is scombrotoxic poisoning (see Table 8.2). It is a chemical intoxication often linked to the presence of large quantities of histamine in the product (Bartholomew *et al.*, 1987). The most prominent symptoms are rash, diarrhoea, flushing, headache, vomiting/nausea and a burning sensation around the mouth. The histamine is derived from histidine found in high concentrations in scombroid fish, i.e. species of tuna, bonito, mackerel, mahimahi, bluefish, etc. of the families Scomberesocidae and Scombridae. Histidine is decarboxylated by enteric bacteria. As the bacteria cannot grow at chill temperatures, some degree of temperature abuse is often implicated in incidents of toxicity. Histamine is heat stable and so survives cooking and canning. Thus, in many instances, there are no bacteria remaining in the suspect material. However, evidence is emerging that the intoxication may not always be due to histamine, although administration of antihistamines relieves the symptoms. Indeed, consumption of large amounts of histamine does not lead to intoxication, and, in many outbreaks, little or no histamine has been found in the remains of the meal (Bartholemew *et al.*, 1987). In addition, there are cases of toxicity involving non-scombroid fish and other foods. The simultaneous presence of mediators of histamine toxicity has been suggested, perhaps low levels of other marine toxins (Gibson, 1992).

There is no doubt, however, that the presence of significant levels (>5 mg 100 g^{-1}) histamine is an indicator that the fishery products have not been kept at chill temperature since harvesting and it is a marker for time–temperature abuse. High levels of histamine in fishery products are undesirable for other reasons. Histamine is an inhibitor of mono- and di-amine oxidases and can interfere with some medical treatments. In the presence of ferrous ions, it is a catalyst of lipid oxidation (rancidity) yielding products which can be mutagenic and perhaps carcinogenic (Ababouch, 1991).

The syndrome has been regarded as an allergy, but, as victims often have a habit of eating fish and subsequently eat the same species of fish without ill effect, this is unlikely.

Aeromonas spp.

While the role of aeromonads in human disease is still controversial, although more positive evidence is emerging, there is no doubt that such organisms are frequently isolated from healthy fish and from fishery products, especially those stored above 20 °C (Liston, 1992). Indeed, aeromonads predominate in the flora of fish stored at 37 °C. There are also reports that *Plesiomonas shigelloides* can be similarly dominant. As these organisms are relatively heat sensitive, they should prove to be no hazard in cooked products, but could cross-contaminate cooked ready-to-eat foods. At present, there are no techniques to differentiate environmental strains including those isolated from fishery products and those isolated from persons ill with gastroenteritis.

Staphylococcus aureus

Staphylococci derived from fish processors due to bad hygienic practices can and do contaminate fishery products. They are commonly found on the skin and cuts or abrasions, the nose, and on hair. Limits for their presence are given in many end product specifications. Incidents of intoxication following fish consumption are not common. The toxins are produced only at temperatures above 15 °C. The risk is greater in some processed fish with increased salt levels as the bacteria are halotolerant and can grow and produce toxin under the conditions, for example, used in the smoking of fish when the growth of other (spoilage) bacteria is inhibited. As the toxins are heat stable, any preformed toxin will survive cooking. As it is an intoxication, the onset of symptoms, viz. nausea, vomiting, diarrhoea, is rapid, and so is recovery.

Escherichia coli and coliforms

Most specifications for fish and shellfish products list these organisms as they are indicators of sewage pollution in growing and harvesting waters and inadequate toilet hygiene by processing staff. Specifications are often stated in terms of 'Most Probable Number' (MPN), a statistical term. Because it is not possible to show the complete absence of an organism, the limits for *E. coli* are given as <3 per gram or 230 per 100 g. As the organisms cannot grow under chill conditions, their presence indicates contamination at source or by workers. As they are relatively easily killed and die off within 2 days in sea water, they are not regarded as the best indicators of sewage pollution. Streptococci survive for longer in aquatic environments. The survival of viruses is not known.

Some specifications quote limits for *E. coli*, others for coliforms and some both, indicating an equivalency between two sets of numbers. It should be noted that all such specifications are dependent on the analytical method used, including parameters such as the bacteriological medium used, the incubation time and temperature. New methods have been developed yielding results in a short time, before rather than after the products have been consumed, as is the case with traditional methods. There can be differences in which bacteria fall into the category of 'coliforms' by such methods. However, it is known that some spoilage bacteria can be detected as coliforms by some long-established conventional methods.

Besides these indicator bacteria, there are some *E. coli* which are pathogenic to humans by way of producing toxins (Doyle, 1990). These strains, referred to as enteropathogenic (EPEC), enterotoxigenic (ETEC), some of which cause 'traveller's diarrhoea', enteroinvasive (EIEC), can be transmitted, but fishery products have not been reported as vehicles so far. The other group is the verotoxin (VTEC)-producing strains, found in animal faeces, which are foodborne, and can cause severe illness. Their toxins are like those of *Shigella*.

Viruses

Sewage contains viruses of human origin which can contaminate fish and shellfish and then infect consumers. As mentioned, specifications for *E. coli* and coliforms are used as indicators for viral contamination, because there are few effective methods for detecting viruses in foods. Current methods require at least 10^6 virus particles to register a positive result but only 100 are needed to cause disease (Iversen *et al.*, 1987). Molecular biological techniques such as those incorporating the polymerase chain reaction hold promise.

Virtually every enterovirus that has been described has been found in sewage. Infection attributed to shellfish consumption is usually due to small round virus (SRV), small round structured virus (SRSV), Norwalk agent, Snow Mountain agent, and various astroviruses and calciviruses (Appleton *et al.*, 1988). They are detected usually in the faeces of victims by electron microscopy. While raw seafood is easily implicated, undercooked shellfish is also a vehicle, and the heating regime necessary to inactivate such viruses is not known. Because they cannot be grown, it is not possible to suggest safe cooking systems. Sewage can also contain the hepatitis A virus (HAV). This is a far more serious infection than viral enteritis, and as it has a longer incubation time, about 6 weeks, it can follow other viral infections. In the UK, such diseases were common until research showed that they could be eliminated by cooking cockles, the incriminated shellfish, at 90°C for 1.5 min. There are some changes in the product with such a heat treatment but these have to be tolerated for the sake of food safety. There are, in many countries, *E. coli* and coliform standards for molluscan shellfish. These standards are imposed to give the consumer protection from all sewage-derived pathogens, but their effectiveness with regard to protection from viral infections is not known. If shellfish fail to meet them, then they can be transferred to other waters.

The relative rates of depuration, or self-cleaning of shellfish with respect to bacteria and viruses, are unknown. Species such as cockles do not survive transfer and so adequate heat treatment is the only means of ensuring a safe product.

The largest outbreak of food poisoning reported occurred in China in 1988. There were nearly 300,000 cases reported, 292,301 to be precise, of HAV attributed to the consumption of sewage-contaminated clams (Anon., 1988).

Other Food Poisoning Agents

In Table 8.2, it can be seen that no agent was found for many incidents of food poisoning attributed to the consumption of fish or shellfish. Even with good laboratory facilities available, about half the reported cases of food poisoning have no known cause or vehicle identified. Sometimes samples of the suspected food are not available. But it is evident that new agents of infection are being discovered. In addition, when the incubation period, that is the time between consumption of the suspected food and the eruption of illness, is more than a few days, then most people cannot accurately remember what they had eaten. Thus foods eaten outside the home are more often implicated because (a) the occasion was more memorable and (b) as more victims are reported, investigation is more likely. More food is being sold in a 'ready-to-eat' format, and/or as mixed commodities rather than as, say fish, alone and the possibilities of cross-contamination between the flora of the different items is significant. Given sufficient time or shelf-life, the contaminant could grow to reach an infectious dose or produce toxin.

Besides the presently unidentified agents of disease, there are reports of bacteria in the marine environment being in a 'viable non-culturable' state (Colwell *et al.*, 1989). This has been well established for some vibrios including pathogenic types, and for enteric bacteria. Essentially, when confronted with adverse growth conditions, the bacteria can shut down their metabolism and remain in a minimal energy state until conditions improve. For the practical microbiologists, this in effect results in a lack of growth and so a negative test result. If the organisms revive before consumption and are not otherwise killed as during cooking, then there is a hazardous situation.

Developments in Fish Technology of Concern for Microbiological Safety

Over the years, researchers have ensured that changes in fish processing technologies have not resulted in any increase in the risk to human health from consuming fish and fishery products. Recent innovations in chilled fishery products include packaging, irradiation and heat pasteurization (*sous vide*). Packaging of fish products with oxygen removed as in vacuum packs, and under modified atmospheres (MAP) have been regarded as potentially hazardous as the conditions in the pack may enhance the growth of *C. botulinum* while inhibiting the growth, and so the protection offered, by the spoilage flora (Conner *et al.*, 1989). Some of the outbreaks of botulism listed in Table 8.4 were attributed to home-produced vacuum packaged products. However, the oxidation–reduction potential in all fish as food is sufficiently low a few millimetres below the surface such that anaerobes can grow. All forms of packaging of chilled fish have been carefully evaluated with respect to the risk of botulism. Packaging does not change the risk. The microbiological safety of foods can be assessed using the techniques of predictive microbiology, which quantify the possibility and kinetics of growth of microbes. Baker and Genigeorgis (1990) have made extensive studies of the time taken for fishery products inoculated with spores of *C. botulinum* to become toxic and showed that packaging had an insignificant effect. The most important factor, accounting for 76% of the experimental variability, was the storage temperature of the products.

Irradiation of fish is permitted in some countries as a means of increasing the shelf-life and to eliminate pathogens. The doses given cause pasteurization rather than sterilization of the product, to avoid giving the product undesirable flavours. At the levels used, the flora is reduced by a factor of 1000. Spoilage and vegetative forms of pathogens are susceptible but toxins already present and spores survive. The effect on viruses is not known. For greatest effectiveness, only very fresh fish and shellfish should be irradiated and the products must be stored/chilled. The possibility of increasing the risk of botulism because of the reduced competition for nutrient and the increased shelf-life has been assessed. At the levels of irradiation used, there is no increase in the risk.

Heat treated or *sous vide* products are becoming popular in some countries. They are portioned processed fish, vacuum packed, then heated, cooled and stored. They are reheated before consumption. The heat treatment, usually at 70 °C or higher for 10–60 min, but sometimes less, kills spoilage and vegetative pathogens but spores can remain. If storage temperatures permit, these could outgrow. As the products are packaged, recontamination is only possible after the packs have been opened. The safety of *sous vide*, vacuum packed and MAP products has been assessed (Mossel and Struijk, 1991; Anon., 1992b). An expert committee in the UK has concluded that the main risk to consumers from such products is botulism and has proposed that they be given a shelf-life of a maximum of 10 days at 10°C, recognizing that lower more desirable chill temperatures may not be achieved. Longer shelf-lives are allowed if the products have been heat treated at 90°C or more for at least 10 min, or have a pH <5 or salt level $>3.5\%$ NaCl in the water phase or water activity below 0.97, or combinations of these factors shown to be effective against *C. botulinum*. These proposals would apply to other ready-to-eat products such as smoked fish, both hot and cold processed products such as smoked trout and salmon. These have traditionally been made with salt levels higher than 3.5% NaCl, but there is a trend towards the production of lower salt (and other preservatives) products.

Because of their high buffering capacity and low carbohydrate content, fermented fish have required added acid either directly or by a co-fermentation. For stability, their pH would be <5. However, there have been outbreaks of botulism due to fermented fish products where the pH has been sufficiently low to inhibit spoilage but not low enough to inhibit the growth of the botulinum organisms. Botulism caused by the consumption of kapchunka, a dried, fermented, whole fish, has been described (Slater *et al.*, 1989). Because of its repeated involvement in incidents, its production is banned in the USA.

Conclusion

The consumption of fish and shellfish has generally led to no untoward increase in the risk of food poisoning except where there is a tradition of eating raw shellfish. As the fish industry becomes more integrated with other foods and less is sold as a single commodity, there is more chance of non-traditional contamination which can survive and grow if products are given extended shelf-lives. In the past, it was assumed that, if the risk of botulism was controlled, then other pathogens would also be controlled. Now, with the emergence of cold-tolerant pathogens which can grow at temperatures below the minimum for *C. botulinum*, and new technologies, risks must be reassessed. With the globalization of fish supplies and of the movement of people, there is an ever-changing risk to the consumer due to the possibility of naturally occurring toxins and pathogens being imported to countries where the risks are not recognized.

References

Ababouch, L. (1991) Histamine food poisoning. An update. *Fish Technology News* 11, 3–11.

Anderson, C. (1991) Cholera epidemic traced to risk miscalculation. *Nature (London)* 354, 255.

Anon. (1986) A primer for parasites. *Seafood Leader* 6, 44–46.

Anon. (1988) Outbreak of hepatitis A – Shanghai. *WHO Weekly Epidemiological Report* 63, 91–92.

Anon. (1990) *The Microbiological Safety of Food*, Part 1. HMSO, London, vii + 147 pp.

Anon. (1991) *The Microbiological Safety of Food*, Part 2. HMSO, London, vii + 220 pp.

Anon. (1992a) *World Health Organization Surveillance Programme for Control of Foodborne Infection and Intoxication in Europe*. Fifth Report 1985–1989. Institute of Veterinary Medicine, Berlin, 235 pp.

Anon. (1992b) *Report on Vacuum Packaging and Associated Processes*. HMSO, London, 69 pp.

Appleton, H., Caul, E.O., Clewley, J.P., Cruickshank, J.P., Cubitt, W.D., Kurtz, J.B., McSwiggan, D.A., Palmer, S.R., Sockett, P. and Riordan, T. (1988) Food-borne viral gastroenteritis. *PHLS Microbiology Digest* 5, 69–75.

Baker, D.A. and Genigeorgis, C. (1990) Predicting the safe storage of fresh fish under modified atmospheres with respect to *Clostridium botulinum* toxigenesis by modelling length of lag phase of growth. *Journal of Food Protection* 53, 131–140.

Baker, M., Brett, M., Short, P. and Thornton, R. (1993) Listeriosis and mussels. *Communicable Disease New Zealand* 93, 13–14.

Barrett, E.L. and Kwan, H.S. (1985) Bacterial reduction of trimethylamine oxide. *Annual Review of Microbiology* 39, 131–149.

Bartholomew, B.A., Berry, P.R., Rodhouse, J.C., Gilbert, R.J. and Murray, C.K. (1987) Scombrotoxic fish poisoning in Britain: features of over 250 suspected incidents from 1976–1986. *Epidemiology and Infectivity* 99, 775–782.

Bryan, F. (1987) Seafood-transmitted infections and intoxications in recent years. In: Kramer, D.E. and Liston, J. (eds) *Seafood Quality Determination*. Elsevier, Amsterdam, pp. 319–339.

Cahill, M.M. (1990) Bacterial flora of fishes: a review. *Microbial Ecology* 19, 21–41.

Castell, C.H. (1946) The effect of trimethylamine oxide on the growth of bacteria. *Journal of the Fisheries Research Board of Canada* 6, 491–497.

Colwell, R.R., Knight, I.T., Somerville, C., Shults, S. and Kaspar, C.W. (1989) Viable but non-culturable phenomenon in relationship to starvation/survival, 'injury', and strategies for survival of bacteria in the environment. In: Hattori, T., Ishida, Y., Maruijama, R.Y., Morita, R.Y. and Uchida, A. (eds) *Recent Advances in Microbial Ecology*. Scientific Societies Press, Tokyo, pp. 85–88.

Conner, D.E., Scott, V.N., Bernard, D.T. and Kautter, D.A. (1989) Potential *Clostridium botulinum* hazards associated with extended shelf-life refrigerated foods: A review. *Journal of Food Safety* 10, 131–153.

Dillon, R.M. and Patel, T.R. (1992) *Listeria* in seafoods: a review. *Journal of Food Protection* 55, 1009–1015.

Doyle, M.P. (1990) Pathogenic *Escherichia coli, Yersinia enterocolitica* and *Vibrio parahaemolyticus*. In: Waites, W.M. and Arbuthnott, J.P. (eds) *Foodborne Illness. A Lancet Review*, pp. 77–85.

Gibson, D.M. (1992) Pathogenic microorganisms of importance in seafood. In: Huss, H.H., Jakobsen, M. and Liston, J. (eds) *Quality Assurance in the Fish Industry*. Elsevier, Amsterdam, pp. 197–209.

Hauschild, A.H.W. (1989) *Clostridium botulinum*. In: Doyle, M.P. (ed.) *Foodborne Bacterial Pathogens*. Marcel Dekker, New York, pp. 111–189.

Hungerford, J.M. (1993) Seafood toxins and seafood products. *Journal of the Association of Official Analytical Chemists International* 76, 120–130.

Huss, H.H. (1981) The ecology of *Clostridium botulinum* type E. In: Roberts, T.A., Hobbs, G., Christian, J.H.B. and Skovgaard, N. (eds) *Psychrotrophic Microrganisms in Spoilage and Pathogenicity*. Academic Press, New York, pp. 463–473.

Huss, H.H., Pedersen, A. and Cann, D.C. (1974) Incidence of *Clostridium botulinum* in Danish trout farms. *Journal of Food Technology* 9, 445–450.

Iversen, A.M., Gill, M., Bartlett, C.L.R., Cubitt, W.D. and McSwiggan, D.A. (1987) Two outbreaks of foodborne gastroenteritis caused by a small round structured virus: evidence for prolonged infectivity in a food handler. *The Lancet*, Sept. 5, 556–558.

Kvenberg, J.E. (1988) Outbreaks of listeriosis/*Listeria*-contaminated foods. *Microbiological Sciences* 5, 355–358.

Liston, J. (1992) Bacterial spoilage of seafood. In: Huss, H.H., Jacobsen, M. and Liston, J. (eds) *Quality Assurance in the Fish industry*. Elsevier, Amsterdam, pp. 93–105.

Lund, B. (1990) The prevention of foodborne listeriosis. *British Food Journal* 92, 13–22.

Mossel, D.A.A. (1991) Food microbiology: an authentic academic discipline with substantial potential benefits for science and society. *Journal of the Association of Official Analytical Chemists* 74, 1–13.

Mossel, D.A.A. and Struijk, C.B. (1991) Public health implication of refrigerated pasteurized ('sous-vide') foods. *International Journal of Food Microbiology* 13, 187–206.

Motohiro, T. (1992) Biotoxins in seafood. In: Huss, H.H., Jakobsen, M. and Liston, J. (eds) *Quality Assurance in the Fish Industry*. Elsevier, Amsterdam, pp. 243–258.

Notermans, S., Dufrenne, J. and Lund, B.M. (1990) Botulism risk of refrigerated, processed foods of extended durability. *Journal of Food Protection* 53, 1020–1024.

Oliver, J.D. (1989) *Vibrio vulnificus*. In: Doyle, M.P. (ed.) *Foodborne Bacterial Pathogens*. Marcel Dekker, New York, pp. 569–600.

Olson, R.E. (1987) Marine parasites of public health importance. In: Kramer, D.E. and Liston, J. (eds) *Seafood Quality Determination*. Elsevier, Amsterdam, pp. 339–355.

Rice, E.W. and Johnson, C.H. (1991) Cholera in Peru. *The Lancet* 338, 455.
Slater, P.E., Addiss, D.G., Cohen, A., Leventhal, A., Chassis, G., Zehavi, H., Bashari, A. and Costin, C. (1989) Foodborne botulism: an international outbreak. *International Journal of Epidemiology* 18, 693–696.
Strøm, A.R., Olafsen, J.A. and Larsen, H. (1979) Trimethylamine oxide: a terminal electron acceptor in anaerobic respiration of bacteria. *Journal of General Microbiology* 112, 315–320.
Twedt, R.M. (1989) *Vibrio parahaemolyticus*. In: Doyle, M.P. (ed.) *Foodborne Bacterial Pathogens.* Marcel Dekker, New York, pp. 543–568.
Weagant, S.D., Sado, P.N., Colburn, K.G., Torkelson, J.D., Stanley, F.A., Krane, M.H., Shields, S.C. and Thayer, C.F. (1988) The incidence of *Listeria* species in frozen seafood products. *Journal of Food Protection* 51, 655–657.
Wright, J.L.C. (1992) Shellfish toxins: a Canadian perspective. In: Bligh, E.G. (ed.) *Seafood Science and Technology*. Fishing News Books, Oxford, pp. 366–375.

Contaminants in Fish 9

Adriaan Ruiter
Department of the Science of Food of Animal Origin, Faculty of Veterinary Medicine, Utrecht University, The Netherlands.

Introduction

Contaminants in food can be defined as undesirable substances that are present in relatively small amounts, as a result of human action. This definition excludes additives which are intentionally introduced with the aim of improving some properties of food products, e.g. colour, flavour, texture or storage life. Compounds resulting from spoilage processes are also excluded, and so are unwanted compounds that might be present naturally (Ruiter, 1985).

This definition can also be applied to contaminants in fish and fishery products. Environmental contaminants account for a large and important section of fish contaminants, and thus, in this chapter, merit the larger part of the attention (in non-cultivated animals such as game and most fish, pollution with environmental contaminants is a larger problem than in domesticated production animals).

Another category of contaminants results from fish treatment, preparation, processing and packaging. As a rule, these contaminants are not characteristic for fishery products. They will also be found, in comparable amounts, in meat products and in some other food products. In this book, some of these will be briefly considered within the scope of a special group of fishery products, i.e. smoked fish, in Chapter 11.

A third category is of particular importance in cultivated fish, namely residues of drugs applied to these fish, mostly to protect them from diseases. The drug residue problem in cultivated fish is not completely comparable to that in warm-blooded animals as, for example, excretion is, on the whole, different from that in slaughter animals traditionally kept in farms. Some words are devoted to this problem here.

The category of biocontaminants may give rise to serious problems. In slaughter animals and milch cows, biocontaminants such as mycotoxins are carried over, for a small part, from the feed to tissues and milk. Comparable contaminations, as a result of mould growth on feedstuffs, might also occur in cultivated fish, but these are not considered here. In shellfish, the occasional presence of some poisonous components originating from parasites presents a serious problem and can be regarded as biocontamination. Because of their direct and dramatic relation to safety (or lack of it), these poisons are discussed in Chapter 8.

Environmental Contaminants

Numerous substances are brought into the environment in some way or another by the actions of man. A distinction can be made between substances normally present in the biosphere but now found, as a result of man-made pollution, in increased amounts (e.g. heavy metals); and substances not originally present in nature but synthetized by man, the so-called *xenobiotics*. Most of these xenobiotics are rapidly broken down in the environment, either as a result of chemical degradations determined by the climate or through attack by living organisms. Some of these, however, are markedly resistant to chemical or biological degradation and will persist somewhere in the environment for a considerable period of time. In the long run, serious problems may arise from the presence of these compounds in the environment, in particular when they are extremely toxic. Special attention must be paid to those that tend to accumulate throughout the food chain (biomagnification) because of their persistency and their lipophilic nature, which allows them to be stored, for a long time, in fatty tissue or fat depots. Another result of this persistency is the slow metabolism (if any), resulting in a low excretion rate from organisms. Heavy metals may also accumulate in fish, but by other mechanisms and not in fatty tissue. It is this category of contaminants which is discussed first.

Heavy metals

Traces of many elements are essential for life. Probably life has adapted itself to the presence of these elements in the environment. It was stated that 'the trace elements are more important than are the vitamins, in the sense that they cannot be synthesized by the living matter, as is the case with vitamins. Thus they are the basic sparkplugs in the chemistry of life' (H.A. Schroeder).

Some of these elements, although essential at low levels, are toxic at higher concentrations. This holds for copper and for fluorine, and it is very striking with respect to elements such as iodine and selenium. Fortunately, hardly any of these elements pollute our environment in such a way that they should be considered here.

A quite different situation exists with regard to heavy metals such as lead, cadmium and mercury. These elements, as far as is known, do not show any essential function in life. They are non-essential and, because of their toxicity, are considered as harmful elements.

Because of this non-essentiality, any concentration of lead, cadmium or mercury in food is, by some people, considered too high. Human beings are accustomed, however, to traces of these elements in their environment. They were always present during the earth's history, and did not prevent the development of either humans or other animals throughout the ages. Nevertheless, these elements merit a somewhat closer inspection, as their amounts have been increased as a result of our activities.

Pollution with heavy metals occurs through effluents which primarily pollute surface water in terrestrial areas and, secondarily, marine areas. For mercury another pollution route, i.e. through the atmosphere, is important as well. It has been supposed that half of the elevated levels observed in pike from southern Swedish lakes is caused by atmospheric deposition (Håkanson *et al.*, 1990).

It is well known that relationships exist between age and size of certain marine organisms, e.g. *Mytilus edulis*, and their heavy metal content (National Academy of Sciences, 1980). With respect to mercury a similar relation exists in freshwater fish such as pike (*Esox lucius*) and pike-perch (*Lucioperca lucioperca*) (Pieters *et al.*, 1993). Such relationships, however, depend on the feeding habits of the organism and its place in the water column. As a consequence, not all marine organisms can be used as indicators for seawater pollution (Law and Singh, 1991). This may also mean that pollution of sea water with heavy metals is not clearly reflected in all organisms living in polluted water. Nevertheless, many such relationships have been established in past years.

Lead and cadmium

As a rule, the levels of lead and cadmium in fish are not higher than those in beef, pork or poultry. In northwest Atlantic cod and bluefin tuna, for example, values in fish muscle are of the order of 0.01 mg kg^{-1} (Hellou *et al.*, 1992a; 1992b). Similar values were frequently reported in cod and other fish from European waters, e.g. by Hagel and Vos (1989), who reported average values of 0.03 mg kg^{-1} for lead and 0.005 mg kg^{-1} for cadmium. Levels in other edible parts such as liver and roe are also low in lead but somewhat higher in cadmium (0.02–0.04 mg kg^{-1} in roe and up to 0.40 mg kg^{-1} in liver).

It has to be noted that, for lead, much higher values were found in a Belgian study on several species from the southern North Sea and from the Irish Sea, i.e. from 0.04 mg kg^{-1} for plaice to 0.3 mg kg^{-1} for lemon sole (De Clerck *et al.*, 1988). Dogfish also shows higher lead contents (Guns *et al.*, 1984). Cadmium levels were reported to be less than 0.01 mg kg^{-1} in this study and in earlier studies in the same areas (Guns *et al.*, 1984; Vyncke *et al.*, 1984).

As for shrimp, lead contents are around 0.1 mg kg^{-1} and cadmium contents around 0.01–0.02 mg kg^{-1} (Hagel and Vos, 1989).

In molluscs, lead and in particular cadmium levels are generally higher as compared to fish. López-Artiguez *et al.* (1989) observed values up to 1.5 mg kg^{-1} for lead and up to 4.1 mg kg^{-1} for cadmium in contaminated molluscs from the Huelva estuary, Spain. Comparable levels were found in Dutch estuaries (Luten *et al.*, 1986), showing a downward gradient for cadmium from inshore locations towards the open sea. Hutagalung (1989) reports cadmium levels from 0.63 to 2.44 mg kg^{-1} in mussels from Jakarta Bay and from 0.08 to 1.24 mg kg^{-1} in other areas. The lead and cadmium levels from the Belgian coast are lower, i.e. 0.15–0.60 mg kg^{-1} for lead and 0.1–0.25 mg kg^{-1} for cadmium (Vyncke *et al.*, 1988).

The uptake of both lead and cadmium by molluscs is directly proportional to the concentrations of these metals in the ambient water (Klumpp and Burdon-Jones, 1982) and, for that reason, is elevated in polluted areas. In bivalves from the Philippines, lead contents were markedly higher than cadmium contents, i.e. 0.64–2.24 mg kg^{-1} and 0.06–0.66 mg kg^{-1}, respectively (Luis *et al.*, 1989), which is related to the pollution of the sea water showing much higher levels for lead than for cadmium. Particularly high levels for lead were found in American crayfish (*Procambarus clarkii*) from a rural area in Spain (Rico *et al.*, 1987), i.e. up to 10 mg kg^{-1}. Cadmium levels were moderately elevated while mercury levels were not.

Lead and cadmium levels in mussels and oysters from sites around the US coasts show some decrease in the 1980s as compared to the 1970s (Lauenstein *et al.*, 1990). The same was observed in a Dutch estuary (Stronkhorst, 1992).

Mercury

The situation with respect to mercury, as a contaminant in fish, is completely different from that regarding lead or cadmium. It has been known for many years that mercury, in particular, tends to accumulate in marine organisms.

The cause of this strong tendency to accumulate in the marine environment is the peculiar property of mercuric ions to be methylated by marine bacteria. The mercuric ion, like other bivalent heavy metal ions, has an absorption rate in the gastrointestinal tract of man and animals of

5 to 10%. The methylmercuric ion (CH_3Hg^+), however, shows lipophilic properties and, for this reason, is much better absorbed by the intestine. Taguchi (1971) found an absorption rate of 6% for mercuric acetate and of 73% for methylmercuric chloride. According to many authors (e.g. Oskarsson, 1992a) the absorption rate of methylmercury may even approach 100%, although, in the body, it is slowly converted to inorganic mercury.

As for fish, the levels of mercury are higher than those of other bivalent metals, despite the comparatively low concentrations in sea water. This also has to be ascribed to the high absorption rate for methylmercury.

Mercury concentrations in fish flesh are usually high in large carnivorous and low in small herbivorous fish (Klemmer *et al.*, 1976). More generally, the mercury level is related to the position of the fish in the food chain, to the length of the fish (Barak and Mason, 1990a; 1990b; Barghigiani and de Ranieri, 1992) and to its age (Krüger, 1990). For these reasons it is difficult to give the 'normal' mercury values of fish. In relatively unpolluted areas, mercury concentrations in fish muscle are in the range of 0.05–0.20 mg kg^{-1} (Shrestha *et al.*, 1988; Hagel and Vos, 1989; Krüger, 1990; Hellou *et al.*, 1992b), but lower values have been reported as well (Gutiérrez-Galindo *et al.*, 1988). Higher values may be considered as caused by pollution (Tariq and Jaffar, 1991). Hagel and Vos (1989) mention an average mercury content of 0.06 mg kg^{-1} for marine fish. Mercury levels in fish liver may be lower than those in muscle (Barak and Mason, 1990b). Others found fish muscle and fish liver showing comparable ability to accumulate mercury (Marcovecchio and Moreno, 1993).

Mercury levels in molluscs are usually in the range of 0.02–0.05 mg kg^{-1}. It frequently happens that mercury levels in molluscs are lower than lead and cadmium levels (e.g. Luten *et al.*, 1986; Luis *et al.*, 1989).

Hutagalung (1989) reports mercury level values ranging from 0.01 to 0.07 mg kg^{-1} for mussels from various areas and values up to 0.85 mg kg^{-1} in mussels from the locally polluted Jakarta Bay.

Mercury levels in polluted areas often exceed 1 mg kg^{-1}, particularly for freshwater fish, and in some exceptional cases even exceed 10 mg kg^{-1} (Francesconi and Lenanton, 1992). Species living in coastal waters such as flounder (*Platichthys flesus*) and molluscs are suitable for monitoring the pollution with mercury as a result of industrial effluents. Values reported from clams, oysters and cockles range from 0.02 to 1.27 mg kg^{-1} (López-Artiguez *et al.*, 1989) and, for flounder, from 0.01 to 1.96 mg kg^{-1} (Leah *et al.*, 1992). In the molluscs examined, other metals show high levels as well (see section on lead and cadmium).

Because of peculiar circumstances such as geological anomalies in local areas, mercury levels in fish may reach very high levels. In the northern Tyrrhenian Sea values are reported exceeding 1 mg kg^{-1} in all species examined and up to 3.2 mg kg^{-1} in hake, *Merluccius merluccius* (Barghigiani and de Ranieri, 1992).

In many industrialized areas the mercury levels in fish show a decreasing trend from the early 1970s to about 1985. As a result of decreased discharge of mercury-containing effluents, the mercury content of freshwater fish such as pike, pike-perch and eel in The Netherlands is considerably lower now (Pieters *et al.*, 1993). The same holds for flatfish caught in Belgian continental shelf waters (Guns *et al.*, 1992a). On the other hand, it is reported from Sweden that methylmercury levels in fish from polluted waters show an increase, despite measures to reduce the mercury pollution. This is probably due to acid precipitation, which facilitates the formation of methylmercury from inorganic mercury in the lakes (Oskarsson, 1992b). It was also reported from Sweden, however, that treatment with sodium selenite incorporated in a rubber matrix is an efficient way to quickly reduce the mercury content of fish living in these lakes (Paulsson and Lundbergh, 1989).

Krüger (1990) points out that not only the mercury levels but also the average weights of the marine fish under consideration have been decreased and is of the opinion that this may be an important reason for reduced mercury levels in fish. Harms (1990) found some increase in the mercury levels for flounder (*Platichthys flesus*) from the German Bight during the period 1984–1989.

It is of importance that, during cooking or processing fish, the mercury content may be considerably reduced (Hernández-García *et al.*, 1988). Roasting sardines, for example, caused a reduction in the mercury level from 0.32 to 0.10 mg kg^{-1}.

The mercury–selenium interaction

Ten years after its recognition as an essential element (see Chapter 6), selenium was found to protect rats against the toxic effects of mercuric chloride (Pařízek and Oštádalová, 1967). It was demonstrated in livers of marine mammals living in areas polluted with mercury that mercury and selenium were present in a 1:1 stoichiometric ratio (Koeman *et al.*, 1973; 1975) and associated with proteins (Lunde, 1972). This 1:1 ratio was also observed in kidney, thyroid, pituitary and brain tissues of miners exposed to mercury (Kosta, *et al.*, 1975). For this reason it has frequently been assumed that detoxification of mercury by selenium can be explained by formation of a 1:1 complex that does not release mercury.

It is well known that selenium may act as a barrier against the poisoning effects of mercury, and that in some fish, e.g. tuna and swordfish, it is able to protect experimental animals against methylmercury toxicity (Ganther *et al.*, 1972; Ganther and Sunde, 1974; Friedman *et al.*, 1978).

As in most marine fish the amount of selenium in muscle tissue is in (stoichiometric) excess to that of mercury (Freeman *et al.*, 1978; Luten *et al.*, 1980), it is frequently thought that consumption of marine fish does not expose the consumer to toxic mercury compounds. If mercury is in excess, which is sometimes the case in heavily polluted fish, the selenium in the body of the consumer may act as a second barrier. As in the miners described by Kosta *et al.* (1975), selenium may accumulate in the body in order to bind mercury compounds that might be ingested. It will be clear that these mechanisms are inadequate if large doses of mercury are administered.

As small amounts of selenite (0.35–6 mg of Se per kg) are able to decrease, to some extent and for some time, the toxic effect of methylmercury (20 mg of Hg per kg) in Japanese quails (El-Begearmi *et al.*, 1977), the conclusion seems justified that selenium is able to decrease methylmercury toxicity in another way than by formation of a 1:1 complex. It should be kept in mind, however, that absorption and distribution of selenium and mercury compounds in the body may follow different pathways, which may result in considerable effects of comparatively low amounts of selenium. In a kinetic study with guinea-pigs it was shown that the nature of the phenomenon is complicated (Komsta-Szumska and Miller, 1984).

Cuvin-Aralar and Furness (1991), stating that the protective effect of selenium against mercury toxicity is not well understood, sum up some possible mechanisms which, in part, may explain this protective effect: (i) formation of a mercury–selenium complex; (ii) mercury redistribution within the organism, i.e. from more sensitive organs to less sensitive sites, promoted by selenium; (iii) competition between mercury and selenium for binding sites, e.g. some receptors located in tissues; (iv) conversion of toxic mercury compounds to less toxic forms, e.g. demethylation of methylmercury; and (v) alleviating of the inhibitory effect of methylmercury on glutathione peroxidase activity. It is known that arsenic compounds and sulfur-containing amino acids may also take part in the mercury–selenium interaction (El-Begearmi *et al.*, 1982).

In contrast to liver, no correlation or only a poor correlation is found between mercury and selenium levels in muscle tissues of marine organisms including fish, crustaceans and cephalopods (Cannizzaro *et al.*, 1986; Yamanobe *et al.*, 1987; 1988; Chvojka *et al.*, 1990; Barghigiani *et al.*, 1991; Guns *et al.*, 1992b). Barghigiani *et al.* (1991) point out that most of the selenium in fish muscle is water-soluble and does not form stable compounds with methylmercury. It is suggested that this compound might be *bis*-methylmercury selenide. Cuvin-Aralar and Furness (1991) mention the formation of $(CH_3Hg)_2Se$ under circumstances where selenium compounds are reduced to selenide. A possible role of this compound, as the result of methylmercury detoxification, can be excluded because of its instability. The protective effect of the mercury–selenium relation has to be searched for in the liver (Barghigiani *et al.*, 1989). In this respect it is worth noting that selenite does not protect mussels against mercury toxicity (Micallef and Tyler, 1990).

It should also be noted that Hansen (1988) is of the opinion that selenium is of no benefit in cases where people are exposed to inorganic mercury or mercury vapours. Apart from the question whether this hypothesis is right or not, this is not contradictory to the observation that selenium compounds from fish muscle are able to protect against toxicity of the methylated mercuric ion present in fish.

Siegel *et al.* (1991), examining five biological systems including tilapia and shrimp, found a protective action of an excess of selenite against mercuric chloride (100 μM of selenite vs. 4 μM of $HgCl_2$), but a comparable protection was observed from sulfite, tellurite, glutathione, DL-methionine, DL-selenomethionine and thiamine added in the same molar 25:1 ratio to $HgCl_2$.

Selenium levels in marine fish are given in Chapter 6.

Arsenic

A short section has to be devoted to the element arsenic, which is reported to be present at high levels in some flatfish species, and particularly in some crustaceans, i.e. in the range of 10–50 mg kg^{-1} and in some cases even higher (Chapman, 1926; LeBlanc and Jackson, 1973; Edmonds and Francesconi, 1977; Edmonds *et al.*, 1977; Luten *et al.*, 1982). These high levels are caused by the ingestion of brown kelp (*Ecklonia radiata*), which is able to bind large quantities of arsenic (Edmonds and Francesconi, 1981). These arsenic compounds are converted by the fish to a practically harmless compound called arsenobetaine (Edmonds *et al.*, 1977). After ingestion by humans or other animals it is excreted without any change within a short time (Yamauchi and Yamamura, 1984) and found to be of no toxicological significance (Kaise *et al.*, 1985).

In freshwater fish only trace amounts of arsenic, i.e. below 0.1 mg kg^{-1}, were observed (Ramelow *et al.*, 1989). In molluscs, arsenic levels were found to be in the range of 0.5–1.5 mg kg^{-1} (Vyncke *et al.*, 1992). Others report 0.2–0.4 mg kg^{-1} for marine fish (Gutenmann *et al.*, 1988). Inorganic arsenic species represent but a minor part of the total amount of arsenic present (Lunde, 1973).

Xenobiotics

Many substances that cause environmental pollution are intentionally produced by man because of their peculiar properties, e.g. pesticides and PCBs, while others are unintentionally formed as a result of man's acts (e.g. dioxins). Artificial radioisotopes, originating from human-induced nuclear reactions, may also be considered to belong to this category.

Organohalogen compounds

Halogenated hydrocarbons such as DDT and the PCBs have been detected in fish all over the world. These compounds, for a large part, enter the environment through wastewater discharges, which are an important route to surface water and the marine environment. Another route is long-range atmospheric transport (Swackhamer and Hites, 1988), which is now considered more important, for some compounds, than was thought before.

Halogenated hydrocarbons accumulate in sediments because of their low solubility in water and their tendency to adhere to particulate matter. They are present in all systems of the aquatic food chain but in particular in the fatty tissues of predator fish, where they accumulate, in some cases, in considerable amounts.

A number of halogenated hydrocarbons have been detected in fish, e.g. polychlorinated biphenyls (PCBs); polychlorinated naphthalenes; tetrachlorobenzyltoluenes; pesticides such as DDT, cyclodienes, toxaphene, hexachlorobenzene (HCB), hexachlorocyclohexanes (HCHs); octachlorostyrene; and not forgetting partly oxygenated components such as polychlorinated dibenzodioxines and dibenzofurans (PCDDs and PCDFs; commonly indicated together as 'dioxins'), etc. In quantitative terms, PCBs are the major contaminants, but, in Atlantic fish caught west of Ireland and Scotland, toxaphene concentrations are sometimes higher than those of PCBs (de Boer and Wester, 1993). It has been stated, however, that more than 90% of the extractable organically bound chlorine in fish is present in compounds which have not yet been identified (Wicklund-Glynn and Slanina, 1992). This is in line with recent findings of Newsome *et al.* (1993), who examined the total organochlorine content of fish from the Great Lakes (USA). They found most of the organic chlorine to reside in compounds of high molecular weight and refer to earlier publications in which it is suggested that humic matter or chlorinated lignins may be responsible for the presence of these compounds in fish lipids. It is also thought that these high molecular weight compounds are degraded to smaller chlorine-containing molecules which accumulate in fish tissue.

DDT has been used as a pesticide for several decades. As this pesticide is banned in most countries now, the environmental load of DDT has been gradually decreased. In many marine areas and even freshwater areas levels are well below 0.01 mg kg^{-1} in whole fish muscle (Lockerbie and Clair, 1988; de Boer and Wester, 1991). (The levels include isomers such as *o,p'*-and *o,o'*-DDT, and metabolites, mainly *p,p'*-DDE.) Fish from rivers and lakes situated in areas where, in the past, DDT was frequently applied, still contain levels in the range of 0.01–0.10 mg kg^{-1} (Arruda *et al.*, 1988; McDougall *et al.*, 1989). Elevated levels of DDT are also found in fish from the Baltic Sea (Kruse and Krüger, 1989).

Traces of cyclodienes are found in fish living in or close to areas where these pesticides were applied in the past. Some of these were epoxides (dieldrin, heptachlor epoxide) while others (aldrin, heptachlor) are metabolized to epoxides. For this reason, cyclodiene residues, with the exception of chlordane, are mostly found as epoxides (Arruda *et al.*, 1988; Swackhamer and Hites, 1988; McDougall *et al.*, 1989). In some individual fish the dieldrin content is unexpectedly high (McDougall *et al.*, 1989), but it is mostly below 0.1 mg kg^{-1}, if at all detectable. In some investigations HCH isomers are found (Rico *et al.*, 1987; Lockerbie and Clair, 1988).

The comparatively high amount of organochlorine compounds in organisms living in the Baltic Sea should be mentioned again, although decreases were observed by 1980 (Moilanen *et al.*, 1982). A further decrease was observed from 1980 onwards (Oehlenschläger and Karl, 1992), but not for HCB. Both HCB and octachlorostyrene are found, at elevated levels, in fish from the River Elbe (Kruse and Krüger, 1989). PCB levels in eel, however, were recently reported to be in the range of only one tenth of the limits laid down in the German directive for tolerable levels of PCBs in eel (Karl and Lehmann, 1993).

Levels of *p,p'*-DDE are significantly correlated with age in lake trout (Gutenmann *et al.*, 1992). This study, however, concerned only one species from one lake. It was observed by Swackhamer and Hites (1988) that the levels of organochlorine compounds, as a function of fish age or size, were not similar between species and inconsistent among compounds. These authors conclude that such differences indicate that physicochemical properties alone do not determine bioaccumulation of these compounds.

The accumulation of toxaphene, i.e. the common name for a mixture of, mainly, polychlorinated terpenes (PCTs), gains more and more attention because of its ubiquitous presence, in comparatively high amounts, in lipid tissues of fish (e.g. Newsome and Andrews, 1993). Being an insecticide, toxaphene was used in the past for the protection of non-food (cotton) crops. Its presence in fish from waters supposed to be unpolluted was first reported, amongst some other organochlorine pollutants, by Zell and Ballschmiter (1980). The presence of PCTs in Canadian east coast fish, in concentrations up to 12 mg kg^{-1} on a fat basis, was demonstrated in 1983 by Musial and Uthe. Later on, van der Valk and Wester (1991) found comparable levels in fish from northwestern Europe. As PCTs have never been used, on a large scale, in this part of the world, transport through the air was suggested (de Boer and Wester, 1993). Because of the supposed high toxicity of PCT mixtures, attention has to be paid to establish tolerance levels for these compounds in fatty fish.

As a result of the diversity of the uses of polychlorinated biphenyls (PCBs) these products are widely spread through the environment. They are ubiquitously present but, with regard to fish, in comparatively higher amounts in some inland waters (e.g. in The Netherlands and in Sweden). In heavily polluted fish, concentrations may reach 6 mg kg^{-1} in fish muscle (Wicklund-Glynn and Slanina, 1992) and even higher. Generally, increased levels of PCBs are found in river fish rather than in marine fish (Kruse and Krüger, 1989).

It is frequently stated that PCBs even pollute organisms in remote areas such as the Antarctic. In sub-Antarctic fish and other animals, however, the levels for PCBs and for many other organochlorine compounds are extremely low, e.g. $3\,\mu g\,kg^{-1}$ for muscle of Argentinian hake (de Boer and Wester, 1991).

It was observed that the *ortho*-unsubstituted (planar) PCBs show a considerable toxicity and, though to a less extent, the mono-*ortho* and some di-*ortho* substituted PCBs as well. The toxic effect of PCBs on organisms can principally be attributed to these groups of congeners (de Voogt *et al.*, 1990). These PCBs, in particular the planar congeners, represent but a minority in the total PCB mixture but contribute more than 90% of the toxicity of PCB preparations.

It should be mentioned that PCB levels in polluted areas, with some exceptions, have hardly decreased up to now. Less chlorinated PCBs, as a rule, are more easily degradable than the more chlorinated congeners; this results in a shift in the composition of PCBs in animal tissues towards these more chlorinated congeners as compared to the PCB mixtures that are originally released in the environment. National regulatory limits for PCBs in fish and shellfish range from 0.5 to $5\,mg\,kg^{-1}$ (Wicklund-Glynn and Slanina, 1992), mostly on a fat basis. Six or seven congeners have now been determined which are considered to be representative of the total amount of PCBs. It should be mentioned that, for this total amount, the interest is decreasing.

Despite their low solubulity in water, PCBs may be leached out from fish by boiling in water. In this way up to 35% of the PCBs present were removed from blue crab muscle. Most of these PCBs were retained in the cooking water (Zabik *et al.*, 1992).

The presence of dibenzo-*p*-dioxins and dibenzofurans, as far as is known at the moment, is mainly a problem of local contamination. Hellou and Payne (1993) state that there is little information available on dioxin levels in fish populations far from local pollution sources. Bergqvist *et al.* (1989) examined the levels of dioxin equivalents (Safe, 1990) in cod and herring from seas around Sweden and found values close to or below the limit of quantification, i.e. 0.2 ng of dioxin equivalents (TEQs) per kg (wet weight base). In herring, the levels range from 1.8 to $9.0\,ng\,kg^{-1}$, the highest levels being found in samples from the Baltic Sea.

Dioxins and dibenzofurans exert their action by binding to the cytosolic *Ah*-receptor, which is thought to control the synthesis of many cytosolic enzymes. Because of the steric similarity of non-*ortho* and mono-*ortho* substituted PCBs it is assumed that these components act in the same way. A competition for binding sites has been suggested as well (Haake *et al.*, 1987; Biegel *et al.*, 1989).

The content of planar PCBs in animals and tissue is often expressed in TEQs nowadays. In this context it should be noted that, in Dutch eel and probably in many other fish as well, the amount of TEQs from planar PCBs by far exceeds that from dioxins (H. Pieters, 1993, personal communication).

Organobromine compounds, e.g. polybrominated phenyl ethers (PBBE), widely used as flame retardants in materials such as textiles and paints, may be present at low levels (up to 0.02 mg kg^{-1}) in fish and mussels from the seashore (Watanabe *et al.*, 1987). It has to be stated that, in earlier times, much higher levels were observed in pike from polluted Swedish waters, and that these levels even exceeded those of PCBs (Andersson and Blomkvist, 1981).

Organotin compounds

Trialkyl and triaryl tin compounds have been increasingly used, from the early 1970s on, as biocides and antifouling paints because of their excellent ability to prevent marine organisms from becoming encrusted on ship bottoms and culturing nets (Suzuki *et al.*, 1992). As a result, however, they have caused pollution of the environment.

As organotin compounds show large bioconcentration factors in marine organisms (Yamada and Takayanagi, 1992), the concentrations that may accumulate in fish should not be neglected. In Japan, bis(tributyltin)oxide was found, in cultured fish and in bay or inshore fish and shellfish, at levels from 0.01 to 0.50 mg kg^{-1} (Takeuchi *et al.*, 1987a; 1987b). It was also observed that levels of tributyl tin (TBT) in fish are gradually increasing and that those of triphenyl tin (TPT) are somewhat decreasing, these trends being consistent with the amounts of these compunds used in antifouling paints (Takeuchi *et al.*, 1991).

As it has been demonstrated that organotin compounds exert adverse effects on the marine fauna even at low concentrations, the use of these compounds has been restricted in the USA in order to reduce levels in coastal marine waters (Short and Sharp, 1989). Uhler *et al.* (1993) found decreases in tissue levels of tributyl tin at US west coast sites while levels at east coast sites fluctuated. In oysters from the Gulf of Mexico, Garcia-Romero *et al.* (1993) found a decline, from 1989 to 1991, in sites where the concentrations were relatively low already, but at concentrations exceeding 400 μg tin per kg hardly any decrease was observed.

Metabolism of TBT involves hydroxylation of alkyl groups (Suzuki *et al.*, 1992) and removal of at least one alkyl group (Uhler *et al.*, 1993). It has been suggested that TBT is rapidly metabolized and that TPT tends to accumulate in tissues because of a much slower rate of metabolism (Mizuishi *et al.*, 1991).

Polycyclic aromatic hydrocarbons

It is questionable to classify these compounds under the xenobiotics as they may be formed as a result of naturally occurring events such as, for example, forest fires. It should be realized, however, that this may even hold for dioxins. However, as the presence of polycyclic aromatic hydrocarbons (PAHs) is usually the result of man's actions, these components may also be regarded as xenobiotic substances.

The major part of PAHs in aquatic systems is associated with particles and they accumulate in sediments but, to a minor extent, are dissolved in water. From there they are taken up by aquatic organisms (Alexander *et al.*, 1992).

There seems to be a diffuse background of PAHs which leads to low levels in aquatic organisms. Local contamination may lead to elevated levels in fish and in particular molluscs, where the content of benzo(a)pyrene frequently exceeds 1 $\mu g\,kg^{-1}$ (Alexander *et al.*, 1992). Such amounts are also observed in traditionally smoked fish (see Chapter 11).

Radionuclides

Human-induced radioactivity in fish may be caused by leakages from nuclear reactors, reprocessing plants and by fall-out resulting from nuclear explosions and by disasters such as the Chernobyl accident in 1986.

The most important radionuclides that may accumulate in the food chain are strontium-90 (^{90}Sr) and caesium-137 (^{137}Cs). As a result of the high ionic strength of sea water the absorption of ions of these isotopes by marine organisms is low. In freshwater fish, however, the uptake will be higher. Fish from lakes in the most northern parts of Europe, which received high amounts of fall-out from the Chernobyl emission, show activities exceeding 10,000 $Bq\,kg^{-1}$, which makes the fish unfit for consumption (Sandberg, 1992). As the emission of ^{90}Sr from Chernobyl was comparatively low, the activity in these fish mainly originates from ^{137}Cs. Compared to warm-blooded animals, the biological half-life of this radionuclide in fish is much longer and increases with decreasing water temperature (Sandberg, 1992), resulting in biological half-lives of 4 to 5 years or perhaps even longer. In contrast, fish from lakes in the southern part of Sweden need not be considered as dangerous.

In fillets from fish caught in the North Sea the ^{137}Cs activity is between 1 and 10 $Bq\,kg^{-1}$ and, in most samples, lower than 2 $Bq\,kg^{-1}$ (Kanisch and Nagel, 1991). The radioactivity in this area is mainly due to emissions from the reprocessing plants in Sellafield (UK) and La Hague (France). As these emissions have largely decreased in the past 15 years, radioactivity in fish is decreasing as well. In 1982, values of 14 $Bq\,kg^{-1}$ were still found in cod, while these values were around 2 $Bq\,kg^{-1}$ in

1991. The role of the Chernobyl accident (1986) in this contamination was insignificant.

In fillets from gadoid fish caught in Nordic areas (Faeroe Islands, North Cape, Bear Island) the ^{137}Cs activity was in the range of 0.3–0.9 Bq kg^{-1} (Kellermann, 1992).

As strontium is mainly incorporated in bones and in the skin, the activity of ^{90}Sr in fillets is low. Values between 0.002 and 0.05 Bq kg^{-1} are observed (Kanisch and Nagel, 1991). These values are dependent on the bone content.

Values for radiation from plutonium isotopes (^{239}Pu and ^{240}Pu) in North Sea fish are also very low, i.e. below 1 mBq kg^{-1} (Kanisch and Nagel, 1991). Alpha radiation from other isotopes was not measured.

Gouvea *et al.* (1987) found surprisingly high levels of polonium-210 radiation in soft tissues from bivalves, i.e. about 1000 Bq per kg dry weight. Due to higher concentrations of radium-226 in the open sea, the values found there were higher than values obtained from bivalves in a bay region. It is difficult to draw conclusions from these observations.

Drug Residues in Cultivated Fish

Because of a high infection pressure, fish diseases are a serious problem in fish reared in ponds, basins or net pens. For this reason a number of drugs are frequently applied either for preventive or for curative purposes. These drugs are mostly mixed with the feed, but treatments to fight ectoparasites are usually carried out by bathing the fish (Yndestad, 1993).

As is the case with many animals kept on farms such as cattle, swine and poultry, these treatments unavoidably lead to drug residues in the edible parts of fish. These residues may be present in the mg per kg range, and the rate of absorption, biotransformation and excretion determines in what time the concentrations of drug residues, including metabolites and conjugates, are reduced to amounts that are of no physiological importance. For the setting of withdrawal periods in order to reach such low concentrations it has to be realized that there is a large variation in pharmacokinetic parameters between fish species (Kleinow *et al.*, 1992). Furthermore, external parameters such as salinity, pH and, in particular, water temperature influence absorption, metabolism and excretion of drugs (Yndestad, 1993). It was stated that, in poikilothermic animals, a temperature increase of 10°C results in a doubling of the metabolic rate (Ellis *et al.*, 1978). Later on it was demonstrated, from several studies, that elimination half-lives are longer at low than at high water temperatures (Schmid, 1980; Borgan *et al.*, 1981; Salte and Liestøl, 1983; Schmidt *et al.*, 1986).

As for the depletion of drugs (or disinfectants) from fish, large differences exist between the various compounds. To mention two extremes as examples, the ectoparasiticide praziquantel disappears completely after not more than 7 days from the end of treatment, at 10°C. Malachite green, on the other hand, has a half-life of 40 days at 9.7°C in fish with a high fat content (von Bauer *et al.*, 1988).

A drug such as furazolidone, which rapidly breaks down in terrestrial production animals, is rapidly degraded in fish as well (Wölwer-Rieck, 1990; Lu and Fang, 1992)). The same holds for virginiamycin (Cravedi, 1990). Mebendazole can be detected for up to 14 days in eels cultivated at 25°C but disappears, for the greater part, within 3 days (Mellergaard *et al.*, 1990). Residues of quinolones seem to be bound to skin and/or bone, which results in prolonged half-lives (Steffenak *et al.*, 1991). These residues are not degraded during cooking of the fish (Steffenak *et al.*, 1993). For oxytetracycline, which also binds to bone, long elimination periods were found for a variety of species (Grondel *et al.*, 1987; Jacobsen, 1989; Chikova, 1992; Uno *et al.*, 1992; Nouws *et al.*, 1993; Wölwer-Rieck *et al.*, 1993). At 30–32°C, however, this period is considerably shorter (Chai-anan *et al.*, 1993).

There is a great need for further pharmacokinetic and depletion studies (Kleinow *et al.*, 1992). Such studies should be performed under varying conditions, in different species of fish, with different drugs, and at least at one low and one high water temperature. There is also a great need for setting maximum residue levels (MRLs) on a scientific basis, which includes the evaluation of toxicological and other relevant factors. This work is in progress in both the Codex Alimentarius and the European Union (Yndestad, 1993).

Contamination of non-treated fish in adjacent net pens, or of wild fish swimming around the nets, should not be left out of consideration (Samuelson *et al.*, 1993). An additional problem is that some of the medicated fish die. These are collected and, for example, ensilaged. Some residues will persist in the silage and might hinder its exploitation.

Conclusions

Many types and kinds of contaminants can be found in fish and some of these require special attention, e.g. PCBs in freshwater fish, toxaphene in marine fish and, in local areas, some other organochlorine compounds such as hexachlorobenzene and octachlorostyrene. This particularly holds for fatty fish, which are able to accumulate considerable amounts of these lipophilic compounds.

As a rule, heavy metals in fish do not give rise to much concern, with the exception of mercury in waters heavily contaminated with mercury compounds. Further elucidation of the mercury–selenium interaction is necessary for a better understanding of the alleviating or even protective effect of selenium on exposure to mercury.

It has to be emphasized that residues of veterinary drugs and related compounds in cultivated fish should not be neglected.

Most fish, however, contain very low levels of contaminants. Therefore it can be agreed that 'this, together with the fish consumption habits, implies that the vast majority of consumers do not need to restrict their consumption. On the contrary, many people should instead eat more fish' (Slorach, 1992).

References

Alexander, J., Becher, G. and Dybing, E. (1992) Polycyclic aromatic hydrocarbons (PAH) in fish. In: *Fish as Food*. Nordiske Seminar- og Arbejdsrapporter 1992:568. Nordic Council of Ministers, Copenhagen, 1992, pp. 181–194.

Andersson, Ö. and Blomkvist, G. (1981) Polybrominated aromatic pollutants found in fish in Sweden. *Chemosphere* 10, 1051–1060.

Arruda, J.A., Cringan, M.S., Layher, W.G., Kersh, G. and Bever, Ch. (1988) Pesticides in fish tissue and water from Tuttle Creek Lake, Kansas. *Bulletin of Environmental Contamination and Toxicology* 41, 617–624.

Barak, N.A.-E. and Mason, C.F. (1990a) Mercury, cadmium and lead in eels and roach: the effects of size, season and locality on metal concentrations in fish and liver. *The Science of the Total Environment* 92, 249–256.

Barak, N.A.-E. and Mason, C.F. (1990b) Mercury, cadmium and lead concentrations in five species of freshwater fish from Eastern England. *The Science of the Total Environment* 92, 257–263.

Barghigiani, C. and de Ranieri, S. (1992) Mercury content in different size classes of important edible species of the Northern Tyrrhenian Sea. *Marine Pollution Bulletin* 24, 114–116.

Barghigiani, C., Pellegrini, D. and Carpene, E. (1989) Mercury binding proteins in liver and muscle of flat fish from the Northern Tyrrhenian sea. *Comparative Biochemistry and Physiology* 94C, 309–312.

Barghigiani, C., Pellegrini, D., D'Ulivo, A. and Ranieri, S. de (1991) Mercury assessment and its relation to selenium levels in edible species of the Northern Tyrrhenian sea. *Marine Pollution Bulletin* 22, 406–409.

von Bauer, K., Dangschat, H., Knöppler, H.O. and Neudegger, J. (1988) Aufnahme und Ausscheidung von Malachitgrün bei Regenbogenforellen. *Archiv für Lebensmittelhygiene* 39, 97–102.

Bergqvist, P.-A., Bergek, S., Hallbäck, H., Rappe, C. and Slorach, S.A. (1989) Dioxins in cod and herring from the seas around Sweden. *Chemosphere* 19, 513–516.

Biegel, L., Harris, M., Davis, D., Rosengren, R., Safe, L. and Safe, S. (1989) 2,2′,4,4′,5,5′-hexachlorobiphenyl as a 2,3,7,8-tetrachlorodibenzo-*p*-dioxin antagonist in C57BL/6J mice. *Toxicology and Applied Pharmacology* 97, 561–571.

de Boer, J. and Wester, P. (1991) Chlorobiphenyls and organochlorine pesticides in various sub-Antarctic organisms. *Marine Pollution Bulletin* 22, 441–447.

de Boer, J. and Wester, P. (1993) Determination of toxaphene in human milk from Nicaragua and in fish and marine mammals from the Northeastern Atlantic and the North Sea. *Chemosphere* 27, 1879–1890.

Borgan, A., Ødegaard, S. and T. Bergsjø (1981) Temperature related absorption and excretion of sulphadimidine in rainbow trout, *Salmo gairdneri. Acta Veterinaria Scandinavica* 22, 211–217.

Cannizzaro, C., Madarena, G., Dazzi, G., Chizzolini, G. and Campanini, G. (1986) [Mercury and selenium levels in fishing catches from the port of Pescara.] *Annali della Facoltà di Medicina Veterinaria, Università di Parma* 6, 187–204; ref. *Food Science and Technology Abstracts* 20 (1988) 10R2.

Chai-anan, P., Limpoka, M., Sittiphuprasert, U., Vichkovitten, T., Wongsommart, D. and Rongrodejanarak, C. (1993) Absorption, distribution and excretion of sulfamethazine, oxytetracycline and oxolinic acid in giant seaperch (*Lates calcarifer*) after single oral dosing and 8 days on medicated diet. In: Haagsma, N., Ruiter, A. and Czedik-Eysenberg, P.B. (eds) *Proceedings of the EuroResidue II Conference*, Veldhoven, The Netherlands, pp. 231–235.

Chapman, A.C. (1926) On the presence of compounds of arsenic in marine crustaceans and shell fish. *The Analyst* 51, 548–563.

Chikova, V. (1992) [Antibiotic residues in fish.] *Khranitelna Promishlenost* 40(2), 32–33; ref. *Food Science and Technology Abstracts* 24 (1992) 11R11.

Chvojka, R., Williams, R.J. and Fredrickson, S. (1990) Methyl mercury, total mercury, and selenium in snapper from two areas of the New South Wales coast, Australia. *Marine Pollution Bulletin* 21, 570–573.

Cravedi, J.-P. (1990) Residue pharmacokinetics of virginiamycin in rainbow trout. In: Haagsma, N., Ruiter, A. and Czedik-Eysenberg, P.B. (eds) *Proceedings of the EuroResidue Conference*, Noordwijkerhout, The Netherlands, pp. 139–143.

Cuvin-Aralar, M.L.A. and Furness, R.W. (1991) Mercury and selenium interaction: a review. *Ecotoxicology and Environmental Safety* 21, 348–364.

De Clerck, R., Guns, M., Vyncke, W. and Hoeyweghen, P. Van (1988) Zware metalen in mariene organismen uit de lerse Zee (Heavy metals in marine species from the Irish sea). *Landbouwtijdschrift* 41, 211–217.

Edmonds, J.S. and Francesconi, K.A. (1977) Methylated arsenic from marine fauna. *Nature* 265, 436.

Edmonds, J.S. and Francesconi, K.A. (1981) Arseno-sugars from brown kelp (*Ecklonia radiata*) as intermediates in cycling of arsenic in a marine ecosystem. *Nature* 289, 602–604.

Edmonds, J.S., Francesconi, K.A., Cannon, J.R., Raston, C.L., Skelton, B.W. and White, A.H. (1977) Isolation, crystal structure and synthesis of arsenobetaine, the arsenical constituent of the Western rock lobster *Panulirus longipes cygnus* George. *Tetrahedron Letters* 18, 1543–1546.

El-Begearmi, M.M., Sunde, M.L. and Ganther, H.E. (1977) A mutual protective effect of mercury and selenium in Japanese quail. *Poultry Science* 56, 313–322.

El-Begearmi, M.M., Ganther, H.E. and Sunde, M.L. (1982) Dietary interaction between methylmercury, selenium, arsenic and sulfur amino acids in Japanese quail. *Poultry Science* 61, 272–279.

Ellis, A.E., Roberts, R.J. and Tytler, P. (1978) The anatomy and physiology of teleosts. In: Roberts, R.J. (ed.) *Fish Pathology*. Baillière Tindall, London, pp. 13–54.

Francesconi, K.A. and Lenanton, R.C.J. (1992) Mercury contamination in a semi-enclosed marine embayment: organic and inorganic mercury content of biota, and factors influencing mercury levels in fish. *Marine Environmental Research* 33, 189–212.

Freeman, H.C., Shum, G. and Uthe, J.F. (1978) The selenium content in swordfish (*Xiphias gladius*) in relation to total mercury content. *Journal of Environmental Science and Health A* 13, 235–240.

Friedman, M.A., Eaton, L.R. and Carter, W.H. (1978) Protective effects of freeze dried swordfish on methylmercury chloride in rats. *Bulletin of Environmental Contamination and Toxicology* 19, 436–443.

Ganther, H.E. and Sunde, M.L. (1974) Effect of tuna fish and selenium on the toxicity of methylmercury: A progress report. *Journal of Food Science* 39, 1–5.

Ganther, H.E., Goudie, C., Sunde, M.L., Kopecky, M.J., Wagner, P., Sang-Hwan Oh and Hoekstra, W.G. (1972) Selenium: Relation to decreased toxicity of methylmercury added to diets containing tuna. *Science* 175, 1122–1124.

Garcia-Romero, B., Wade, T.L., Salata, G.G. and Brooks, J.M. (1993) Butyltin concentrations in oysters from the Gulf of Mexico from 1989 to 1991. *Environmental Pollution* 81, 103–111.

Gouvea, R.C., Santos, P.L. and Gouvea, V.A. (1987) Contribution to the study of radioactivity in marine organisms: dosage of ^{210}Po in *Perna perna*, L. *The Science of the Total Environment* 61, 117–120.

Grondel, J.L., Nouws, J.F.M., Jong, M. de, Schutte, A.R. and Driessens, F. (1987) Pharmacokinetics and tissue distribution of oxytetracycline in carp, *Cyprinus carpio* L., following different routes of administration. *Journal of Fish Diseases* 10, 153–163.

Guns, M., Clerck, R. De, Vyncke, W. and Hoeyweghen, P. Van (1984) Verdere studie van het gehalte aan zware metalen in mariene organismen uit de Noordzee (Further studies on the heavy metal content of marine species from the North Sea). *Landbouwtijdschrift* 37, 315–322.

Guns, M., Hoeyweghen, P. Van, Vyncke, W. and Clerck, R. De (1992a) Selenium assessment and its relation to mercury levels in fish, shrimp and mussels from Belgian continental shelf waters. *Landbouwtijdschrift/Revue de l'Agriculture* 45, 731–739.

Guns, M., Hoeyweghen, P. Van, Vyncke, W. and Clerck, R. De (1992b) Mercury concentrations in plaice, flounder and dab from Belgian continental shelf waters. *Landbouwtijdschrift/Revue de l'Agriculture* 45, 959–964.

Gutenmann, W.H., Bache, C.A., McCahan, J.B. and Lisk, D.J. (1988) Heavy metals and chlorinated hydrocarbons in marine fish products. *Nutrition Reports International* 38, 1157–1161.

Gutenmann, W.H., Ebel, J.G., Jr, Kuntz, H.T., Yourstone, K.S. and Lisk, D.J. (1992) Residues of *p,p'*-DDE and mercury in lake trout as a function of age. *Archives of Environmental Contamination and Toxicology* 22, 452–455.

Gutiérrez-Galindo, E.A., Muñoz, G.F. and Flores, A.A. (1988) Mercury in freshwater fish and clams from the Cerro Prieto geothermal field of Baja California, México. *Bulletin of Environmental Contamination and Toxicology* 41, 201–207.

Haake, J.M., Safe, S., Mayura, K. and Phillips, T.D. (1987) Aroclor 1254 as an antagonist of the teratogenicity of 2,3,7,8-tetrachlorodibenzo-*p*-dioxin. *Toxicology Letters* 38, 299–306.

Hagel, P. and Vos, G. (1989) Zware metalen in visserijprodukten (Heavy metals in fishery products). *Landbouwkundig Tijdschrift* 101(2), 39–42.

Håkanson, L., Andersson, T. and Nilsson, Å. (1990) Mercury in fish in Swedish lakes – Linkages to domestic and European sources of emission. *Water, Air and Soil Pollution* 50, 171–191.

Hansen, J.C. (1988) Has selenium a beneficial role in human exposure to inorganic mercury? *Medical Hypotheses* 25, 45–53.

Harms, U. (1990) Untersuchungen zur zeitlichen Veränderung der Quecksilbergehalte in Flundern aus der inneren Deutschen Bucht. *Informationen für die Fischwirtschaft* 37(3), 126–129.

Hellou, J. and Payne, J.F. (1993) Polychlorinated dibenzo-*p*-dioxins and dibenzofurans in cod (*Gadus morhua*) from the Northwest Atlantic. *Marine Environmental Research* 36, 117–128.

Hellou, J., Fancey, L.L. and Payne, J.T. (1992a) Concentrations of twenty-four elements in bluefin tuna, *Thynnus thynnus*, from the Northwest Atlantic. *Chemosphere* 24, 211–218.

Hellou, J., Warren, W.G., Payne, J.F., Belkhode, S. and Lobel, P. (1992b) Heavy metals and other elements in three tissues of cod, *Gadus morhua*, from the Northwest Atlantic. *Marine Pollution Bulletin* 24, 452–458.

Hernández-García, M.T., Martínez Para, M.C. and Masoud, T.A. (1988) Variación en la cantidad de mercurio en muestras de pescado sometidas a diversos procesos culinarios (Changes in the content of mercury in fish samples subjected to different cooking processes). *Anales de Bromatología* 40, 291–297.

Hutagalung, H.P. (1989) Mercury and cadmium content in green mussel, *Mytilus viridis* L., from Onrust waters, Jakarta Bay. *Bulletin of Environmental Contamination and Toxicology* 42, 814–820.

Jacobsen, M.D. (1989) Withdrawal times of freshwater rainbow trout, *Salmo gairdneri* Richardson, after treatment with oxolinic acid, oxytetracycline and trimetoprim. *Journal of Fish Diseases* 12, 29–36.

Kaise, T., Watanabe, S. and Itoh, K. (1985) The acute toxicity of arsenobetaine. *Chemosphere* 14, 1327–1332.

Kanisch, G. and Nagel, G. (1991) Radioaktivität in Fischen aus der Nordsee. *Informationen für die Fischwirtschaft* 38(4), 32–39.

Karl, H. and Lehmann, I. (1993) Organochlorine residues in the edible part of eels of different origins. *Zeitschrift für Lebensmittel-Untersuchung und -Forschung* 197, 385–388.

Kellermann, H.-J. (1992) Radioaktivität in Fischen aus dem Eingang zur Barentssee. *Informationen für die Fischwirtschaft* 39(3), 134–136.

Kleinow, K.M., James, M.O. and Lech, J.J. (1992) Drug pharmacokinetics and metabolism in food-producing fish and crustaceans – Methods and examples. In: Hutson, D.H., Hawkins, D.R., Paulson, G.D. and Struble, C.B. (eds) *Xenobiotics and Food-Producing Animals – Metabolism and Residues*. ACS Synposium Series No. 503. American Chemical Society, Washington, DC, pp. 98–130.

Klemmer, H.W., Unninayer, C.S. and Okubo, W.I. (1976) Mercury content of biota in coastal waters in Hawaii. *Bulletin of Environmental Contamination and Toxicology* 15, 454–457.

Klumpp, D.W. and Burdon-Jones, C. (1982) Investigations of the potential of bivalve molluscs as indicators of heavy metal levels in tropical marine waters. *Australian Journal of Marine Freshwater Research* 33, 285–300.

Koeman, J.H., Peeters, W., Koudstaal-Hol, C., Tjioe, P.S. and Haaften, J.L. van (1973) Mercury–selenium correlations in marine mammals. *Nature (London)* 245, 285–286.

Koeman, J.H., Van de Ven, W.S.M., De Goeij, J.J.M., Tjioe, P.S. and Haaften, J.L. van (1975) Mercury and selenium in marine mammals and birds. *Science of the Total Environment* 3, 279–287.

Komsta-Szumska, E. and Miller, D.R. (1984) A kinetic analysis of the interaction between methylmercury and selenium. *Toxicology* 33, 229–238.

Kosta, L., Byrne, A.R. and Zelenko, V. (1975) Correlation between selenium and mercury in man following exposure to inorganic mercury. *Nature (London)* 254, 238–239.

Krüger, K.-E. (1990) Der Quecksilbergehalt der Seefische: Erste Ergebnisse einer Trendstudie. *31. Tagung des Arbeitsgebietes 'Lebensmittelhygiene'*, Garmisch-Partenkirchen, 2–5 October 1990, pp. 292–300.

Kruse, R. and Krüger, K.-E. (1989) Kongenere polychlorierte Biphenyle (PCB's) und chlorierte Kohlenwasserstoffe (CKW's) in Fischen, Krusten-, Schalen- und Weichtieren und daraus hergestellten Erzeugnissen aus Nordatlantik, Nordsee, Ostsee und deutschen Binnengewässern. *Archiv für Lebensmittelhygiene* 40, 99–104.

Lauenstein, G.G., Robertson, A. and O'Connor, Th.P. (1990) Comparison of trace metal data in mussels and oysters from a mussel watch programme of the 1970s with those from a 1980s programme. *Marine Pollution Bulletin* 21, 440–447.

Law, A.T. and Singh, A. (1991) Relationships between heavy metal content and body weight of fish from the Kelang estuary, Malaysia. *Marine Pollution Bulletin* 22, 86–88.

Leah, R.T., Evans, S.J. and Johnson, M.S. (1992) Mercury in flounder (*Platichthys flesus* L.) from estuaries and coastal waters of the north-east Irish Sea. *Environmental Pollution* 75, 317–322.

LeBlanc, P.J. and Jackson, A.L. (1973) Arsenic in marine fish and invertebrates. *Marine Pollution Bulletin* 4(6), 88–90.

Lockerbie, D.M. and Clair, T.A. (1988) Organic contaminants in isolated lakes of Southern Labrador, Canada. *Bulletin of Environmental Contamination and Toxicology* 41, 625–632.

López-Artiguez, M., Soria, M.L. and Repetto, M. (1989) Heavy metals in bivalve molluscs in the Huelva estuary. *Bulletin of Environmental Contamination and Toxicology* 42, 634–642.

Lu, C.-F. and Fang, L.-S. (1992) [Furazolidone residues in bass, as determined by high performance liquid chromatography]. *Taiwan Journal of Veterinary Medicine and Animal Husbandry* No. 60, 21–26; ref. *Food Science and Technology Abstracts* 25 (1993) 12R27.

Luis, E.S., Balagot, A.H., Villaflor, A.C., Sanchez, F.C. and Develles, E.N. (1989) Pb, Cd and Hg contents of bivalves collected during the different seasons of the year. *Food Chemistry* 32, 239–255.

Lunde, G. (1972) Location of lipid-soluble selenium in marine fish to the lipoproteins. *Journal of the Science of Food and Agriculture* 23, 987–994.

Lunde, G. (1973) Separation and analysis of organic-bound and inorganic arsenic in marine organisms. *Journal of the Science of Food and Agriculture* 24, 1021–1027.

Luten, J.B., Ruiter, A., Ritskes, T.M., Rauchbaar, A.B. and Riekwel-Booy, G. (1980) Mercury and selenium in marine and freshwater fish. *Journal of Food Science* 45, 416–419.

Luten, J.B., Riekwel-Booy, G. and Rauchbaar, A.B. (1982) Occurrence of arsenic in plaice (*Pleuronectes platessa*), nature of organo-arsenic compound present and its excretion by man. *Environmental Health Perspectives* 45, 165–170.

Luten, J.B., Bouquet, W., Burggraaf, M.M., Rauchbaar, A.B. and Rus, J. (1986) Trace metals in mussels (*Mytilus edulis*) from the Waddenzee, coastal North sea and the estuaries of Ems, Western and Eastern Scheldt. *Bulletin of Environmental Contamination and Toxicology* 36, 770–777.

Marcovecchio, J.E. and Moreno, V.J. (1993) Cadmium, zinc and total mercury levels in the tissues of several fish species from La Plata river estuary, Argentina. *Environmental Monitoring and Assessment* 25, 119–130; ref. *Food Science and Technology Abstracts* 25 (1993) 8R19.

McDougall, K.W., Ahmad, N., Harris, C.R. and Higginson, F.R. (1989) Organochlorine insecticide residues in fish and birds from three river systems on the north coast region of New South Wales. *Bulletin of Environmental Contamination and Toxicology* 42, 884–890.

Mellergaard, S., Buchmann, K. and Bjerregaard, J. (1990) Mebendazole residues in eels after anthelmintic treatment in intensive eel culture systems. In: Haagsma, N., Ruiter, A. and Czedik-Eysenberg, P.B. (eds), *Proceedings of the EuroResidue Conference*, Noordwijkerhout, The Netherlands, pp. 272–275.

Micallef, S. and Tyler, P.A. (1990) Effect of mercury and selenium on the gill function of *Mytilus edulis*. *Marine Pollution Bulletin* 21, 288–292.

Mizuishi, K., Takeuchi, M., Yamanobe, H. and Nakamura, H. (1991) Hygienic chemical studies on organotin compounds. VIII. Distribution of tributyltin and triphenyl tin compounds in fish. *Annual Report of Tokyo Metropolitan Research Laboratory of Public Health* 42, 86–89; ref. *Food Science and Technology Abstracts* 24 (1992) 12R17.

Moilanen, R., Pyysalo, H., Wickström, K. and Linko, R. (1982) Time trends of chlordane, DDT, and PCB concentrations in pike (*Esox lucius*) and Baltic herring (*Clupea harengus*) in the Turku archipelago, Northern Baltic sea for the period 1971–1982. *Bulletin of Environmental Contamination and Toxicology* 29, 334–340.

Musial, C.J. and Uthe, J.F. (1983) Widespread occurrence of the pesticide toxaphene in Canadian East coast marine fish. *International Journal of Environmental Analytical Chemistry* 14, 117–126.

National Academy of Sciences (1980) *The International Mussel Watch*. Report of a workshop sponsored by the Environmental Studies Board Commission on Natural Resources. National Research Council, National Academy of Sciences, Washington, DC.

Newsome, W.H. and Andrews, P. (1993) Organochlorine pesticides and polychlorinated biphenyl congeners in commercial fish from the Great Lakes. *Journal of AOAC International* 76, 707–710.

Newsome, W.H., Andrews, P., Conacher, H.B.S., Rao, R.R. and Chatt, A. (1993) Total organochlorine content of fish from the Great Lakes. *Journal of AOAC International* 76, 703–706.

Nouws, J.F.M., Boon, J.H., Driessens, F., Mengelers, M.J.B., Booms, G.H.R. and Heijden, M.H.T. van der (1993) Residue depletion of oxytetracycline in European eel. In: Haagsma, N., Ruiter, A. and Czedik-Eysenberg, P.B. (eds) *Proceedings of the EuroResidue II Conference*, Veldhoven, The Netherlands, pp. 514–517.

Oehlenschläger, J. and Karl, H. (1992) Zur Belastung von Ostseefischen und anderen Meerestieren mit anorganischen und organischen Rückständen. *Deutsche Lebensmittel-Rundschau* 88, 115–116.

Oskarsson, A. (1992a) Methylmercury. In: *Fish as Food*. Nordiske Seminar-og Arbejdsrapporter 1992:568. Nordic Council of Ministers, Copenhagen, 1992, pp. 73–79.

Oskarsson, A. (1992b) Hälsoaspekter av metylkvicksilver i fisk. (Adverse health effects from methylmercury in fish.) *Vår Föda* 44, 148–154.

Pařízek, J. and Oštádalová, I. (1967) The protective effects of small amounts of selenite in sublimate intoxication. *Experientia* 23, 142–143.

Paulsson, K. and Lundbergh, K. (1989) The selenium method for treatment of lakes for elevated levels of mercury in fish. *The Science of the Total Environment* 87/88, 495–507.

Pieters, H., Hagel, P. and Ruiter, A. (1993) Kwik in milieu en voeding in Nederland. (Mercury in the environment and in food in The Netherlands.) Report, Commissie voor Metingen in het Milieu (Committee on Environmental Measurements; 'CCRX'). National Institute of Public Health and Environmental Protection (RIVM), Bilthoven, The Netherlands, 211 pp.

Ramelow, G.J., Webre, C.L., Mueller, C.S., Beck, J.N., Young, J.C. and Langley, M.P. (1989) Variations of heavy metals and arsenic in fish and other organisms from the Calcasieu river and Lake Louisiana. *Archives of Environmental Contamination and Toxicology* 18, 804–818.

Rico, M.C., Hernández, L.M., González, M.J., Fernández, M.A. and Montero, M.C. (1987) Organochlorine and metal pollution in aquatic organisms sampled in the Donaña National Park during the period 1983–1986. *Bulletin of Environmental Contamination and Toxicology* 39, 1076–1083.

Ruiter, A. (1985) Contaminants in meat and meat products. In: Lawrie, R. (ed.) *Developments in Meat Science – 3*. Elsevier Applied Science Publishers, London/New York, pp. 185–220.

Safe, S. (1990) Polychlorinated biphenyls (PCBs), dibenzo-*p*-dioxins (PCDDs), dibenzofurans (PCDFs), and related compounds: environmental and mechanistic considerations which support the development of toxic equivalency factors (TEFs). *Critical Reviews in Toxicology* 21, 51–88.

Salte, R. and Liestøl, K. (1983) Drug withdrawal from farmed fish – Depletion of oxytetracycline, sulfadiazine, and trimethoprim from muscular tissue of rainbow trout (*Salmo gairdneri*). *Acta Veterinaria Scandinavica* 24, 418–430.

Samuelsen, O.B., Lunestad, B.T., Thorsen, B., Eriksen, V., Ervik, A. and Solheim, E. (1993) Residues and effects of oxolinic acid in wild fauna following medication in fish farms. In: Haagsma, N., Ruiter, A. and Czedik-Eysenberg, P.B. (eds) *Proceedings of the EuroResidue II Conference*, Veldhoven, The Netherlands, pp. 606–610.

Sandberg, E. (1992) Radionuclides. In: *Fish as Food.* Nordiske Seminar-og Arbejdsrapporter 1992:568. Nordic Council of Ministers, Copenhagen, 1992, pp. 129–134.

Schmid, A. (1980) Arzneimittelrückstände bei Fischen. *Tierärztliche Praxis* 8, 237–244.

Schmidt, T., Richter, O. and Büning-Pfaue, H. (1986) Auswertung von Rückstandskurven mit biphasischer Kinetik. *Zeitschrift für Lebensmittel-Untersuchung und -Forschung* 183, 325–330.

Short, J.W. and Sharp, J.L. (1989) Tributyltin in bay mussels (*Mytilus edulis*) of the Pacific coast of the United States. *Environmental Science and Technology* 23, 740–743.

Shrestha, K.P., Noguera, R., Chopite, J. and Sosa, P. (1988) Mercury content of some marine fish from the southern Caribbean sea. *The Science of the Total Environment* 73, 181–187.

Siegel, B.Z., Siegel, S.M., Correa, T., Dagan, C., Galves, G., LeeLoy, L., Padua, A. and Yaeger, E. (1991) The protection of invertebrates, fish, and vascular plants against inorganic mercury poisoning by sulfur and selenium derivatives. *Archives of Environmental Contamination and Toxicology* 20, 241–246.

Slorach, S. (1992) Kvicksilver och andre främmande ämnen i fisk – åtgärder för att begränsa hälsoriskerna (Mercury and other foreign substances in fish – measures to reduce health risks). *Vår Föda* 44, 163–170.

Steffenak, I., Hormazabal, V. and Yndestad, M. (1991) Reservoir of quinolone residues in fish. *Food Additives and Contaminants* 8, 777–780.

Steffenak, I., Hormazabal, V. and Yndestad, M. (1993) Residues of quinolones in fish and the effect of cooking on the residues in the fish. In: Haagsma, N., Ruiter, A. and Czedik-Eysenberg, P.B. (eds) *Proceedings of the EuroResidue II Conference*, Veldhoven, The Netherlands, pp. 646–649.

Stronkhorst, J. (1992) Trends in pollutants in blue mussel *Mytilus edulis* and flounder *Platichthys flesus* from two Dutch estuaries, 1985–1990. *Marine Pollution Bulletin* 24, 250–258.

Suzuki, T., Matsuda, R. and Saito, Y. (1992) Molecular species of tri-*n*-butyltin compounds in marine products. *Journal of Agricultural and Food Chemistry* 40, 1437–1443.

Swackhamer, D.L. and Hites, R.A. (1988) Occurrence and bioaccumulation of organochlorine compounds in fishes from Siskiwit Lake, Isle Royale, and Lake Superior. *Environmental Science and Technology* 22, 543–548.

Taguchi, Y. (1971) *Nippon Eiseigaku Zasshi* 25, 563. Ref. Underwood, E.J. (1977) *Trace Elements in Human and Animal Nutrition*, 4th edn. Academic Press, New York/San Francisco/London, p. 378.

Takeuchi, M., Mizuishi, K., Yamanobe, H. and Watanabe, Y. (1987a) Survey of pollution with bis(tributyltin)oxide (TBTO) in fish and shellfish. II. TBTO content in cultured and inshore fish and shellfish. *Annual Report of Tokyo Metropolitan Research Laboratory of Public Health* 38, 155–159; ref. *Food Science and Technology Abstracts* 20 (1988) 5R7.

Takeuchi, M., Mizuishi, K., Yamanobe, H. and Watanabe, Y. (1987b) Survey of pollution with bis(tributyltin)oxide (TBTO) in fish and shellfish. III. Monitoring TBTO pollution in Tokyo Bay. *Annual Report of Tokyo Metropolitan Research Laboratory of Public Health* 38, 160–166; ref. *Food Science and Technology Abstracts* 20 (1988) 5R8.

Takeuchi, M., Mizuishi, K., Yamanobe, H. and Nakamura, H. (1991) Hygienic chemical studies on organotin compounds. VII. Content of tributyltin and triphenyl tin compounds in fish and shellfish (1988–1990). *Annual Report of Tokyo Metropolitan Research Laboratory of Public Health* 42, 77–85; ref. *Food Science and Technology Abstracts* 24 (1992) 12R16.

Tariq, J. and Jaffar, M. (1991) Mercury concentrations in ten fish species from the Arabian Sea, Pakistan. *Toxicology and Environmental Chemistry* 34, 57–63.

Uhler, A.D., Durell, G.S., Steinhauer, W.G. and Spellacy, A.M. (1993) Tributyltin levels in bivalve molluscs from the East and West coasts of the United States: Results from the 1988–1990 National Status and Trends Mussel Watch Project. *Environmental Toxicology and Chemistry* 12, 139–153.

Uno, K., Aoki, T. and Ueno, R. (1992) Pharmacokinetic study of oxytetracycline in cultured rainbow trout, amago salmon, and yellowtail. *Bulletin of the Japanese Society of Scientific Fisheries* 58, 1151–1156.

van der Valk, F., and Wester, P.G. (1991) Determination of toxaphene in fish from Northern Europe. *Chemosphere* 22, 57–66.

de Voogt, P., Wells, D.E., Reutergaardh, L. and Brinkman, U.A.Th. (1990) Biological activity, determination and occurrence of planar, mono and di-ortho PCBs. *International Journal of Environmental Analytical Chemistry* 40, 1–46.

Vyncke, W., Guns, M., Clerck, R. De and Hoeyweghen, P. Van (1984) Het gehalte aan zware metalen in tong uit de Noordzee en de Ierse Zee: 10 jaar onderzoek (1973–1982) (Heavy metals in sole (*Solea solea*) from the North Sea and the Irish Sea: 10 years of monitoring (1973–1982)). *Landbouwtijdschrift* 37, 1229–1234.

Vyncke, W., Guns, M., Baeteman, M. and Hoeyweghen, P. Van (1988) Het gehalte aan zware metalen en PCB's in mosselen van de Belgische kust (Heavy metals and PCBs in mussels from the Belgian coast). *Landbouwtijdschrift* 41, 713–719.

Vyncke, W., Guns, M., Hoenig, M. and Hoeyweghen, P. Van (1992) Concentration of arsenic in blue mussels (*Mytilus edulis*) from the Belgian coast. *Landbouwtijdschrift/Revue de l'Agriculture* 45, 741–746.

Watanabe, I., Kashimoto, T. and Tatsukuwa, R. (1987) Polybrominated biphenyl ethers in marine fish, shellfish and river and marine sediments in Japan. *Chemosphere* 16, 2389–2396.

Wicklund-Glynn, A. and Slanina, P. (1992) Halogenated hydrocarbons: DDT, polychlorinated biphenyls, dioxins and dibenzofurans. In: *Fish as Food.* Nordiske Seminar- og Arbejdsrapporter 1992:568. Nordic Council of Ministers, Copenhagen, 1992, pp. 95–119.

Wölwer-Rieck, U. (1990) Investigations about the furazolidone metabolite 3-(cyano-2-oxobutylideneamino)-2-oxazolidone in trout. In: Haagsma, N., Ruiter, A. and Czedik-Eysenberg, P.B. (eds) *Proceedings of the EuroResidue Conference*, Noordwijkerhout, The Netherlands, pp. 420–423.

Wölwer-Rieck, U., Blankvoort, E., Haagsma, N. and Büning-Pfaue, H. (1993) Improved assessment of the residual behaviour of oxytetracycline in trout. In: Haagsma, N., Ruiter, A. and Czedik-Eysenberg, P.B. (eds) *Proceedings of the EuroResidue II Conference*, Veldhoven, The Netherlands, pp. 685–689.

Yamada, H. and Takayanagi, K. (1992) Bioconcentration and elimination of bis(tributyltin)oxide (TBTO) and triphenyltin chloride (TPTC) in several marine fish species. *Water Research* 26, 1589–1595.

Yamanobe, H., Takeuchi, M., Mizuishi, K., Watanabe, Y. (1987) Mercury and selenium contents in fish and shellfish. *Annual Report of Tokyo Metropolitan Research Laboratory of Public Health* 38, 167–172; ref. *Food Science and Technology Abstracts* 20 (1988) 5R9.

Yamanobe, H., Takeuchi, M., Mizuishi, K. and Watanabe, Y. (1988) Mercury and selenium contents in sea bass from Tokyo Bay. *Annual Report of Tokyo Metropolitan Research Laboratory of Public Health* 39, 101–103; ref. *Food Science and Technology Abstracts* 21 (1989) 12R16.

Yamauchi, H. and Yamamura, Y. (1984) Metabolism and excretion of orally ingested trimethylarsenic in man. *Bulletin of Environmental Contamination and Toxicology* 32, 682–687.

Yndestad, M. (1993) Problems with drug residues in farmed fish. In: Haagsma, N., Ruiter, A. and Czedik-Eysenberg, P.B. (eds) *Proceedings of the EuroResidue II Conference*, Veldhoven, The Netherlands, pp. 115–124.

Zabik, M.E., Harte, J.B., Zabik, M.J. and Dickman, G. (1992) Effect of preparation and cooking on contaminant distribution in crustaceans: PCBs in blue crab. *Journal of Agricultural and Food Chemistry* 40, 1197–1203.

Zell, M. and Ballschmiter, K. (1980) Baseline studies of the global pollution. II. Global occurrences of hexachlorobenzene (HCB) and polychlorocamphenes (Toxaphene®) (PCC) in biological samples. *Fresenius' Zeitschrift für Analytische Chemie* 300, 387–402.

Freezing and Chilling of Fish 10

CEES KEIZER
Netherlands Institute for Fisheries Research, RIVO-DLO, IJmuiden, The Netherlands.

Introduction

Temperature lowering, e.g. frozen storage, retains the properties of fresh fish more than any other preservative process.

The consumer expects from frozen fish that, after thawing, the product is identical to freshly caught fish. Although consumer acceptability (or shelf-life) can be maintained for a storage period of more than one year provided proper pre-processing and adequate frozen storage conditions are obeyed, the sensory properties of the fish gradually deteriorate and, to a minor extent, the nutritional properties as well.

The quality of seafood and safety standards are still under development, e.g. following the Hazard Analysis Critical Control Point (HACCP) system. This system essentially aims at microbial conditions that can be easily determined. The definition of quality in terms of sensory properties will be difficult, though required to eliminate confusion.

For both the fish industry and consumer it is important to know for what time frozen fish can be stored before quality loss exceeds a certain predetermined level. This means that definitions of shelf-life and methods should be clear, consistent and applicable to all products, packaging methods and storage conditions.

To a certain extent a well-trained expert panel is able to reproducibly establish sensory characteristics, while the average consumer may not observe changes in loss of quality. These incompatible observations by experts and consumers also play a role in arriving at definitions on loss of quality. The introduction of the description High Quality Life (HQL), being the time in which a taste panel does not notice statistically significant differences from the initial high quality, does not show an adequate relation with the quality perception by consumers. As a consequence, this description cannot be applied in practice. Nevertheless, product quality is often assessed by sensory panels, e.g. to determine Practical Storage Life (PSL) or the time during which the stored product is still acceptable to the consumer.

It is also difficult to objectively measure sensory properties. Microbiological and physicochemical characteristics can be measured but an incontestable indication of quality is still lacking. Jul (1985) puts forward the question whether consumer preferences will coincide with results of objective tests, as marketing experience and studies of consumer behaviour point out that what may be considered as very fresh food is not always most favoured by the consumer.

The Freezing Process

Depending on species, fish contain 60–80% of water and, as a consequence, the freezing process mainly concerns the phase transition from liquid to solid H_2O (removal of latent heat) and secondarily the removal of sensible heat (reduction of temperature), both with water still being in the liquid phase and in the solid state. The larger part of heat to be removed from the fish is required for the formation of ice.

As removal of heat proceeds during ice formation, the concentration of solutes, i.e. salts and other components, in the remaining liquid water gradually increases. Due to these continuously increasing concentrations the temperature at which the remaining water is converted into ice gradually decreases. As a result, the water becomes frozen over a temperature range.

Most of the water is turned into ice between 0 and $-5\,^{\circ}C$. This range is called the *freezing zone* or *critical zone*. At lower temperatures more water is frozen, although a rather low temperature, i.e. about $-60\,^{\circ}C$, is required to completely convert all the liquid water, as far as it is not chemically bound, into ice.

The percentage of water frozen, at different temperatures, is given in Fig. 10.1. The enthalpy of fish is related to the water fraction and temperature. For calculation of thermal processes such as freezing and thawing, and for equipment engineering, differences in enthalpy are essential. As an example, a graph is shown in Fig. 10.2 which allows establishment of the enthalphy of lean fish under given circumstances.

Model simulation

Theories on freezing behaviour at the cellular level have been developed and discussed in the past, but have not yet resulted in full understanding of this behaviour.

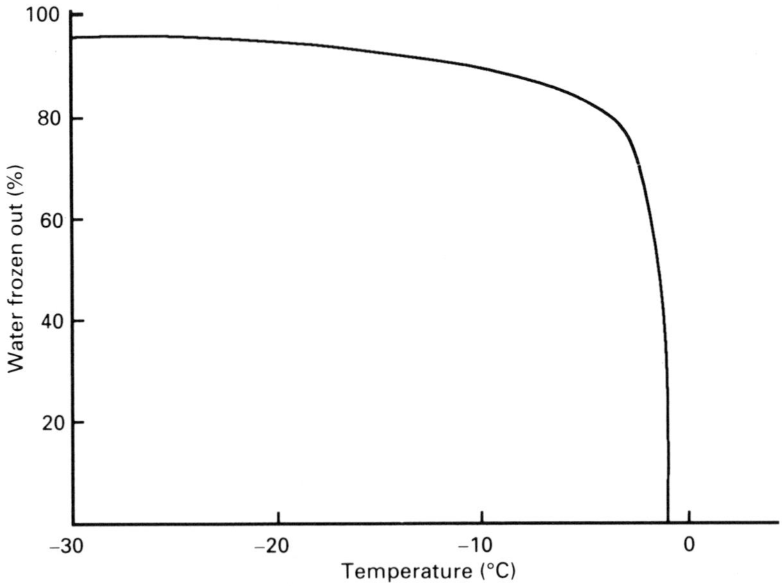

Fig. 10.1. Percentage of total water in fish muscle, frozen out as ice, related to end temperature.

Recently, however, attention has been paid to detailed mathematical simulation models (Gao *et al.*, 1991). In these models, heat transfer and thermodynamics are combined with kinetics of water loss from cells and other processes which are thought to be of importance. Hayashi and Momose (1991) performed such simulations considering equilibrium as well as non-equilibrium conditions and assuming sub-cooling for intracellular nucleation of ice crystals.

Non-equilibrium conditions cause strong fluctuations in the (decreasing) temperature, due to the local disappearance of sub-cooling at the moment of actual nucleation. Changes of cell volume due to water migration, being dependent on the rate of nucleation, determine the successively and/or simultaneously occurring physicochemical processes as well. It may be concluded, therefore, that a further elaboration of these models will be essential to the understanding of freezing at micro level and its effects on the physical damage to biological materials.

Freezing time

Freezing food involves a situation of unsteady heat transfer. Calculation of the time required to reduce the temperature of a substance is rather easy if simple boundary conditions apply. Such conditions are satisfied if the shape of the substance can be considered to be regular in one, two or three

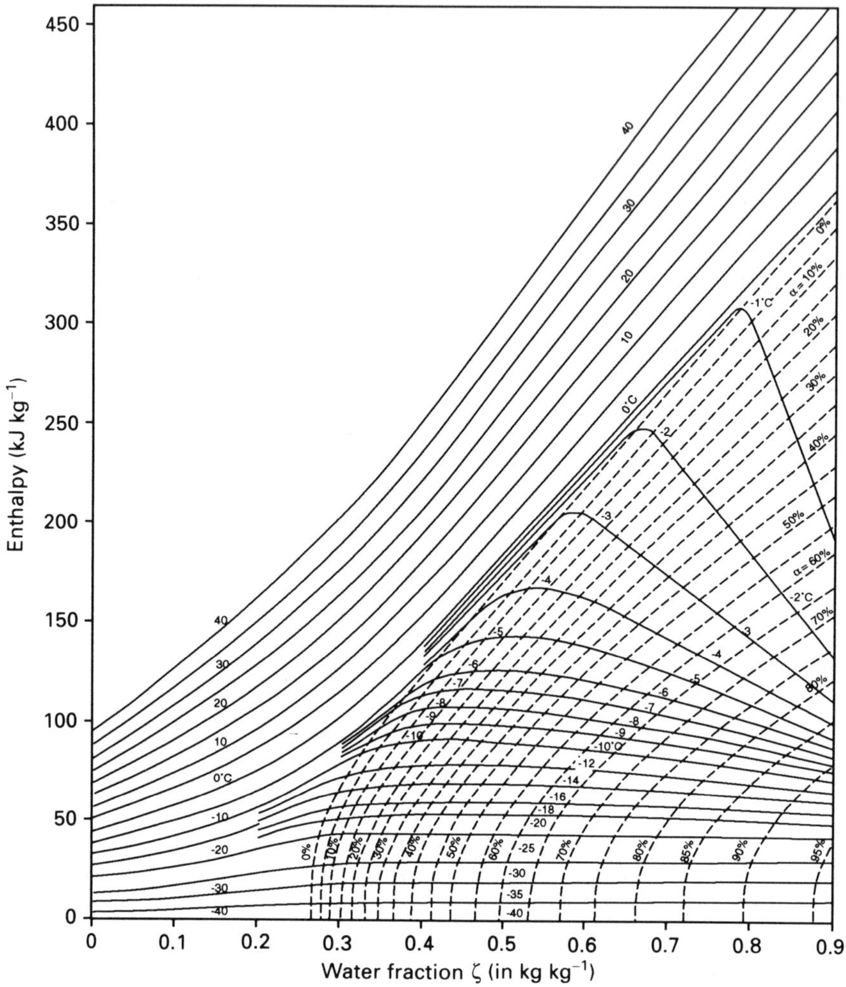

Fig. 10.2. Relation between the enthalpy of lean fish and the water fraction. α = water frozen out as % of the initial water fraction.

dimensions, i.e. can be considered as a slab, a cylinder and a sphere or a cube, respectively. Apart from this, it has to be assumed that the material should be homogeneous. If not only temperature reduction but also a change of state from the liquid to the solid state is involved, the freezing time can also be analytically calculated if a single freezing point, i.e. not a freezing zone, is assumed.

Departing from these simple boundary conditions and assuming that heat transfer is relatively slow, which means that this can be considered as a quasi-steady state, a basic equation to calculate the time t (seconds) required to attain the core temperature, i.e. the highest temperature in the product, has been put forward by Plank (1960) as:

$$t = \frac{\Delta H}{T_f - T_r} C \left(\frac{L}{h} + \frac{L^2}{4k_{solid}} \right), \text{ where}$$

ΔH = latent heat to be removed for conversion of the liquid to the solid state ($J\,m^{-3}$),
T_f = initial freezing point of the substance (°C),
T_r = temperature of the refrigerant (°C),
C = constant (1/6 for a sphere, 1/4 for a cylinder, 1/2 for a slab),
L = characteristic dimension of the substance (in m), e.g. thickness of a slab, diameter of a cylinder or a sphere,
h = surface heat transfer coefficient ($W\,m^{-2}\,K^{-1}$), and
k_{solid} = thermal conductivity of the frozen substance ($W\,m^{-1}\,K^{-1}$).

As can be noted, the characteristic dimension of the product is important with respect to the freezing time due to its quadratic effect. If the linear effect of the dimension is neglected, doubling or tripling the dimensions requires a freezing time of four and nine times as long, respectively.

In general, prediction of freezing time by Plank's equation underestimates the actual freezing time. This holds particularly in those cases where thick products are involved, the deviations amounting up to 50% from this equation.

Both the sensible heat for precooling from the initial temperature to the freezing point and for sub-cooling from the freezing point to the final core temperature, however, are not accounted for in the calculation of freezing time. The thermal conductivity of the liquid phase, which is about a quarter of that of ice, is not considered either. The effect of food composition, e.g. the amount of fat, is only expressed by the latent heat to be removed and not in the values for thermal conductivity of either frozen or unfrozen food.

Nevertheless Plank's equation remains a popular analytical approach for predicting freezing time, but can be used only for a rough estimation of freezing time and is not accurate enough for predicting the freezing time of foodstuffs.

Food usually does not satisfy the boundary conditions as assumed by Plank. Most often the food shows an irregular shape. Furthermore, the (thermo)physical properties, as a rule, differ locally, dependent on the composition of the food, and are subject to changes during the freezing process. Several modifications, e.g. accounting for removal of sensible heat by insertion of experimentally obtained coefficients in Plank's equation, or in similar equations either expressed or not expressed in dimensionless numbers, have been derived. More accurate calculations can be obtained by applying mathematical models which can be numerically solved. An advantage of these models is that, apart from predicting the average temperature, the temperature can be predicted at any location of the product, e.g. in the core or at the surface. For an accurate prediction of the freezing time, sufficient and exact information should be available to simulate the model.

One of the relatively simple methods to improve the prediction of freezing time for elementary configurations, by introducing analytical approximations, has been proposed by Pham (1984). According to this method, which can be considered as an extension of Plank's equation, sensible heat effects are taken into account and the process of heat transfer is divided into: (i) a precooling period; (ii) the actual freezing period, i.e. the period where the process of phase change occurs; and (iii) a sub-cooling period. Apart from a more accurate calculation of separate and total cooling times, this subdivision allows better designs of freezing as well as thawing equipment.

After rearranging Plank's equation, Pham obtains the following equations for each of the periods to be distinguished:

$$t = \frac{Q}{A\Delta T}\left\{\frac{1}{h}\left(1 + \frac{Bi}{C}\right)\right\}, \text{ where}$$

Q = the heat (in J) to be removed in each period: $Q_1 = c_l(T_i - T_f)V$, $Q_2 = \Delta H \cdot V$ and $Q_3 = c_s(T_f - T_r)\ V$ (V = volume in m^3),
A = heat transfer surface (m^2),
C = the volumetric specific heat ($J\,m^3\,K^{-1}$) for either the liquid or the solid phase, i.e. l or s, respectively,
T = the temperature (in K) of the cooling medium (T_r), the initial temperature (T_i), the freezing temperature (T_f) and the final core or mean temperature (T_m),
ΔT = temperature differences for the distinguished periods:

$$\Delta T_1 = (T_i - T_f)/\ln\{(T_i - T_r)/(T_f - T_r)\},$$

$$\Delta T_2 = T_f - T_r,$$

$$\Delta T_3 = (T_f - T_m)/\ln\{(T_f - T_r)/(T_m - T_r)\},$$

Bi = the Biot number ($h\,D\,k^{-1}$) for the liquid (Bi_l) or solid (Bi_s) phase:

$$Bi_1 = \tfrac{1}{2}(Bi_l + Bi_s),$$

$$Bi_2 = Bi_3 = Bi_s, \text{ and}$$

C = a constant, being 6 during periods (i) and (iii), and 4 in period (ii).

These equations are more meaningful due to the introduction of the *Biot number*, indicating the ratio of the internal resistance to heat transfer, inversely proportional to the thermal conductivity (k) and the external resistance, governed by the heat transfer coefficient (h) at the product surface. The value for h amounts to approximately 20 in the case of applying an air-blast freezer and about 200–300 for a plate freezer.

The heat transfer for air-blast freezing can be increased by imposing higher air velocities, though at the expense of increased fan power. An increase of air speed requires an increase of energy consumption proportional to the third power. When cryogenic refrigerants are applied, h can assume comparatively high values, e.g. 10,000 ($Bi \rightarrow \infty$), when the liquid is boiling. In that case the heat conductivity of the product is governing the resistance to heat transfer. The fast freezing rate in the case of cryogenic freezing is actually determined by the large difference between the temperatures of the product and the refrigerant.

In cases where the food is contained in packaging material, Pham's equation may be adapted as

$$t = \frac{Q}{A\Delta T}\left\{\frac{1}{h}\left(1 + \frac{Bi}{C}\right) + \frac{x}{k_1}\right\}, \text{ where}$$

x = the thickness of the packaging material (in m), and
k_1 = the thermal conductivity of the packaging material ($W\,m^{-1}K^{-1}$).

It should be noted that, in practice, affirmation of actual freezing time remains necessary, in particular if the properties of the product and/or the packaging method are not known.

Freezing and Quality Aspects

Deterioration during freezing and frozen storage

Spoilage as a result of microbial activity can be stopped, and deterioration as a result of (bio)chemical and physical changes can be retarded, at frozen storage conditions. The water activity is also reduced. Although the effects of freezing on microbial and biochemical processes are well understood, physical deterioration during freezing and frozen storage has always been a topic yielding confusion in the literature up till now. This mainly concerns phenomena occurring during ice formation at freezing and changes of ice crystals during frozen storage.

Loss of quality during freezing

The first theory on quality loss of fish during freezing, still adhered to by many, is based on the assumption that cell walls are damaged by ice crystals. At a higher freezing rate the number of ice crystals will be higher and the dimensions of these crystals will be smaller than at low freezing rates. The relatively large crystals formed under the latter conditions are thought to penetrate the cell walls and, as a consequence, cause damage to these cell walls. This damage could occur during frozen storage as well, in particular at large temperature fluctuations, as can be explained by the thermodynamic principle that larger ice crystals will be formed at the expense of smaller ones. This explanation, being determined by the freezing rate and by storage conditions, might be considered appropriate when visually observing the coarse appearance of slowly frozen tissue.

Deterioration during the freezing process itself and subsequent thawing could not be confirmed by organoleptic testing. It may be concluded that the singular advantage of freezing is the possibility it offers to store highly perishable commodities for considerable lengths of time which are virtually indistinguishable, sensorially as well as nutritionally, from the starting material (Lavety, 1991). In practice, however, there are reasons why this favourable situation is not obtained. Moreover, it might be stated that freezing, particularly frozen storage, imposes excellent conditions for denaturation of fish muscle proteins. This latter phenomenon is considered nowadays as the explanation for deterioration of frozen food.

Recent theories on deterioration of fish during freezing or frozen storage involve protein denaturation which, in part, may be considered as an agglomeration of protein molecules, resulting in changes in both physical and chemical properties of the protein. The conditions for protein denaturation are created by solution effects and by the difference between intracellular and intercellular ice formation.

Freezing starts with intercellular ice formation occurring at equilibrium conditions, while intracellular ice formation is retarded due to non-equilibrium behaviour, i.e. believed to occur at sub-cooling conditions. As a consequence, water from the cells will diffuse through the cell walls to maintain osmotic equilibrium. This cell dehydration creates the best conditions for protein agglomeration. Denaturation is particularly enhanced by a prolonged exposure of the cells to the concentrated solution, which cannot or can only partially be restored and will result in textural changes and drip at thawing.

Actually two oppositely acting factors can be distinguished with respect to the denaturation rate: (i) temperature decrease; and (ii) increasing concentration of enzymes and other compounds such as salts. The combined effect of these two factors results in a maximum denaturation rate at a temperature of −1 to −2°C.

As the optimal temperature coincides with the first ice formation, it might be concluded that a further concentration increase of the compounds dissolved, as a result of further temperature reduction, is not decisive in the combination of these two effects and/or the effect of temperature on denaturation is non-linear.

Obviously the combination of the two factors is governed by time. As a consequence, slow freezing results in a fish product inferior to that obtained by fast freezing. This means that freezing damage, according to this concept, may be at least partly considered as a consequence of ice formation but is not directly affected by the dimensions of the ice crystals.

An additional remark may be made on the older explanation which considers penetration of cell walls by ice crystals to be the cause of damage, i.e. the flexibility of animal tissue is sufficient to avoid cell rupture or penetration by ice crystals. With respect to the degree of deterioration during freezing the opinions are different. Jul (1984) states that the effect of freezing rate has been exaggerated in the past. Lavety (1991) is of the opinion that irreversible denaturation will occur only at long periods of frozen storage and if the storage temperature is unsuitable.

According to Lavety a stable temperature of −30 °C is required to avoid serious deterioration of most fish. A temperature of −28 °C, for frozen storage of fish at the producer and wholesale level, is recommended. For lean fish such as cod and plaice the temperature might be higher (−20 °C) and for fatty fish such as herring and mackerel the temperature might be even lower (−30 °C), while in Japan fish is commercially stored at temperatures as low as −50 °C.

At lower storage temperatures both the vapour saturation pressure and the vapour pressure over the product will be lower. The difference between these pressures will be smaller as these pressures decrease non-linearly with temperature decrease. The rate of dehydration is approximately proportional with this pressure difference and will therefore be reduced at a low storage temperature. As desiccation of the product surface accelerates fat oxidation, the latter is also reduced by decreasing the temperature.

A higher relative humidity, the ratio between the actual and the maximum vapour pressure of the air, will evidently be an advantage in avoiding dehydration and oxidation. Most cold stores, however, are not equipped with devices to control the relative humidity.

According to Waddell (1979), the need for lower storage temperatures has been substantially demonstrated, e.g. by the work on time–temperature tolerances for frozen foods. For every frozen product a relationship exists between storage temperature and the time needed, at this temperature, for the product to undergo a certain amount of quality change. At different temperatures during storage and distribution these changes are cumulative.

At adequate frozen storage conditions most water will diffuse back into the cells. In that case the protein will partly adopt its original state and its former water-binding capacity again once the product is thawed.

Apart from this, however, freezing rate may affect quality, despite application of excellent techniques such as air-blast freezing at low temperature. This may be due to dense packaging of the product, or packaging in large-sized units. As a consequence the product is very slowly cooled, as the blast air cannot flow along the product due to either dense storage or short-circuiting of the blast air circulating over the evaporator only.

Microbial and biochemical deterioration may occur if the product has been kept for a too long time at a too high temperature. As a result, quality loss during freezing, in general, is due to bad conditions before freezing and to bad frozen storage conditions. Deterioration during frozen storage, even if the conditions are suitable, will always go on as a result of biochemical reactions which cannot completely be stopped by lowering the temperature. Apart from redistribution of water, due to the heterogeneous ice formation and the corresponding concentration of solutes, surface dehydration (freezer burn) as a result of a low humidity in the cold store, contributes to texture deterioration by denaturation.

Development of rancidity, however, is mainly due to lipid oxidation, accompanied by other reactions such as lipid hydrolysis and protein denaturation. Hydrolysis may even occur at a faster rate below than above the freezing point (Hanaoka and Toyomizu, 1979).

If fish is frozen pre-rigor, rapid thawing has to be prevented in order to avoid a fast rigor mortis and to allow rehydration with the protein returning to its original gel condition. If thawing proceeds too fast, 'gaping' may occur or thaw rigor may be induced, the latter causing tough flesh and a significant drip.

Lindeløv (1978) reported that the relation between product temperature after freezing and storage temperature is important for salted meat, shelf-life being shorter for bacon frozen to −40°C and stored at −5°C and −12°C than frozen to and stored at the latter temperature. This may be due to the effect of salting and/or recrystallization at higher storage temperatures.

Confusion still exists on the effect of temperature fluctuations. Jul (1984) stresses that deterioration due to temperature fluctuations has been exaggerated. Garthwaith (1992) mentions the large effect of even small fluctuations. One can imagine that the air temperature, as a consequence of temperature control of the cold store, intermittently assumes a higher and a lower value than the surface temperature of the product. At air temperatures lower than the product temperature, dehydration will occur while, at a higher air temperature, water vapour will condense and freeze onto the product.

At frozen storage air humidity is normally less than the saturation pressure. This leads to dehydration, finally resulting in 'freezer-burn' on the surface of the frozen fish. Dehydration, however, always occurs, though at a lower rate if the cold-store temperature is controlled more accurately, as the removal of heat from the cold store, commonly performed by air coolers, is accompanied by withdrawal of moisture from the air, which results in a low air humidity.

Prevention of quality loss

In order to limit quality loss, fish should be frozen within a short time after catching. The freezing rate should be fast enough to prevent both microbiological deterioration and dehydration. There may also be logistic arguments for enabling a fast freezing rate, as this determines the production capacity. Some sources indicate that at a slow freezing rate there is more protein denaturation, as the percentage of drip is higher. A freezing rate of at least $0.5\,\mathrm{cm\,h^{-1}}$ is often recommended (Dyer, 1971; Heldman, 1983), though no quality differences between freezing rates of 0.15 and $5\,\mathrm{cm\,h^{-1}}$ are noted.

As far as the storage condition itself is concerned, one should preferably take care to keep a constant temperature in order to minimize the rate of (bio)chemical and physical changes such as dehydration and oxidation. This macro climate might be controlled as far as the gaseous composition is concerned, particularly for fatty fish. Evidently the design of a cold store should involve a homogeneous distribution of the air, resulting in a uniform temperature throughout the store. Proper insulation and therefore a minimal heat transmission directly limit the required refrigerating capacity and also the rate of dehydration. Due to the minimum rate of air circulation there is an indirect effect as well.

Glazing

A method to limit deteriorative effects even occurring at optimal conditions is glazing, i.e. the application of a thin layer of ice on the already frozen product, providing protection against dehydration but also oxidation of lipids in fatty fish.

Because of the high surface-to-volume ratio of small products such as shrimps, glazing is an adequate means to restrict dehydration in these cases.

A layer of ice can be applied by dipping the product into a glaze solution, i.e. water either with or without an additive, or by spraying a glaze solution onto the product.

A drawback of glazing is a temperature increase of the product. As a result, water present in the product which has been converted into ice will melt again because the cold required to transform the glaze solution into the solid state is withdrawn from the cold already applied to the product earlier.

The amount of glaze added to the product will cause approximately the same amount of water already in the solid state to melt again. This may have a negative effect on product quality.

Large amounts of glaze, added in one step to relatively small products, may particularly result in a temperature increase to values well within the critical zone, primarily near to the surface but in thin products up to the core as well. Decreasing the product temperature to values below −40 °C in order to avoid these high end temperatures, on the other hand, may result in cracking of the glaze.

It should be stressed that the product has to be frozen again after glazing. Control of the process of glazing is important to attain a sufficient and a uniform coverage of the surface of the product.

Packaging

Proper packaging of frozen fish prevents the moisture from escaping and avoids admission of oxygen to the product. Prevention of water loss does not always mean that the product is not dehydrated. If the product is not tightly packed, or if there is an air gap of only a few millimetres between the product and the package, the product will dehydrate and the moisture will deposit as ice crystals at the inner side of the package.

The product will also lose water, even when tightly packaged, if the packaging material is permeable to water vapour. Evidently dehydration will depend on product temperature or temperature fluctuations in the cold store. At a lower cold-store temperature dehydration will be less. Where the temperature changes and exceeds the temperature of the product, water vapour from the air in the cold store will deposit as frost on both the product and the packaging material. Reversible diffusion/migration of moisture to the product may be excluded.

Freezing Methods and Equipment

Freezing equipment may be classified according to different criteria, such as: (i) the method of generating the refrigerating effect; and (ii) the method of applying the refrigerating effect, i.e. by convection or by conduction.

Convective heat transfer may involve a gaseous cooling medium (such as air) or a liquid cooling medium which may or may not be boiling (such as a brine or a cryogenic refrigerant, respectively). In case of conductive heat transfer the product to be frozen is in contact with a metal surface which is directly cooled by a refrigerant or secondarily cooled by another medium, e.g. a brine which is cooled by a refrigerant.

The selection of any specific freezer equipment depends on requirements characteristic for the product(s) to be frozen as well as for the manufacturing conditions. Apart from the quality of the product, the following characteristics have to be taken into account:

- the rate of freezing;
- the shape and size of the product;

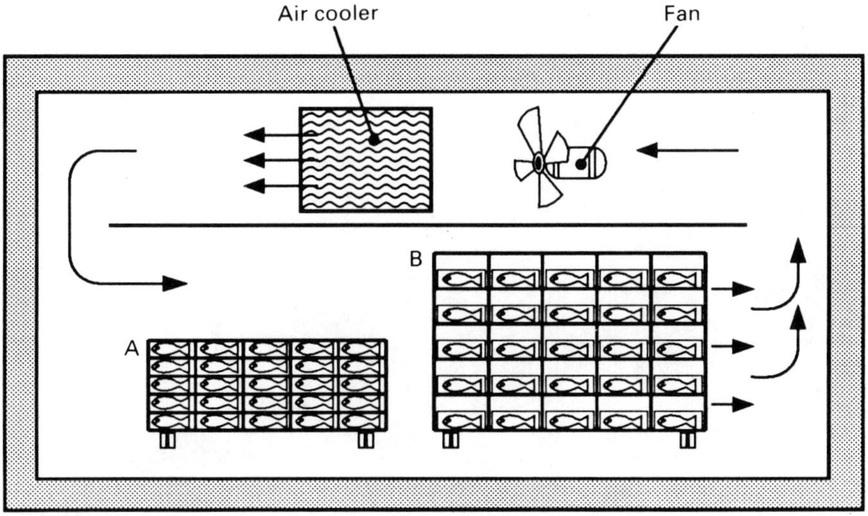

Fig. 10.3. Air-blast freezer. A = stowage too compact. B = proper stowage density.

- the rate of weight loss;
- exclusion of oxygen during freezing;
- freezing packaged or unpackaged products;
- capacity of refrigerating effect versus frequency of operation;
- batchwise or continuous operation;
- reliability of the supply of refrigeration;
- flexibility to freeze different products and to integrate these in the production process;
- ease of cleaning the equipment;
- labour and maintenance required for operation;
- energy consumption;
- costs of investment and exploitation.

Air-blast freezers

The simplest type of blast freezer consists of a stationary blast tunnel, i.e. an insulated enclosure in which air is circulated by fans over the product stacked on trays to cool these and successively over air coolers to cool the air again in order to repeat this sequence (Fig. 10.3). The velocity of the air should be high enough, i.e. about 5 m s^{-1}, to allow an acceptable heat transfer resulting in an adequate freezing time as well as to limit the temperature increase of the air. It may be noted that higher air velocities hardly lead to a better heat transfer.

Special attention should be paid to ensure the air speed is constant and

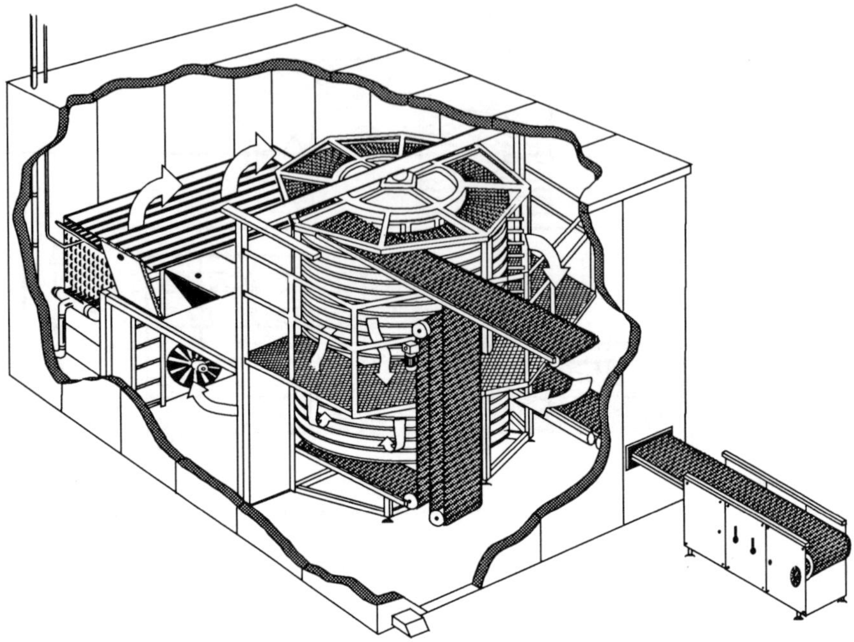

Fig. 10.4. Spiral freezer. (Reproduced courtesy of Triphase Ltd, England.)

uniform throughout the cross-section of the tunnel. As has been indicated before, this latter condition is often not fulfilled in practice. The product should be evenly distributed over the total height and width of the tunnel. A too compact stowage (Fig. 10.3; A) has to be avoided. Proper storage (B) allows a uniform air speed and temperature throughout any cross-section of the tunnel.

The capacity of the air coolers, i.e. the difference between the air temperature and that of the refrigerant, ought to be as small as (economically) possible in order to limit both dehydration and the need for defrosting the air coolers. On the other hand, in spite of batch operation, the blast tunnel offers much flexibility with regard to freezing of products of different dimensions and shapes.

Belt freezers

Continuous operation of air-blast freezing can be provided by belt freezers. This type of freezer is equipped with one or more linear belts, mostly made of stainless steel wire mesh, moving the fish through the freezer. The spiral freezer (Fig. 10.4) is provided with spiral belts.

Belt freezers are commonly applied to freeze individual, unpacked and individually quick-frozen (IQF) products and allow an easy integration with the processing line.

The remark on incorrect operation of blast tunnels by imposing a non-uniform air flow holds for belt freezers as well. The product should be spreaded uniformly on the belt to avoid bypassing (channelling) of the air, both in the case of series flow and particularly if cross-flow air circulation is applied.

The velocity of the air is mostly between 10 and 15 $m\,s^{-1}$, which is economically feasible due to the limited size of the product. The higher fan power requirement is compensated by the smaller freezer, which means lower costs of investment. A possible drawback of wire mesh belts may be that the belt presses its form in the product. Special attention is required to prevent sticking of the product to the belt and for removal of the product from the belt. These problems may be avoided by integration of a belt washer with the freezer and/or choosing a flat belt of plastic (in the first part of the belt) or of stainless steel, as applied in the Torry freezer.

Fluidized bed freezers

Fluidized bed freezers are air-blast freezers where the air has an additional function to fluidize the product. This means that air is blown from below through the product, fast enough to lift it. Lifting and floating of the product will be obtained if the individual particles have approximately the same shape and dimension. Fluidized bed freezing offers an adequate method for IQF freezing. Particularly wet products, which usually stick together, can be easily frozen and kept separated.

In general one should take care to obtain sufficiently low air temperatures and, in the case of air-blast freezing, adequate air velocities to limit weight losses of unpackaged products. Although fluidized bed freezing offers excellent opportunities for freezing wet products, the water attached to the product will be easily withdrawn by the dry air and deposited at the air cooler.

Air coolers for fluidized bed freezers, as a consequence, ought to satisfy severe specifications regarding fin spacing. If fin spacing is small, frost accumulating in the air cooler will soon lead to higher pressure losses in the air flow, yielding a lower air velocity and a reduced heat transfer and, as a consequence, higher air temperatures. Moisture removal from the product, comprising both the water attached to the product and water as one of the constituents of the product, will be attenuated due to this effect.

Plate freezers

Plate freezers are contact freezers. Metal plates are in direct contact with the product or its package. With the aid of hydraulic pressure the plates can be moved to obtain a compact product and a good contact to ascertain good heat transfer by conduction. The plates may be cooled by a brine or a refrigerant, which is usually pumped through the plates. The geometries of the plates may be horizontally or vertically arranged. Generally the former are used to freeze whole fish, mainly at sea, and for bulk freezing. The horizontal plate freezer is used both for freezing laminated fish blocks and for freezing consumer packages.

A principal advantage of plate freezers is that the original shape of the product or its package can be maintained. This advantage may change into a drawback because of the limitations imposed on the geometry of the product. With respect to flexibility of shape and dimensions, air-blast devices are preferable to plate freezers. Incorrect operation due to void spaces should be prevented: consumer packages should be compactly filled and when freezing whole fish one might consider the addition of water to improve contact for heat transfer, i.e. to avoid the presence of air gaps separating the fish, and to prevent oxidation and dehydration during frozen storage, in particular if fatty fish is involved. Correct operation of plate freezers also involves defrosting of the plates to facilitate the removal of the product and to avoid damage to both the product and the plates.

Cryogenic freezers

In cryogenic freezers the product is brought into direct contact with the refrigerant, which is nowadays usually liquid nitrogen or liquid carbon dioxide, and sometimes solid carbon dioxide. The fish is moved on a conveyer belt counter-current to the flow of cryogenic gas, which is generated by injection of the cryogenic liquid and successive extraction of heat from the fish due to simultaneous evaporation of the liquid refrigerant. This counter-current flow is of particular importance with respect to energy cost if liquid nitrogen is applied, because of its low boiling temperature (−196°C at atmospheric pressure).

As a result of this very low temperature, cooling of the fish will be so rapid that cracking of the product may occur. Also for that reason, the fish is moved counter-current to the nitrogen gas, which precools the fish. At the same time this allows the use of the refrigerating effect available due to the sensible heat required to increase the temperature of the gas from −196°C to a higher temperature, preferably to that of the fish when entering the freezer. This heat effect is approximately equal to the heat required for evaporation of the nitrogen.

Although the sublimation temperature of carbon dioxide is higher (−78°C), application of counter-current flow is also adequate here.

The fast rate of freezing is often mentioned as a main advantage of cryogenic freezing, e.g. with respect to weight losses. Rates of weight loss during cryogenic freezing are claimed to amount to 0.5%, while rates with respect to losses during air-blast freezing are indicated to be 1%. The latter, however, is related to a large extent to the design and, more importantly, the operation of the freezer, and can adopt rates of up to 5%. In general, claims regarding lower weight losses in the case of cryogenic freezing compared to adequate air-blast freezing ought to be judged for their significance. Even if the statement is generally true, the total weight loss during freezing and frozen storage (dehydration as well as drip losses) is an important consideration. The losses during frozen storage, after cryogenic freezing, might be higher, as physical damage to the fish may more easily occur than in the case of proper air-blast freezing. Evidence for this comparison is not abundant. Additional research on the opportunities provided by fast cryogenic freezing might yield interesting results if conversion to the glassy state could be applied in practice or if adequate cryoprotectants are developed to stabilize, for example, a mince-based product such as surimi.

An additional advantage of nitrogen and carbon dioxide may be that oxygen can be excluded and, for carbon dioxide, that it shows bacteriostatic properties. Another advantage of cryogenic refrigeration may be the relatively small floor space required.

Disadvantages of cryogenics are the high variable costs of exploitation, mainly related to the consumption and the costs per kilogram of these cryogenics. The consumption amounts to between 1 and 2 kg per kg of product to be frozen, the amount of carbon dioxide being slightly higher than that of nitrogen.

In practice the consumption rates will not differ much, as storage losses of liquid nitrogen at −196°C may amount to 0.5% per day, apart from losses during transport to the vessel at the site of storage and use. Carbon dioxide can be stored in vessels at a moderate pressure at ambient temperatures without significant losses.

The high variable costs of exploitation are not always a drawback. In particular in the case of seasonal production with high demands for refrigerating capacity and low running hours the capital costs, being relatively low for cryogenic freezers compared with those for mechanical refrigeration, might be decisive. In practice one should think of a running time of 500 to 1000 hours. At shorter times cryogenic freezing is economically advantageous in comparison with mechanical refrigeration, but mechanical refrigeration ought to be favoured for longer running times and a preferably constant demand for refrigeration capacity.

Recently combinations of cryogenic and conventional freezers, i.e. compression-type refrigeration units, have been introduced (e.g. Eek, 1991). In a first freezing step the product surface is frozen with liquid nitrogen while, in a second step, freezing is conventionally completed. Compared with complete cryogenic freezing the demand for liquid nitrogen is lower. Consequently, operating costs of cryo-mech freezing can be less. Apart from this, dehydration losses will be lower and total freezing time will be less. A practical advantage is that fewer defrosting operations are required.

The combination of freezing steps offers the opportunity to freeze low viscosity products such as fish soups and minced fish, in order to obtain stable products in the initial, cryogenic freezing step before further conventional freezing.

Chilling

Not all fish is immediately frozen aboard ship. Often the fish is chilled first or chilled only, the latter mainly aboard smaller vessels. A main advantage of chilling fish, i.e. cooling fish down to a temperature just above the freezing point, is that chilling causes minimal changes to sensory properties and appeals more to the consumer.

Ice is still often applied as a means to chill and to preserve fish aboard, either mixed as plain ice with fish or as chilled sea water, i.e. sea water mixed with ice. Ice is also often used in retail shops, restaurants, etc. Apart from providing refrigeration the melting ice removes slime from the surface of the fish which is polluted with bacteria.

Instead of chilling sea water by adding ice the same results can be obtained by mechanical refrigeration of sea water.

Because of the higher rate of spoilage as compared to freezing, chilled fish has to be consumed shortly after catching the fish. If not, chilling has to be combined with another method of preservation or pre-processing or with the increasingly popular method of modified air packaging. Evidently, the supply of chilled foods, in particular highly perishable fish, is to a large extent dependent on the adequate operation of the cold chain.

Icing

Icing fish aboard ship provides an excellent method to preserve fish whether or not it is eviscerated and washed.

Icing, however, requires proper sanitary conditions. Nowadays the most common type of ice used aboard is flake ice because of its adherence to the fish surface, providing optimal contact for heat transfer and causing a minimum of mechanical damage to the fish. Apart from these advantages flake ice, as compared with other types of ice, shows less tendency to fuse and is comparatively easy to mix with fish. It should be stressed that mixing is essential to achieve fast and uniform cooling of the total amount of fish in a box or container.

Flake ice is mainly produced from potable water. At places where the infrastructure is limited, crushed block ice is still popular because of the robust equipment which can be used. Another frequently applied method of chilling and storing fish aboard, though mainly used by larger vessels, is salt-water icing. The use of sea water or potable water to which 3% of sodium chloride is added enables icing temperatures of about −1 °C or lower at higher concentrations. Due to the lower ice temperature and the fine ice particles, the mixture of brine and ice looks like wet snow. The cooling rate is higher and the final temperature of the product will be lower.

Salt water or slush ice is attracting renewed attention as, apart from the advantages mentioned, it offers additional benefits due to reduced handling both in icing and at offloading, as slush ice alone or mixed with fish can easily be pumped.

A possible disadvantage of salt-water icing is the salt uptake (the level depending on the species), though in practice methods might be adopted to prevent this. A similar drawback holds for refrigerated sea water. Consequently, such methods should be applied for short-term periods only.

It is not yet clear whether salt-water icing at freezing temperatures below −1 °C, which allows partial freezing of the fish, is favourable or disadvantageous with respect to quality.

Ice is also used in retail displays to keep fish chilled and because of commercial arguments related to an attractive appearance. Fish should be stored *in* ice and not on top of a layer of ice, as the temperature of the fish, due to abundant illumination, may rise to even more than 10 °C at the upper surface.

Predicting the shelf-life of chilled fish

Microbial spoilage is the main cause of deterioration of chilled fishery products and, as temperature and time are the major parameters determining bacterial growth, the rate of spoilage is largely dependent on these parameters. Fishing techniques, contamination due to gutting and filleting, or the method of packaging, also affect this rate.

The growth of bacteria is non-linear with temperature and often assumed to behave according to an exponential Arrhenius type of relation. Olley and Ratkowsky (1973) laid the basis for the establishment of a relation enabling the comparison of shelf-life days on ice with those at higher temperatures, which, in a simplified form, is given in Chapter 7.

Modified atmosphere packaging

Another method to extend the preservative effect of chilling is storage in an atmosphere other than ordinary air. An increase of the carbon dioxide concentration reduces microbial growth. Carbon dioxide has long been known to inhibit the growth of many psychrotropic spoilage bacteria. One of the mechanisms thought to be involved is a pH decrease in the product, due to the dissolution of carbon dioxide in the water present in the fish. Carbon dioxide may also inhibit certain enzyme systems, though details on this mechanism are still lacking. As the solubility of carbon dioxide in water increases with decreasing temperature, the effect of this gas is more pronounced at low temperatures.

In the lag phase growth inhibition is most pronounced. Successful modified atmosphere packaging therefore requires high initial product quality and high hygienic standards. As temperature essentially governs microbial growth, a strict temperature regime through the cold chain must be maintained.

The term MAP (Modified Air Packaging) is used where a fixed composition of a gaseous mixture is added to a package. During the time for which the fish remains in this package, both gas composition and gas pressure change as a result of the carbon dioxide absorption. For this reason inert nitrogen is commonly added to the package to avoid a pseudo vacuum and package collapse. Normally a fish to gas ratio of 1:3 is applied.

In some cases, especially if white fish is concerned, oxygen should be present in the gas in order to minimize anaerobic spoilage and to prevent excessive drip and colour changes. If fatty fish species or smoked fish are involved, however, vacuum packaging may be preferred to exclude oxygen and, in this way, to prevent oxidative rancidity. Recently, research and development have focused on the permeability of packaging material. Impermeability or selective permeability with respect to components of the gas in the package is the main issue.

Another method is controlled atmosphere packaging (CAP), in which the atmosphere initially introduced to the package is maintained during storage of the product. Due to the higher costs this method is only applied at a larger scale than that of consumer packs.

Regenstein and Regenstein (1991) point out that claiming extension of shelf-life may not be justified due to improper experimental conditions. Extension of shelf-life by storage in a modified atmosphere is decreased or even nullified by the negative effect of a higher storage temperature during shelf-life, due to a larger period required for cooling the product. Application of modified atmosphere packaging, therefore, should be preceded by utmost attention to realization of optimal product temperature. If this condition is fulfilled, modified atmosphere packaging may be a valuable tool in extending shelf-life. However, it should certainly not be considered as a substitute for chilling.

Partial Freezing

Partial freezing means that the fish is brought to a temperature of about −2 to −3°C. As a result, approximately half of the amount of the unbound water present in the fish is converted to the solid phase.

Partial freezing or superchilling gained much attention in the 1960s, but has not been successful. Recently, however, it has received renewed interest.

It has been shown that storage life could be doubled by partial freezing, i.e. from 14 days storage on ice to 28 days at partially frozen conditions. A problem is the loss of sensory quality after about 5 days. This quality loss is thought to be caused by temperature fluctuations. A temperature increase or decrease of less than 0.5°C may cause intermittent melting, or, respectively, freezing, of 5–10% of the water in the fish. This may have a marked effect on the process of ice formation and, as a result, on protein denaturation.

A pH decrease post mortem, as a result of glycolysis, is also thought to be responsible for protein denaturation, as this results in a lower water binding capacity of the proteins. Depending on species, season, condition of the fish, the catching ground and the method of catching, the muscle pH may decrease, in some cases to 5.5 (Huss, 1988), i.e. close to the isoelectric point of fish muscle proteins where the water binding capacity is minimal. In later stages basic components are formed and the pH may increase again.

Others point at the maximum rate of post-mortem biochemical effects and the hydrolysis of lipids occurring at these temperatures, although opinions on the effect of partial freezing on oxidation are contradictory.

At these temperatures bacterial growth in packaged fishery products proceeds remarkably faster than in unpacked fish. Storage life of partially frozen fish which has subsequently been frozen was found to be extremely short, due to dehydration of the surface, and this is relevant to further processing.

Maintaining a very constant temperature is one of the aims in the attempts to improve partial freezing. Another aspect concerns freezing with a minimum moisture loss after earlier partial freezing.

Thawing

The thawing process in fish is mainly governed by heat conduction. It can be considered, to a large extent, as the opposite of the freezing process, though with some essential differences.

The rate of thawing is less important with respect to the physical changes than is the rate of freezing, which determines the number and size

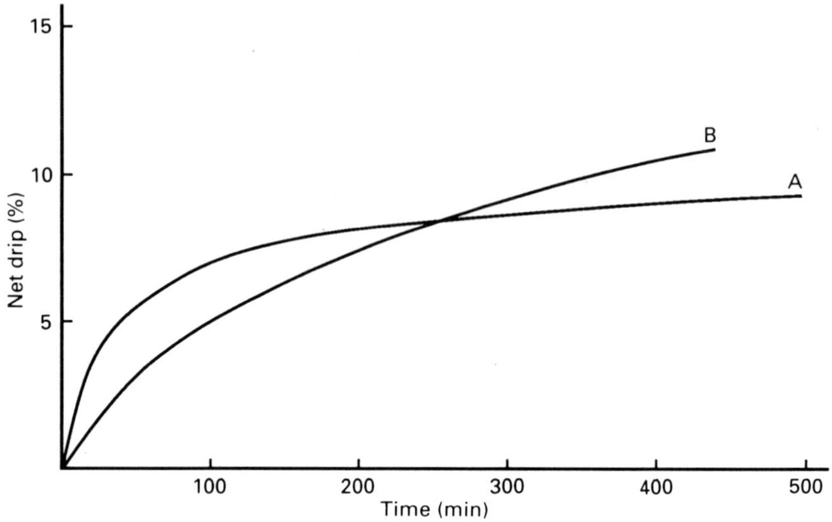

Fig. 10.5. Effect of cooling and thawing on net drip loss. A = cooling from 0 to −5°C. B = thawing from −5 to 0°C.

of the ice crystals. Another difference between freezing and thawing is related to the difference in heat conductivity between liquid and solid H_2O, the latter being about four times that of the first. As a result, the thawing time will substantially exceed the freezing time under identical conditions and if heat is supplied to the surface of the product.

The thawing rate should be slow enough to allow the water to move to its original position to avoid drip loss. On the other hand, the thawing rate should not be too slow, as time–temperature effects play a similar role related to spoilage as in chilled fish.

In Fig. 10.5 the effects of freezing and cooling rates on drip are shown. These data indicate that drip is lower at rapid freezing and standardized thawing, or rapid thawing after freezing at a constant rate. During commercial freezing drip rate is practically independent of freezing rate (Jul, 1984), though generally higher at low freezing rates. Fig. 10.5 shows that slow thawing results in the highest amount of drip.

During thawing specific attention is required: (i) to prevent overheating and even cooking of the product; (ii) to avoid dehydration; and (iii) to optimize thawing time, i.e. to avoid excessive drip loss due to too fast thawing and to limit bacterial spoilage due to too slow thawing.

Thawing methods

The methods available for thawing can mainly be distinguished regarding the way of supplying heat: (i) at the surface of the product with subsequent heat transfer into the product by conduction; and (ii) within the product.

The first group of methods, in particular those using air, water or brine as a heating medium, are most common. It is generally assumed that the temperature of the heating medium should not exceed 20 °C. At least the surface temperature of the fish should not exceed this value and should preferably be kept lower. In the case of fatty fish, recrystallization of the oils might occur and produce texture changes.

The heat flux is not only determined by heat conductivity but is also dependent on the heat transfer coefficient at the surface, i.e. the resistance to heat transfer from the heating medium to the product as well as the temperature difference between these.

The temperature difference is restricted by the maximum temperature of the heating medium as explained before. The only way to improve heat transfer consists of increasing the heat transfer coefficient at the surface of the product, e.g. by increasing the speed of the medium flowing around the product, although the effect which can be attained is limited.

If water is applied as the heating medium, the surface heat transfer coefficient is higher than if air is applied. Consequently, the thawing time will be longer for air than for water. Apart from the flow speed, surface heat transfer will depend on the relative humidity of the air. Spraying water in the flowing air increases the heat flux. This holds in particular for air saturated with water vapour, as condensation of water vapour goes along with a high heat transfer. At the same time a high relative humidity prevents the product from dehydration.

In practice, these conditions cannot be realized by humidification with liquid water. Injection of live steam may be applied, but care has to be taken to obtain a homogeneous distribution of steam in the air flow. An oversaturated mixture, i.e. an unstable fog, is preferred to avoid excessive temperatures. Difficulties in realizing this mean that humidification using steam has not been feasible up to now, but recently new efforts have been made.

An advantage of moisture-saturated air is that oxidation of fatty fish is not accelerated by dehydration.

Vacuum thawing may be considered as an alternative to air-blast thawing with injection of steam. In applying vacuum thawing, water vapour is generated by boiling water under vacuum. The vapour condenses on the cold surface of the product with simultaneous release of heat, though in absence of air. Thus, oxidation of fatty fish can be excluded by vacuum thawing.

Evidently, the need for a proper distribution of the air in a thawing tunnel also holds for air thawing. Additionally, attention should be paid to the humidity gradient as well, in order to obtain uniform thawing, when the air flows in series with the product over a long route. The shape and dimension of the product, as a matter of fact, have an effect on the thawing period, as is the case in freezing. Dehydration does not occur when water is applied as the heating medium, and heat transfer is commonly faster when using air unless live steam is applied for heating and humidification, resulting in condensation of water on the product.

A drawback of using water as the heating medium, in particular when products without skin such as fillets are being thawed, is leaching of taste and flavour components, and, for whole fish, negative effects on its appearance.

The second group of thawing methods departs from heat supply to the interior of a product by application of an electric current and/or changing electric fields. One possibility is electrical resistance heating, where the product is brought into good contact with two electrodes imposing an alternating electrical current to the product. As a result, heat is generated in the product. Another possibility is dielectric heating, in which a high frequency field is imposed on the product. In practice two frequencies are applied, i.e. approximately 27 MHz and 2450 MHz, which are defined as high frequency or radio wave heating and microwave heating, respectively. The heating effect is obtained by internal friction in products due to oscillating dipoles and movements of ions in the frequently changing electric fields.

The composition of the product, in particular the presence and distribution of water and salts, is essential for all these methods, as heat is mainly generated where the water is located.

In practice it is extremely difficult to achieve a uniform temperature throughout the product and to avoid edge effects through local overheating. Intermittent heating or combinations with thawing methods using air or water as a heating medium might yield technically and technologically feasible concepts for practical application. Up to now, internal heating methods are rather seldom applied, due to the lack of practically feasible solutions and because of the relatively high costs of investment and exploitation.

The Cold Chain

Maintaining the cold chain is of paramount importance in determining shelf-life of a perishable product such as fish. It has been pointed out that the production of food worldwide is twice the consumption (Stolk, 1988), indicating that facilities of proper distribution throughout the cold chain are lacking.

It is often stated that refrigerated retail cabinets form the critical link in the cold chain. Jul (1984) points out that the critical points in the cold chain concern the transfer of refrigerated products, (i) at (un)loading at docks, (ii) at loading refrigerated cabinets in retail shops and (iii) during transport from the store to the consumer's home.

Although transfer of products in the cold chain may be responsible for the main loss of product quality due to temperature increase, it is not very important whether the product is immediately cooled down on delivery. Serious quality losses may occur in retail freezer cabinets if products are stored for longer times at a temperature higher than prevailing in the preceding part of the cold chain (Spiess, 1988). It is not uncommon that products adopt temperatures of up to −5°C though mean temperatures of −15 to −18°C are most common. In particular, the products at the top layer in the display may show higher temperatures due to abundant illumination and exposure to the temperature in the shop.

As a matter of fact, most important to the final quality of the product is the initial quality of the product entering the cold chain. Van Hiele (1988) stresses that standardization of procedures and techniques implied in cold chain operations should not be exaggerated, indicating the ATP regulations (1970) for refrigerated trucks, but leaving the way open to innovative approaches in order to arrive at a product quality which can be well defined, thus allowing a check of this quality.

Adequate logistic organization and quick handling by skilled labour are essential to a strong cold chain. Apart from this, attention to hygienic handling and packaging is of utmost importance.

References and Further Reading

Aguilera, J.M. and Stanley, D.W. (1988) *Microstructural Principles of Food Processing and Engineering*. Elsevier Applied Science, London/New York, 331 pp.

ASHRAE Equipment Handbook (1983) Chapter 41: Automatic ice makers, pp. 41.1–41.8.

ASHRAE Refrigeration Handbook (1986) Chapter 14: Fishery products, pp. 14.1–14.12.

ATP (1970) *Agreement on the International Transport of Perishable Foodstuffs.* UNO/ECE, 1 September 1970, Geneva.

Burgess, G.H.O., Cutting, C.J., Lovern, J.A. and Waterman, J.J. (1965) *Fish Handling and Processing*. Ministry of Technology, Torry Research Station. HMSO, Edinburgh, 390 pp.

Dyer, W.J. (1971) Speed of freezing and quality of frozen fish. In: Kreuzer, R. (ed.) *Fish Inspection and Quality Control.* FAO/Fishing News Books, pp. 75–81.

Eek, L. (1991) A convenience born of necessity: the growth of the modern food freezing industry. In: Bald, W.B. (ed.) *Food Freezing: Today and Tomorrow.* Springer-Verlag, London, pp. 143–155.

FAO Fisheries Circular 145 (Rev. 2) (1977) *Code of Practice for Frozen Fish*, 49 pp.
Fellows, P. (1988) *Food Processing Technology*. Ellis Horwood Series in Food Science and Technology/VCH Publishers, Chapters 18 and 19, pp. 359–400.
Gao, D.Y., Kornblatt, J.A. and Lin, S. (1991) Osmotic injury to cells during freezing preservation. *Proceedings of the 18th Congress of the International Institute of Refrigeration, Montreal*, pp. 1692–1695.
Garthwaith, G.A. (1992) Chilling and freezing of fish. In: Hall, G.M. (ed.) *Fish Processing Technology*. Chapman and Hall, London, pp. 89–113.
Haard, N.F. (1992) Biochemical reactions in fish muscle during frozen storage. In: Bligh, E.G. (ed.) *Seafood Science and Technology*. Proceedings International Conference Seafood 2000, Halifax (Canada), 1990. Fishing News Books, Oxford, pp. 176–209.
Hanaoka, K. and Toyomizu, M. (1979) The effects of dehydration and concentration of components on phospholipid decomposition in frozen stored fish. *Scientific Bulletin of the Faculty of Agriculture, Kyushu University* 4, 225.
Hayashi, Y. and Momose, N. (1991) Simulation of micro-freezing of biological substances. *Proceedings of the 18th Congress of the International Institute of Refrigeration*, Montreal, pp. 1657–1661.
Heldman, D.R. (1983) Factors influencing food freezing rates. *Food Technology* 37(4), 103–109.
van Hiele, T. (1988) *Commission Meeting*, International Institute of Refrigeration, Wageningen, p. 25.
Huss, H.H. (ed.) (1988) Fresh fish quality and quality changes. *FAO Fisheries Series* 29, 54.
Jason, A.C. Thawing of frozen fish. *Torry Advisory Note* No. 25 (revised). HMSO, Edinburgh, 11 pp.
Jul, M. (1984) *The Quality of Frozen Foods*. Academic Press Inc., London, pp. 28–31.
Jul, M. (1985) *Commission Meeting*, International Institute of Refrigeration, Aberdeen, pp. 23–28.
Kramer, D.E. and Liston, J. (1987) Seafood quality determination. *Proceedings of the International Symposium, Anchorage* (1986). Elsevier, Amsterdam, 677 pp.
Lavety, J. (1991) Physicochemical problems associated with fish freezing. In: Bald, W.B. (ed.) *Food Freezing: Today and Tomorrow*. Springer-Verlag, London, pp. 123–131.
Lindeløv, F. (1978) *Freezer Storage of Salted Products*. Danmarks Tekniske Højskole, Copenhagen.
Olley, J. and Ratkowsky, D.A. (1973) The role of temperature function integration in monitoring of fish spoilage. *Food Technology in New Zealand* 25, 8, 13, 15 and 17.
Pham, Q.T. (1984) Extension to Plank's equation for predicting freezing times of foodstuffs of simple shapes. *International Journal of Refrigeration* 7, 377–383.
Plank, R. (ed.) (1960) *Handbuch der Kältetechnik*. Band X: Die Anwendung der Kälte in der Lebensmittelindustrie. Springer-Verlag, Berlin, pp. 22–31, 234–236.

Regenstein, J.M. and Regenstein, C.E. (1991) *Introduction to Fish Technology*. Van Nostrand Reinhold, New York, pp. 104–119.

Spiess, W.E.L. (1988) *Commission Meeting*, International Institute of Refrigeration, Wageningen, pp. 105–120.

Stolk, A.L. (1988) *Commission Meeting*, International Institute of Refrigeration, Wageningen, pp. 12–17.

Waddell, I. (1979) Low temperature freezing maintains high quality – prevents deterioration. *Australian Fisheries* 38(2), 34–36.

Fish Products 11

Zdzisław E. Sikorski[1], Asbjørn Gildberg[2] and Adriaan Ruiter[3]

[1]*Technical University Polytechnika Gdańska, Department of Food Preservation, Gdańsk, Poland;* [2]*Norwegian Institute of Fisheries and Aquaculture, Tromsø, Norway;* [3]*Department of the Science of Food of Animal Origin, Faculty of Veterinary Medicine, Utrecht University, The Netherlands.*

Introduction

A large part of the world fish catch is used in the form of a number of different preserved products, prepared by the food industry or at home. Some of these products have no longer shelf-life than refrigerated fish, others are resistant to spoilage for months, while canned fish is shelf-stable for years. Their nutritive properties, sensory quality and stability depend on the species and initial quality of the raw material, on the method of preservation and on the processing parameters. The typical sensory properties of the products develop as a result of changes in lipids and nitrogenous components, caused by chemical and enzymatic processes controlled by temperature, pH and the presence of salts and of other reactive components (Pan, 1988; Doe and Olley, 1990; Miler and Sikorski, 1990; Naczk and Artyukhova, 1990; Shenderyuk and Bykowski, 1990; Lanier, 1994).

The seafood consumption patterns in many countries are changing. The main trend is more emphasis on high quality than on price. Large groups of consumers have become more health-conscious and are interested in convenience foods. The changing pattern of life and the increasing number of small households have an impact on the market demands. Many consumers look for ready-to-eat products in small unit volumes. Generally the product is expected to be not salty, sour, dry or tough, to retain all nutrients present in the raw material, to contain no chemical additives, contaminants and hazardous microorganisms, and to be free of parasites. Developments in technology of seafood processing can help to satisfy these consumer demands of sensory quality, food safety and socioecologic value. This can be illustrated by the increase in production of many seafood

Table 11.1 Change in the annual production of fish commodities from 1989 to 1990 in Germany (Keller, 1992).

Commodity	Percentage change
Fresh fish	−13.7
Fresh	−28.4
Frozen battered fillets	+26.4
Smoked herring and sprats	+45.5
Other smoked fish	+31.1
Marinades	+33.3
Fish salads	+2.1
Canned fish	+16.1

products suitable for being served after a minimum of preparation at home (Table 11.1).

Dried, Salted and Smoked Fish

Dried fish

Drying in the sun was possibly the first method used for preserving seafoods (Cutting, 1955). It is still applied to a large part of the world catch of fish and squid.

In tropical climates, under humid conditions, heavy infestation of unsalted dried fish by blowflies may cause up to 30% loss of the product (Wood, 1981). The solar tent drier has made it possible to rationalize this old preservation treatment in developing countries and to reduce larval infestation during drying (Doe *et al.*, 1977). In other areas sun drying has been largely replaced by operations in driers with circulating hot air of controlled temperature and humidity, or by freeze drying for expensive products (Doe and Olley, 1990). Small fish may be dried whole, while larger fish are gutted, split, filleted or minced before drying.

There are, at least, the following categories of dried seafoods: plain-dried, boiled and dried, broiled and dried, salted and dried, frozen and dried, and boiled, roasted and dried.

The principle of preservation by drying involves a decrease in water activity (a_w) by removal of water, in order to arrest the activity of microorganisms. Most bacteria do not grow and multiply at a_w values below 0.95. Growth and multiplication of some species, however, is possible at lower a_w values, as is the case with yeasts and moulds. In order to prevent rapid spoilage the a_w should be brought below 0.95 as fast as possible, but for obtaining a microbiologically stable product it is necessary to decrease the a_w to around 0.7, which, in lean fish, implies a moisture content of about 15%. Other deteriorative processes in the fish are changed with changing water content, reaching their minimum at different levels of a_w. In salted and dried seafoods a_w is affected both by loss of water and by the amount of added salt.

Stockfish is naturally dried unsalted fish produced in the temperate windy climate as occurring at the North Atlantic coast (Pedersen, 1981a). It is the oldest preserved fish product in Scandinavia, and was probably the most important provision for the Vikings during their ravagings around Europe more than a thousand years ago.

Cod, haddock and saithe are the main species used for stockfish production. The fish can be dried either as unheaded and eviscerated round fish, or as split fish with the anterior two thirds of the backbone removed and just connected at the tail. The fish is hanged on drying racks in open air. The ideal drying temperature is 8–10°C. For round fish hanged in the spring (May) the drying takes 2–3 months, whereas split fish dries faster. The drying process comprises both some kind of ripening and preservation. The final product contains about 15% water and 75–80% protein, and the nutritional value is similar to fresh fish, as calculated on a moisture-free basis.

The quality of the dried fish is related to the final a_w. At low a_w values, uptake of water proceeds more easily. In properly dried fish the water uptake is complete in not much more than 3 minutes. The speed of water uptake is regarded as a quality criterion. If it is more than 15 minutes the quality of the fish is considered to be questionable. Protein denaturation may be the cause of a decreased ability to rehydrate.

Stockfish has a strong but natural fish flavour, which is very much appreciated particularly in southern Europe. It is either consumed as it is, or serves as a raw material in the preparation of *bacalao*, which is a casserole with fish, potatoes, tomatoes, onion and red pepper in oil.

In Norway a traditional dish, called *lutefisk*, is made after treatment of the stockfish in a potash lye.

Quality of the products

The quality of dried seafoods depends on the properties of the raw material, the applied procedure of curing/drying, the barrier properties of the packaging, and on the time and conditions of storage. The consumer preference regarding the quality of dried seafoods is very diversified. In some countries dried products are unknown, while in other areas, e.g. many tropical countries, they are staple foods. Some dried seafoods are even regarded as a delicacy. Depending on the colour, texture and flavour, the different assortments of dried seafoods may be used either as material for preparing specific dishes such as fish balls, stews and soups, may be consumed as they come, e.g. the dried Caspian roach (Zaitsev *et al.*, 1969), as a snack compatible with beer, or the dried abalone as an esoteric appetizer (Olley and Thrower, 1977), or else they are mainly applied as flavouring additives such as dried bonito.

The sensory properties of dried products may deteriorate during storage due to oxidation of lipids (Bligh *et al.*, 1988), browning, and infestation by insects. Drying slightly decreases the nutritional properties of seafood proteins, depending on the severity of heat treatment, oxidation, and cross-linking between the amino acid residues. Maillard-type non-enzymatic browning reactions may involve NH_2 groups in bound as well as in free amino acids. These browning reactions have their maximal rate at a_w values between 0.6 and 0.7.

Excessive heating, which also promotes reactions at NH_2 groups, may lead to losses in free amino acids and in lysine residues in proteins. Opstvedt (1988) recommends that, whenever possible, drying should be conducted at maximally 70 to 80°C in order to ensure maintenance of protein quality. Reduction in temperature should have priority over a reduction in processing time, and drying at temperatures above 115°C should be avoided.

Storage under abuse conditions may lead to formation of small amounts of undesirable compounds, e.g. histamine (Pan, 1988). Generally, however, more severe loss is caused by insect infestation in unsuitably packaged dried products during storage at abuse temperature and air humidity (Olley *et al.*, 1988).

Shelf-life

The high-quality shelf-life of dried fish is usually several months at room temperature. It depends on a_w, air humidity, temperature, exposure to oxygen, the presence of preservatives, and infestation by insects. The a_w values inhibiting spoilage microorganisms are not low enough to inactivate the insects in dried fish. The total storage losses caused by beetle infestation may be as high as 50% during several months, depending upon the salt content, the rate of drying and the conditions of storage (Wood, 1981). In the tropics dried fish containing even as much as 21% salt are heavily infested by insects of different species (Indriati *et al.*, 1985). Drying conditions unfavourable for infestation and tight packaging of the product are an effective, although expensive, method of preventing losses of dried fish caused by insects. An alternative way to control the infestation and growth of blowflies and dermestid larvae would be to apply approved insecticides.

Salted fish

The salting process

In fish processing, salt is almost invariably used in one form or another.

The suitability of salt for any particular process depends on several factors, the most important of which are the chemical composition, the bacteriological conditions and the physical properties.

Commercial salts can be divided into three groups:

1. *Solar salts*, prepared by natural evaporation of sea water. These salts are often less pure than others, the main impurities being chlorides and sulfates of calcium and magnesium, and insoluble matter such as dust and sand grains. Solar salts usually contain more bacteria than the other groups. These are mostly halophiles. The counts may go up to 10^5 per gram of salt.
2. *Rock salts*, i.e. the natural deposit of salt. Rock salt is ground to various finenesses without any purification. It is generally of higher purity than solar salt and contains few bacteria.
3. *Vacuum salts*, prepared by artificial evaporation, under vacuum, of purified brine. These contain only traces of impurities and, in principle, do not contain bacteria.

Fish can be salted in different ways. As a pretreatment, the fish may be immersed in a 3 to 5% brine in order to remove blood and slime and to improve flavour. As for the final product, in many cases a brine is added to the final pack just before sterilizing.

Salting fish with a saturated brine, for preserving purposes, is used with fatty fish, as immersion in a salt solution keeps the air out, thus preventing development of oxidative rancidity in lipids.

In dry salting the rate of sodium chloride penetration is critical and should be as fast as possible. It depends on several factors such as the amount of calcium and magnesium salts and on the grain size.

Salt with appreciable amounts of Ca^{2+} and Mg^{2+} may have a penetration rate inadequate to inhibit bacterial action, as was mentioned as early as 1920 by Tressler, but by many others as well. This, however, seems to hold only if the salt has to pass a fish skin before it can penetrate into the fish muscle. This may suggest that the retarding effect of Ca^{2+} is a function of skin thickness.

The salt should not be too finely grained. The reason for this is that a fine-grained salt rapidly dissolves in the fish muscle fluids, causing a too rapid withdrawal of moisture from the surface tissue. As a consequence, rapid protein denaturation and coagulation occur, preventing further penetration of the salt into the fish, and giving rise to a condition known in the trade as 'salt burn'. This demonstrates that fish proteins, like many others, can be denatured by the action of salt, and that some phenomena quite resemble those that may occur during fish drying. In salted fish this denaturation always occurs, but normally denaturation proceeds more slowly than salt penetration (Tülsner, 1978).

The cause of salt denaturation, however, differs from that of heat denaturation, as the Na^+ and Cl^- ions act as counter ions towards negatively and positively charged groups, respectively. These groups play an important part in maintaining the native structure of these proteins. In this way fish proteins can be made edible by the action of salt ('salt curing'). Denaturation by salt, like denaturation by heat, results in a decreased protein extractability.

Another effect of salt has to do with the ion-exchanging properties of proteins. In canned fish, for example, Ca^{2+} and Mg^{2+} ions present in the muscle may be replaced by Na^{+} ions from the brine. This causes a soft fillet, giving rise to a 'mushy' texture. As this process is reversible, Ca^{2+} and Mg^{2+} from impure salt may accumulate in the flesh, resulting in a tough texture (Sikorski and Ruiter, 1994).

A high content of calcium and magnesium salts should be avoided for other reasons as well. In particular magnesium sulfate may cause a bitter flavour. An excess of Mg^{2+} may lead to the formation of 'glass crystals' (struvite, NH_4MgPO_4) in canned lobster and crab.

Even chemical reactions leading to formation or cleavage of covalent bonds may be influenced by the addition of salt. For example, proteins will become more accessible to reactions such as these. Catalytic effects of salt cannot be excluded either, but these mostly result from impurities present in the salt. Traces of heavy metals may accelerate Maillard-type reactions.

The sensory characteristics of salted fish result from enzymatic changes in proteins, lipids and carbohydrates, from different interactions of the products of these enzymatic reactions, and from the added spices. Lipids affect directly the texture, while the products of lipid hydrolysis and oxidation contribute to the flavour, directly and by interacting with nitrogenous compounds. If present in excessive amounts the oxidation products impart undesirable rancid sensory properties to the product. Oxidative rancidity is controlled mainly by the fat content and enzymatic properties of the raw material, concentration of salt, temperature, pH, and access of oxygen. In fish salted with a mixture of salt and spices the aroma compounds, mainly terpenoids and phenols, not only have distinct flavour but also a significant antioxidant activity.

When salted fish is stored for longer periods it undergoes partial proteolysis. This is due to the activity of muscle proteases which, in the living animal, play an important role in muscle protein metabolism (Tülsner, 1978; Shenderyuk and Bykowski, 1990). The activity of these muscle proteases, however, is very small as compared to the proteolytic activity of the gastrointestinal tract (Tülsner, 1978). In order to restrict excessive proteolysis, salted fish has to be stored at low temperatures, i.e. from about 5 °C to the cryoscopic temperature for this kind of production (Shenderyuk and Bykowski, 1990).

In contrast, some salted products are kept for long periods before use, allowing a maturation or ripening process. Salted anchovies are an example of this kind of product. The reactions which give rise to the characteristic sensory properties are not well known (Filsinger *et al.*, 1987) but probably start with enzymic proteolysis.

The fish proteases have different pH optima for activity. In whole Baltic sprats the highest proteolytic activity on the muscle proteins is in the pH range 3–5 (Kowalewski, 1989).

Generally the autolytic activity in fish is greatly affected by seasonal changes in feeding intensity. Due to the hydrolytic processes a desirable, tender texture of the fish develops. Prolonged storage of ripened salted fish, however, even under refrigeration, may lead to excessive softening. Overripe salted fish may have on the surface a white 'bloom' of crystallized peptides and amino acids (Levieva, 1956). These deposits on salted Baltic sprats consist of about 50% of tyrosine, 20% of phenylalanine and 10% of glutamic acid (Kowalewski, 1989).

Heavily salted fish lose much water, are tough, and have poor flavour.

Halophilic bacteria from solar salt are the cause of a peculiar type of spoilage. These bacteria are strictly aerobic, proteolytic, and show an optimal growth rate at about 37°C and a salt concentration of 16 to 25% (w/w). The serious spoilage phenomena in salted fish are nearly always accompanied by a red discoloration of the fish surface, as a result of the production of red or reddish carotenoid pigments.

The control of red discoloration and spoilage in salted fish is more difficult than that of other types of spoilage. Although solar salt is the primary source of this trouble, the use of bacteria-free salt will not solve this problem unless a thorough decontamination of the fish salting plant accompanies this change. If present, red halophiles grow wherever concentrated brine solutions and suitable temperatures prevail. In order to suppress fish spoilage by halophiles, low temperatures should be maintained at all stages (Wood, 1963). Fish immersed in brine does not spoil, as halophiles are strictly aerobic.

Salted fish products

Lightly salted sturgeon, salmon, fatty young herring or anchovy have a very tender texture, which is that of a visco-elastic body. The flavour is very characteristic with a predominantly fishy and salty note, slightly meaty, cheesy, greasy, and possibly very slightly rancid.

Klipfish is traditionally salted split fish dried by the sun and wind on slopes of bare rock along the seashore to about 65% dryness. The product has been made in northern and western Europe since the 15th century (Pedersen, 1981b). At present, artificial warm air driers are used. The species used for klipfish production are cod, haddock, saithe, ling and torsk, and the product is popular in many countries in Europe, Africa and Latin America. Most of the klipfish is used for *bacalao* preparation.

A fast ripening occurs in Dutch '*maatjes*' made from immature fatty herring, caught in or at the end of the feeding period. Before salting the fish is gibbed, whereby the pyloric appendages of very high proteolytic activity are not removed from the belly cavity. These participate in ripening, causing the characteristic texture and flavour (Luijpen, 1959; Meyer, 1965; Ruiter, 1972). It should not be excluded that components from the feed also contribute to the characteristic flavour, either directly or by participating in some reactions with the muscle components.

Fish fillets or gutted fish are less suitable for salt ripening than whole fish. However, ripening may be induced by adding proteolytic enzymes to the brine. Rapidase B-500, in amounts of 0.5–0.7 $g\,kg^{-1}$, was used for salting gutted fish and fillets of Atlantic herring and mackerel by Góra (1972). At 10% salt in the fish the ripening of headed and gutted mackerel lasts 1.5 to 2 months at −1 to −4°C and 1 month at 0–5°C. The technology of salting with the use of added enzymes must provide for control of overripening during storage of the product.

Typical Scandinavian sugar-salted herring (or sprat) products such as *kryddersild, tidbits* and *gaffelbiter* are prepared with addition of nitrate and some other ingredients, at low temperatures during up to 18 months. The pickle is low in salt and contains various spices such as pepper, allspice, clove, cinnamon, nutmeg, ginger and Spanish hop (Pedersen, 1983). Microbial activity in ripening is generally low (Knøchel and Huss, 1984a), although nitrate is partially reduced by bacteria to nitrite. Part of this nitrite reacts with myoglobin. The nitrosomyoglobin formed gives a bright-reddish colour to the muscle (Fichtl and Schreiber, 1987). The product does not contain significant amounts of nitrosodimethylamine (Knøchel and Huss, 1984b). These semipreserved products are used as taste sharpeners in appetizers and egg sandwiches.

One of the major delicatessen-type salted fishery products is caviar. Black caviar is prepared from the eggs of sturgeon, red caviar from Salmonidae. Other less expensive caviar is made from the eggs of several other fish, i.e. cod, catfish, herring, capelin, lumpfish and some freshwater species. The preparation of caviar starts with separation of the fish eggs from the ovaries. These should be taken from the fish at the moment of slaughter. The eggs are immediately washed with cold water, salted with dry, fine salt, 3–5% of the egg weight, or brined for 8–18 min in saturated brine at egg:brine ratio 1:3, drained, and packed in cans or jars. Some assortments of caviar are pasteurized in sealed cans at 60–70°C for 2.5–3 h. The shelf-life of pasteurized caviar at 10–18°C is 3–4 months and at −2°C 6–8 months (Shenderyuk and Bykowski, 1990). The quality and stability of caviar is greatly affected by the freshness of the raw material and the concentration of salt and preservatives in the product. Caviar may also be produced in the form of pressed blocks, containing less water and more salt.

Smoked fish

The principles of smoke curing

Three types of smoked products are generally recognized, i.e. cold smoked at temperatures not exceeding about 35 °C, hot smoked in conditions causing thermal denaturation of the muscle proteins, and hot smoke-dried fish (Poulter, 1988). For heavily smoked products preservation is the primary objective of processing, while for soft-cured products the organoleptic properties are the main reason for smoking the fish.

In Europe, fish (mainly herring) is brined prior to cold smoking in order to make the fish edible by means of salt denaturation of the proteins. Most of the salt is leached out subsequently. The fish is dried, to some extent, before smoking starts. Loss of moisture goes on during the smoking step, which may take several hours. During cold smoking the temperature should not exceed 30–35 °C in order to prevent coagulation of proteins with concomitant structure loss.

Traditional African smoked products are rarely brined, but some may be sun-dried before smoking starts (Poulter, 1988).

Smoke curing is a typical combination treatment. Its preservation effect depends on several factors. For preserving fish for months, heavy smoking is applied to cause a substantial loss of moisture and to impregnate the product with a large amount of smoke constituents. Some smoked seafoods may contain up to 0.5% of smoke compounds. Many of these compounds have a distinct antimicrobial action, i.e. formic and acetic acid, carbonyl compounds such as formaldehyde and methylglyoxal, and phenols. Vegetative forms of bacteria are most susceptible to smoke, while spores and moulds are more resistant.

Fish lipids may undergo some oxidative changes during hot smoking, as reflected in slight increase of the peroxide number and thiobarbituric acid (TBA) value, as well as loss in the contents of eicosapentaenoic and docosahexaenoic acids (Beltran and Moral, 1989). The strong antioxidant activity of smoke phenols, however, protects the lipids during storage (Miler and Sikorski, 1990). This activity, which was observed by Lea as early as 1933, allows fatty fish to be smoke-dried without becoming rancid.

Salting of fish prior to smoking contributes to the decrease in a_w. In hot smoking the temperature in the central part of the fish may be, in some cases, as high as 80–90 °C, thus contributing to a pasteurizing effect. Lightly smoked products, containing less salt and smoke components but more moisture, must be kept under refrigeration to extend their storage life. Vacuum packaging, and in some assortments also sodium nitrite, may add to the preservation effect.

Fish containing about 5.5% salt in the water phase and smoked for at least 6–7 h at about 35 °C, have a shelf-life of about 2 months at 10 °C.

Lightly hot-smoked fish keeps not more than 2 weeks at 4°C. Presently the main reason for smoking is often not so much preservation as altering and improving the sensory properties of the product.

The most important contribution to the characteristic flavour of smoked fish is due to the phenols and to some carbonyl compounds. Reactive carbonyl components such as formaldehyde and methylglyoxal, possessing an unpleasant smell, react with surface proteins within a short time and do not contribute to the overall flavour except shortly after smoking.

Smoke curing also affects the sensory attributes of fish by inducing desirable protein changes due to salt curing and heating. In some assortments of smoked herring distinct ripening occurs due to enzymatic protein changes, just like in salted fish. The colour of the smoked fish surface is generated by reactions of carbonyl components from the smoke with proteins through their amino groups (Ruiter, 1979), which is a type of reaction showing a great similarity with the well-known Maillard reaction between amino groups and reducing sugars. High smoke temperatures favour the deposition of higher-boiling compounds and increase the rate of chemical interactions.

The gloss associated with smoked fish products is thought to be caused by extraction and setting of some of the proteins at the surface (Reay, 1933).

Nutritional aspects

The treatment of proteinaceous foods with wood smoke may bring about a substantial reduction in the contents of amino groups (Krylova *et al.*, 1962) and of available lysine due to reactivity of carbonyl compounds present in smoke (Dvořák and Vognarová, 1965; Chen and Issenberg, 1972; Opstvedt, 1988). In a hot-smoked Atlantic mackerel fillet the maximum loss of lysine and tryptophan has been found in the outer layer (Bhuiyan *et al.*, 1986). Regardless of the reactivity of smoke components towards the functional groups in protein amino acid residues, there is no significant decrease in the nutritional value of whole smoked fish, as many products in the market are rather lightly smoked and a high proportion of the total mass of smoke components penetrates hardly deeper than one millimetre under the skin of the fish.

Safety aspects

Smoked fish belong to the group of foods which are known to be contaminated with polycyclic aromatic hydrocarbons (PAH). Many of these compounds are carcinogenic. Benzo(a)pyrene (BaP) is regarded as an indicator of carcinogenicity in smoked foods. Its content in the whole edible parts of hot smoked fish is in the range of 0.5 to 3.5 $\mu g\,kg^{-1}$, depending on the size and preparation of the fish and on the conditions of smoking. The external parts of the fish, exposed to the smoke, especially the skin, may contain up to five times more PAH. Furthermore, the concentration of BaP in smoked fish has been found to decrease significantly during storage of the product (Simko, 1991). Many other foods contain more BaP than smoked fish (Sikorski, 1988).

The conditions of smoking may also favour the formation of N-nitroso compounds from nitrogen oxide with, mainly, secondary amines. Fish products are rich in nitrosatable amines, while nitrogen oxides are present in wood smoke. The content of N-nitroso compounds in the edible parts of smoked fish is, however, rather lower than in smoked meats and in dark malt (Sikorski, 1988).

The bacterial contamination of hot smoked fish just out of the smokehouse is usually low, i.e. below $10^3\,g^{-1}$. However, secondary infections in low-hygiene plants, or during distribution, may cause high bacterial counts in smoked fish in markets remote from the producers, if the product is kept under abuse temperature conditions.

Lightly salted hot-smoked fish may cause botulism because of the presence of *Clostridium botulinum* type E toxins, if the raw material is contaminated (cf. Chapter 8), if the temperature in the thickest part of the fish does not destroy the toxin and inactivate the spores, and if the conditions after smoking are not inhibiting for the growth of *C. botulinum* and toxin production. The main factors controlling the growth and toxin production of *C. botulinum* are the storage temperature and salt content. At 3°C and 3.5% salt in the water phase of the smoked fish, *C. botulinum* hazard can be practically excluded. The recommended good manufacturing practice for hot smoking is 30 min at 82°C in the centre of the thickest part of the fish containing 3.5% NaCl in the water phase. According to Kosak and Toledo (1980), smoking of mackerel fillets in conditions of initial dry and wet bulb temperature 71 and 60°C, respectively, and heating at a rate of 10° per 10 min till 90°C in the centre of the fillet, brings about a decrease in the number of viable spores of *C. botulinum* type E by 12 log cycles, whereby the product has good sensory properties. Hot-smoked fish, vacuum-packaged after smoking, can be heat-pasteurized for 55 min at 92°C, which is sufficient to prevent *C. botulinum* type B and E toxin production during 21 days at 25°C (Eklund *et al.*, 1988). Modified atmosphere packaging as well as proliferation of aerobic organisms consuming oxygen from the atmosphere surrounding the product may in some cases increase the hazard of toxigenicity, although the effect of packaging is only one of several other factors and has been recently shown to be rather insignificant. In ordinary packs the anaerobes can, in any case, grow under the surface of smoked fish. A high oxidation–reduction potential decreases the salt tolerance of *C. botulinum* type E (Huss, 1981). In the presence of $NaNO_2$ the salt concentration in hot-smoked salmon required to prevent toxin formation by *C. botulinum* types A and E during storage at 25°C may be reduced (Pelroy *et al.*, 1982).

It has recently been shown that smoked salmon can support the growth of *Listeria monocytogenes* during 4 weeks at 4°C (Rørvik *et al.*, 1991). This shows the importance of measures aimed at preventing contamination during smoking and handling of the product.

One more health concern, stressed recently, is the possible survival of parasites in cold-smoked fish. To kill *Anisakis* larvae at least 1 min heating at 60°C is required. According to Gardiner (1990) salmon fillets salted for 15 h in a salt and sugar mixture and smoked for 12 h at 26°C contained live larvae after 27 days of storage. Generally an effective measure, eliminating the risk of parasites in uncooked fishery products, is freezing and storage at −20°C for at least 3 days.

Marinated Fish

Marinating is the soaking of meat or fish in a water solution of a variety of compounds such as spices and herbs, salt, organic acids, and others. In a more strict sense, and applied to fish, it means preserving by a solution of sodium chloride and, mostly, acetic acid.

Preservation of fish by marinating goes back to the 7th century BC (Shenderyuk and Bykowski, 1990). Nowadays it is particularly herring that is marinated but, mainly in The Netherlands, mussels as well. As for herring, the usual procedure consists of leaving the raw fish in a comparatively strong solution of salt and acetic acid (the marinating bath) for some weeks and subsequent replacement of this bath by a solution containing much lower amounts of these components (the cover brine).

Unlike mussels, herring are usually marinated in the raw state. In some procedures herring (and other fish as well) are cooked or fried before marinating. The cover brine sometimes contains gelatine which forms a gel in the final product.

The marinating process causes denaturation of fish proteins by sodium chloride but by the acid as well. Denaturation by acetic acid starts at a pH value of about 5 and proceeds faster at lower pH values. The acetate ion binds to the basic groups of the proteins and in this way changes the charge on the protein molecules.

Protein hydrolysis by acetic acid is improbable. If the acetic acid concentration is high, the skin of the herring may come off the flesh. This also happens if the marinated herring is stored for too long a time.

At pH values of 4.0 to 4.2 which normally occur in marinades, however, the proteolytic enzymes present in muscle are optimally active. This implies that marinated herring may be subjected to proteolysis, which, in extreme cases, amounts to 50% of the total protein mass (Tülsner, 1978). Disintegration of the Z lines in the myofibrils as well as myosin breakdown was demonstrated (Rodger *et al.*, 1984). The protein changes result in a desirable modification in the rheological properties of the flesh and development of the characteristic flavour. Proteolysis can be controlled by temperature and concentration of salt.

In some countries hydrogen peroxide may be added to the marinating bath in order to bleach the herring. It may, however, oxidize the SH groups in proteins to sulfoxides and decrease the proteolytic activity.

High acidity of fresh marinated fish guards against the growth of most spoilage bacteria. The heterofermentative *Betabacterium büchneri* and *Betabacterium breve*, however, may proliferate and decarboxylate amino acids liberated from proteins and peptides during ripening, especially arginine, lysine, histidine, tyrosine and glutamic acid. The decarboxylation leads to an increase in pH of the medium and, consequently, to the development of spoilage microflora during prolonged storage of the product under abuse temperature conditions. For this reason, proteolysis should be restricted. On the other hand, amino acids and decomposition products of these contribute to the taste and flavour of marinated herring (Tülsner, 1978). It requires some skill to prepare a sufficiently stable marinade having a pleasant aroma as well.

Marinades containing onion cuttings may develop a brown discoloration if the cut onions are left aside for some time (e.g. a weekend) before addition to the marinade. Bacteria such as *Erwinia herbicola* (De Smedt and De Ley, 1979), naturally present on the onion surface, are able to oxidize glucose present in the onions to reactive metabolites which combine with NH_2 groups and give rise to brown reaction products. This reaction even proceeds at the pH of the marinade, but is greatly accelerated at higher pH values and temperatures.

Canned Seafood

Principles of preparation

Several kinds of fish, i.e. mainly the fatty species, but also crustaceans and molluscs, are especially suitable for canning. Some are known to most consumers mainly as canned products, e.g. salmon, tuna, mackerel, sardine, herring, sprat, shrimp, king crab, scallop, oyster, or clam. The cans may contain whole, deheaded small fish, trunks, fillets, steaks, boneless flesh parts or mince.

In canning, the raw material usually undergoes short brining and some kind of heat treatment prior to packing and sealing. Some assortments of canned seafoods, however, are prepared as ‘natural products’ which are packed into the cans only with salt and sometimes spices. The same holds for liver from some species such as the Gadidae. The heat treatment, if applied, should reduce the moisture content by 10–30%, facilitate peeling, shucking or removal of dark meat, or develop special desirable sensory properties in the product. Different methods are used, mainly steaming, cooking, blanching in brine, frying, pre-drying or hot-smoking. Steaming is often carried out in open cans. Depending on the type of product, different sauces, dressings, brines or oil is added. The cans or jars are sealed under conditions causing reduced pressure in the container. Some cans are prepared as dry pack, without any added liquid. The sealed cans are sterilized in stationary or rotating retorts in time–temperature conditions required by the given product and can dimensions to achieve storage stability at ambient temperature and bacteriological safety. As most of the canned fishery products belong to the low acid food category, the botulism hazard must be totally eliminated by sterilization values high enough to reduce the number of *C. botulinum* spores eventually present by 12 log cycles. Spores from thermophilic spore-forming spoilage organisms of the genera *Bacillus* and *Clostridium*, as a rule, are more resistant to heat than the spores of *C. botulinum*. Thus, if commercially required stability of the product is assured by processing, the safety requirements are also fulfilled (Naczk and Artyukhova, 1990). The sterilized cans are cooled, washed and dried under conditions preventing bacterial recontamination via microleaks eventually present or formed in the can seams due to transient (thermal) deformations of the container.

Quality and stability of canned seafoods

Heat sterilization affects not only the fish microflora but also the sensory properties and nutritional value of the product. In canned sprat and other small fish the bony parts become softened due to collagen breakdown, and can be digested to be used as a source of nutritionally valuable calcium. Fish collagen is not as heat-stable as that of terrestrial animals and denatures at lower temperatures (see Chapter 3). Severe heat treatment may have an adverse impact on the colour, texture and flavour of canned fish due to degradation and chemical interactions of proteins, sugars and lipids. Reactions such as these may take place within the fish, but also as a result of the presence of other ingredients, e.g. sauces added to the product. Components present in tomato sauce, for example, may be oxidized by trimethylamine oxide (TMAO) from the fish during heat sterilization, the oxidation products giving rise to a dark discoloration of the sauce and to a 'burnt' taste.

The nutritive value may decrease as the result of partial thermal destruction of essential amino acids and vitamins, in particular thiamine. The losses of this vitamin may be as high as 80%. Decomposition of the nutritionally important sulfur amino acids also results in the generation of objectionable volatile compounds. In canned seafoods prepared in large containers these undesirable changes may be significantly larger than in home-prepared dishes. In order to reduce these effects high-temperature/short-time sterilization of canned fish is used, whenever the size of the cans and the character of heat exchange inside the can make it possible.

Canned seafoods are shelf-stable at ambient temperature, provided the cans are protected from corrosion and mechanical damage and the can material is resistant to the food product. Careful handling during transportation and storage is of particular importance when modern containers made of laminates and flexible pouches are used. During the first few weeks of storage the typical sensory quality of the products develops as a result of chemical reactions between the components of the pack.

Fish Minces

Raw material and principles of preparation

Fish of prime quality of the most valuable species is usually sold fresh, as frozen fillets, or else as smoked, salted, marinated or canned products. There are, however, many species which have been traditionally processed into different ball-shaped products made from minced flesh. This category of produce from minced fish has been extended during the last 30 years, beyond just traditional items. The raw material, in the past mainly white fish, now also includes many pelagic fatty species, a large group of different underutilized resources, and remnants of flesh which can be recovered from filleting offal (Grantham, 1981).

The mince is produced on board fishing vessels or in shore plants by separating the flesh in a comminuted form from the bones, skin, scales and fins by screening through a perforated sheet, mainly in drum-type separators. The effectiveness of separation of the non-flesh material and the extent of comminution of the flesh, as well as the sensory and functional properties of the mince, depend upon the species of fish and the initial preparation of the raw material, the diameter of the perforations of the screen, and the shear rate developed in the separator. In a multi-pass system high quality mince is produced at low pressure, while higher yields of products of inferior quality are obtained in the next stages, at lower diameters of the screen holes and higher pressure. The mince is either immediately processed into different products or may by frozen after stabilization by cryoprotectants.

Generally, however, the mince is further refined by leaching the soluble fraction with water, straining to remove any remaining connective tissue elements, separating of lipids, dewatering and stabilization, to be used as frozen material known as *surimi*. This bland, preferably white myofibrillar protein concentrate, is used for manufacturing different gelled or emulsion-type fishery products or as an ingredient in processing other foods. Although leaching removes about a third of the initial content of protein from the minced fish flesh, which is generally lost and creates pollution problems, it significantly improves the functional properties of the product.

Improvement of functional properties can be also achieved by the presence of Ca^{2+} and Mg^{2+} ions, in concentrations of about 10 mM, in the wash water (Solberg *et al.*, 1992). It is now possible to produce surimi in a continuous system (Whittle and Hardy, 1992).

Stability

The functional properties of frozen fish minces deteriorate during frozen storage due to denaturation of proteins, involving loss in solubility, water retention, gelling ability and lipid emulsifying capacity. Freeze denaturation may be caused by ions and ice crystals, chemical interactions of proteins with formaldehyde generated from TMAO in gadoid fish, binding of fatty acids and lipid oxidation products, and formation of disulfide bonds (Sikorski and Pan, 1991). These protein changes can be minimized or prevented by applying a very low and constant temperature of storage, by protective packaging, by inhibiting oxidation of lipids through eliminating from the raw material various tissues very rich in haematin compounds and the addition of antioxidants, by removing the water-soluble fraction containing both TMAO and the factors catalysing the formation of formaldehyde, and addition of cryoprotectants. Another proposed approach to improve the structure of gelled products consists of adding reductants to the fish meat during the early stages of grinding to liberate the SH groups and, after adjustment of pH to neutral, re-formation of S—S bonds by adding oxidants during the final stage of grinding (Jiang *et al.*, 1986).

The following compounds are mainly used as cryoprotectants: saccharose (5–10%), sorbitol (about 4%), NaCl (about 2.5%), citrates (1–1.5%), polyphosphates (0.2–0.4%), soyabean proteins (4%), milk proteins (5%), sodium alginate (0.5%) or ascorbate (0.6%). The quantities and proportions of stabilizers depend on the properties of the raw materials and final use of the surimi as well as on the food law regulations. Washed hake mince, stabilized by 4% saccharose + 4% lactose + 0.2% polyphosphate, has a high quality life of 11 months at −20°C, while the value for whiting mince produced during the spawning season, stabilized by the same cryoprotectants, is only 3 months (Kołodziejska *et al.*, 1989). Fish minces containing 5% saccharose have a distinct sweet taste and may not be accepted by different consumers. Some cryoprotectants, e.g.

sorbitol, although very effective, are not permitted in fishery products in all countries.

The microbiological stability of fish minces can be increased by the addition of a preservative in combination with sodium chloride and acetic acid. These additions may be combined with a heat treatment (80°C, 2 min) of the washed mince (Aguilera and Figueroa, 1992).

Mince products

Blocks of frozen fish mince are used for producing battered fingers, sticks, steaks and cakes, which are distributed in the frozen state, in manufacturing of fish balls, loaves, patties, and sausages, as well as different textured and extruded commodities. Some of the mince products are canned. By far the largest application of surimi is in manufacturing of various types of Japanese heat-gelled products such as *itatsuki*, *kamaboko*, *chikuwa*, *hanpen* and *satsuma-age*. The characteristic rheological properties of the traditional Japanese products are based on the gel-forming ability of the fish myofibrillar proteins. The quality of these products is highly affected by the species-determined properties of proteins and by the parameters of manufacturing. The latter also takes into account the activity of the endogenous heat-stable alkaline protease, which may have a gel-softening effect.

A growing field of application of surimi is the manufacture of extruded shellfish analogue products. According to the fabrication process and structural characteristics of these products they may be categorized as moulded, fiberized, composite-moulded or emulsified (Lee, 1984). Recently surimi has been regarded as a valuable, functional proteinaceous ingredient, just like soyabean proteins, in a variety of food products. According to Lanier (1994) many fatty fish species could become the 'soybeans of the sea', valued mainly for the functional qualities of their refined oils and proteins.

Special Fishery Products and By-products

Fish sauce

Fish sauce is a food product made by storing salt-preserved fish material at tropical temperatures until it is solubilized by endogenous enzymes. Small fish or minced fish material is mixed with salt and stored in a closed container. During storage the digestive enzymes slowly hydrolyse the fish protein, producing the clear amber-coloured fish sauce which can be used as a condiment or dressing on rice or vegetable dishes.

Fish sauce was produced in ancient Greece and Rome (Corcoran, 1963). In Europe this tradition vanished, but in South East Asia fish sauce is an important food ingredient used daily by major parts of the population (Amano, 1962; Lee, 1990). The annual production in this region is estimated to be about 250,000 tonnes (Stefansson and Steingrimsdottir, 1990).

To obtain a stable product the sauce must contain at least 20% of salt. Hence, about three parts of small fish or minced by-products are mixed with one part of salt.

Traditionally, solar salt is used. The tryptic enzymes are of major importance for the solubilization process (Orejana and Liston, 1982). The activity of these enzymes is reduced due to the high salt content and endogenous enzyme inhibitors. An acceptable yield (above 50%) of fish sauce is only obtained after at least 6 months of storage.

Several methods have been proposed to reduce the production time. These include raising the initial temperature (Hamm and Clague, 1950; Mabesa *et al.*, 1990), addition of plant enzymes (Beddows and Ardeshir, 1979a), addition of acid combined with reduced salt content (Beddows and Ardeshir, 1979b; Gildberg *et al.*, 1984) and initial alkalization at low salt concentration (Gildberg, 1989). An improved hydrolytic rate has been achieved by all these methods but inferior flavour, i.e. too weak or bitter, is a problem in most cases when the process is accelerated. Addition of the pineapple enzyme bromelain has given promising results, and there are reasons to believe that part of the commercial fish sauce production in Thailand is based on enzyme addition (Thongthai, pers. comm.). In China a method based on initial acidification is apparently used by some fish sauce manufacturers (Shi, pers. comm.).

First-grade fish sauce, which is an undiluted filtrate of the fish autolysate, has a high nutritional value being rich in the essential amino acids lysine and methionine which are usually low in vegetable proteins. However, the recommended daily intake is limited due to the high salt content. A common daily per capita consumption is about 20 g, corresponding to a salt intake of 5 g. This is normally not harmful in tropical regions where the rate of perspiration is high. Health risk may occur, however, when people change from rural to urban life style without changing the food habits. Work is being done to find new safe recipes with a lower salt content, but this is not easy since some harmful bacteria, e.g. *Staphylococcus aureus*, may grow even at high salt concentrations (Owens and Mendoza, 1985).

Second and third-grade qualities of fish sauce are obtained by brine extraction of the residue after harvesting of the first-grade fish sauce. These products are of course cheaper and more available to poor people. The nutritional value, however, is very much inferior to the first-grade product.

At present fish sauce is an important product only in South East Asia, but emigration and internationalization of food habits may very well make the tasty fish sauce a common food item also in the Western world. As a result, a fish sauce factory has recently been established in Canada.

Fish silage

The product of the process of preserving and storing wet biological material in a silo is called silage. The traditional use of the word has been in conjunction with green forage preserved by acid addition and lactic acid production during anaerobic fermentation. Fish silage has been adopted for analogous products of whole fish or parts of fish (Tatterson and Windsor, 1974; Raa and Gildberg, 1982).

Fish silage was first developed in Sweden around 1940 (Edin, 1940) and recognized as an alternative to fish meal in animal feeding. The advantage with fish silage production is the simple technology which can be adapted to any production scale. In principle only three units are necessary to set up a production line, i.e. a mincing device, a dispenser pump for acid addition and a storage tank. Hence, the capital investments are minor as compared with fish meal production. The main disadvantages are the high water content (70–80%), resulting in high transportation costs, variable storage stability and variable chemical composition.

The first significant commercial production of fish silage was established in Denmark in 1948 (Petersen, 1953). The production was based on by-catch fish and fish waste, and formic acid, or a mixture of formic and sulfuric acid, was used for preservation. The silage was used as a protein supply mainly for pigs and poultry. During the 1970s the annual production in Denmark reached a level of about 60,000 tonnes a year (Wignall and Tatterson, 1976). Since then pilot-scale and small-industrial-scale production has been introduced in many countries without approaching the same order of magnitude as the fish meal industry. Increasing demand for standardized feeds in modern husbandry prevented further growth in silage production until methods for silage fractionation and processing had been developed.

During the 1980s industrial-scale silage processing equipment was developed in Norway. Fish silage preserved with about 2% formic acid and antioxidant is produced at fish processing plants or at fish farming slaughterhouses and transported to a central silage processing plant. The main option of the processing plant is to separate the oil and the protein liquid fraction and to concentrate the protein liquid. This is achieved by a processing line as illustrated in Fig. 11.1.

To obtain a good separation it is necessary to solubilize most of the fish tissue. This is achieved by utilizing the digestive enzymes present in the raw material. The acidity of the silage (pH = 4) is optimal for protein digestion by the fish pepsins (Raa and Gildberg, 1976). By moderate heating to 30–40°C for about 2 days most of the protein tissue is

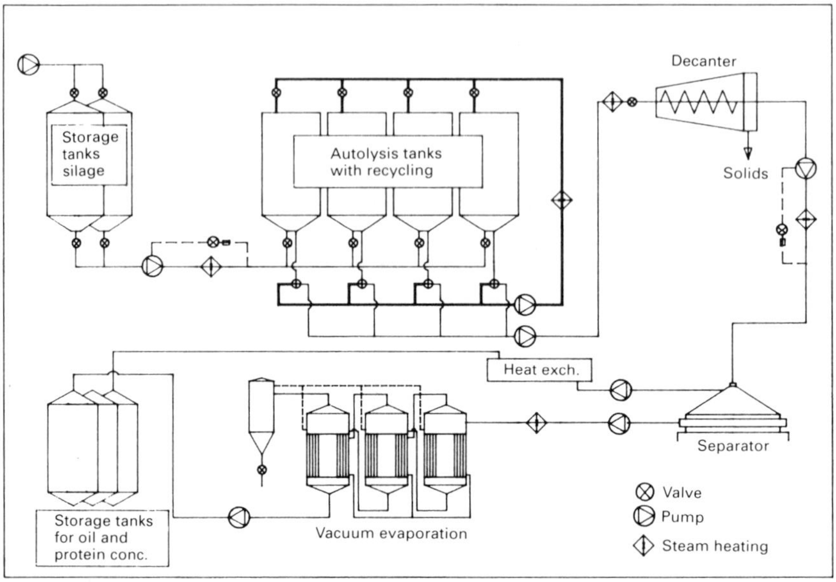

Fig. 11.1. Fish silage processing plant.

solubilized. The solids are removed by a decanter centrifuge, and the oil is separated from the protein liquid after heating to 90°C. Finally the protein liquid is vacuum evaporated to 40–50% dryness.

The protein concentrate is then sold to feed factories, which include certain amounts of fish silage concentrate in standardized feed products for the commercial market. Properly stabilized oil can be applied both in food and feed production.

The silage concentrate is a highly digested protein hydrolysate which is convenient as a protein supply for weaning calves and pigs (Aas, 1982; Opstvedt, 1982) as well as poultry (Krogdahl, 1985a; 1985b). When high amounts of fish silage protein are fed to mature ruminants or fish, the animal production and growth are reduced. This is probably due to adverse effects of highly hydrolysed protein in the metabolism of these animals (Johnsen and Skrede, 1981; Hardy *et al.*, 1984). However, 5–10% of the feed protein may be substituted by silage protein without negative effects. Actually, there are indications that health, fertility and general appearance are improved when some fish silage protein is included.

In Norway the annual silage production at present is about 100,000 t. Fisheries and aquaculture contribute with about equal amounts of raw material. The silage technology is now being recognized as the most useful method for solving the waste problem in fish farming.

Fish protein hydrolysate

Development of fish protein hydrolysates started in Canada in the 1940s (Tarr, 1948; Tarr and Deas, 1949). The first intended application was as an amino nitrogen source in growth media for microorganisms. It was shown that fish flesh hydrolysed by enzymes supported bacterial growth better than hydrolysates prepared by chemical hydrolysis.

Since then it has been proved that fish protein hydrolysates are very good bacterial peptones and comparable with the best meat peptones on the market (Beuchat, 1974; Gildberg *et al.*, 1989; Vecht-Lifshitz *et al.*, 1990). However, so far industrial production of bacterial peptones from fish has not been a commercial success. The major efforts have been made on utilization of fish protein hydrolysates as food or feed ingredients (Mackie, 1982). Normally by-products from the filleting industry, by-catch fish from the trawlers or small underutilized lean fish species have been used as raw material. The general procedure for fish protein hydrolysate preparation is as follows:

- The raw material is minced or comminuted. An equal amount of water is mixed in, and the pH is adjusted to the desired level.
- A commercial preparation of hydrolytic enzymes, usually of plant or microbial origin, is added before the temperature is raised to 50–60°C under stirring. At the end of the incubation period the enzymes are inactivated, and the material is pasteurized by heating to 80–90°C.
- Bones and scales are removed by sieving. Oil, sediment and hydrolysate are subsequently separated by a three-phase separator.
- The hydrolysate is filtered, evaporated and dried by a vacuum or spray-drier.

The average chemical composition of fish protein hydrolysate is about 85% hydrolysed protein, mainly small peptides and free amino acids, 10% inorganic matter and 5% water. The lipid content is normally very low. Hydrolysate yield and degree of hydrolysis depend on factors such as the type of raw material, the hydrolytic power of the enzyme, the temperature and the incubation time.

The choice of enzyme preparation and the process conditions determine the properties of the product. Mild conditions, a short time of hydrolysis and moderate temperature must be chosen to retain emulsifying or gelling properties. Mild properties normally also give rise to a pleasant flavour but to a comparatively low yield and sometimes to an imbalanced amino acid composition, which reduces the nutritional value (Sripathy *et al.*, 1963). If high yield and good solubility are most important, extensive hydrolysis at higher temperatures must be applied. Such conditions are normally chosen if the hydrolysate has to be used as a feed ingredient or as an amino nitrogen source in microbial growth media. Often such hydrolysates are very bitter due to the release of small hydrophobic peptides and, for this reason, are unsuitable as food ingredients. This problem can be eliminated by adding enzymes able to degrade the bitter peptides (In, 1990; Sugiyama *et al.*, 1991).

Today a significant commercial production of fish protein hydrolysate is established in France and Japan (In, 1990; Uchida *et al.*, 1990). The main application in food is as a flavouring in soups and shellfish analogues, while the major feed applications are as milk replacers for calves and weaning pigs and as attractants in fish feed. An interesting market in the near future is application as an appetizer in cat feed (Willard, 1990).

Fish protein hydrolysates have exposed some properties that are very interesting but still poorly investigated. One observation is that domestic animals given fish protein hydrolysate apparently are more resistant to diseases than others. The reason for this is unknown; but *in vitro* experiments have shown that the proliferation of mouse lymphocytes is stimulated by peptides from fish protein hydrolysates (Vinot *et al.*, 1989). Another interesting observation is that spraying fish protein hydrolysate on vegetable crops such as tomatoes reduces stress and improves survival during transplanting (Wyatt and McGourty, 1990).

Work is being done to explore these remarkable properties of fish protein hydrolysates.

Fish enzymes

During the last decades significant efforts have been made to investigate the possibilities for commercial utilization of enzymes from fish and shellfish. These are usually different from the corresponding mammalian enzymes. As most fishes have low habitat temperatures, their enzymes are adapted to function well at low temperatures (Asgeirsson *et al.*, 1989; Simpson *et al.*, 1989). This property may be advantageous in many biotechnological processes involving food and feed processing or preservation.

One such process is caviar production, where enzymes of fish origin are used to achieve a gentle separation of roe from the connective tissue (Raa, 1986). By this process it has been possible to improve the yield from about 70 to 90% in salmon caviar production (Raa, 1990). In Norway and Iceland there is a small commercial production of enzymes obtained from the viscera of Atlantic cod (Almås, 1990; Stefansson and Steingrimsdottir, 1990). Methods for using such enzyme preparations in descaling of haddock and deskinning of skate wings have been developed. In Norway preparations of fish digestive enzymes have also been used in maturation of herring fillets (Opshaug, 1983), but microbial preparations seem to compete favourably and acquire this market.

Exploration of fish pepsin as a digestive aid in fish feed is probably in progress in several countries, but so far the results from this work are not available.

Canadian research has shown that cod pepsin can be used in cheese production and cod trypsin to accomplish cold extraction of the pigments from shrimp waste (Brewer *et al.*, 1984; Cano-Lopez *et al.*, 1987). These methods have not yet been applied in commercial production.

A number of interesting enzymes have been detected in marine invertebrates. A joint venture company between interests in Russia and Liechtenstein (Seatec) produces a collagenase preparation from crab hepatopancreas which may be used in deskinning of squid, caviar production and ripening of salted fish.

A Norwegian company (Marine Biochemicals) produces alkaline phosphatase from shrimp processing waste water. The enzyme is used as a marker in diagnostics and in gene technology (Olsen *et al.*, 1991). A protease preparation from Antarctic krill can promote wound healing by rapid digestion of necrotic tissues (Campbell *et al.*, 1987; Hellgren and Vincent, 1989). In Sweden the preparation is now being tested out for approval in medical application (Hellgren *et al.*, 1991). The preparation is made from an extract of defatted krill. A simple method for bulk production of hydrolytic enzymes from squeezed krill autolysate has also been developed (Karlstam *et al.*, 1991). During autolysis most of the krill protein is digested to free amino acids, which may be recovered as a by-product from the enzyme production (Ellingsen and Mohr, 1979).

Lysozyme from the clam shell *Chlamys islandica* is very active even at 0°C. A method for large-scale extraction of this enzyme from the waste fraction in shell processing is being developed in Norway. Intended application is as a bacteriostatic and prophylactic agent in speciality food and feed (Raa, 1990).

Exploitation of fish enzymes is complicated by variable availability of raw material and expensive production due to comparatively low enzyme concentrations. It seems that profitable processes can only be established when the fish enzymes have unique properties that cannot be imitated by cheaper enzymes of plant or microbial origin. In the future some of these enzymes may very well be produced more profitably by recombinant technology.

Miscellaneous products

By-products from fish and shellfish are rich in interesting tissue fractions and chemical compounds which may be, and to a certain extent already are, subjected to commercial exploitation.

Fish skin and shellfish processing waste are two major fractions that are poorly utilized. Although small quantities of fish skin are traditionally used for leather production (Lewis, 1990), gelatin extracted from the skin may become a more important product (Osborne *et al.*, 1990). The gelatin polymer is an important ingredient in many food items. Although fish gelatin may have a lower gel strength than gelatin produced from mammalians, there is a significant market for kosher gelatin produced from non-mammalian sources. Fish gelatin gels at much lower temperatures than gelatin extracted from mammalian tissues, and this may imply special technical applications. Today fish gelatin is produced commercially in Nova Scotia (Norland, 1990).

Chitin yields about one third of the weight of crustacean shells and is one of the most abundant polymers on earth (Martin and Patel, 1991).

Chitin is insoluble in most solvents, but by alkali treatment it is deacetylated to yield the water-soluble chitosan. Chitosan is a biodegradable cationic polysaccharide with a great application potential ranging from flocculant in waste water treatment and additive in foods and cosmetics to numerous technical and medical applications (Lower, 1984; Knorr, 1991). Although the application potential is large, chitosan has not yet obtained a large part of the polymer market, and today industrial production of chitosan from shellfish waste is only taking place in Japan and the USA (Skaugrud and Sargent, 1990). In Poland, shellfish chitosan is produced for research and commercial purposes (Sikorski and Bykowski, 1994).

Carotenoprotein can be extracted as a by-product in chitosan production from shellfish waste and used as a pigment and protein source in salmon feed (Ya *et al.*, 1991).

Nucleic acids are a third polymer group which can be obtained from fish by-products. Various preparations of nucleic acids are made from fish milt (Murata *et al.*, 1991; Richardsen, 1991). These products are used, particularly in France and Japan, in speciality foods and for medical purposes.

Many chemical compounds with antibacterial, antivirus or antitumour effect have been isolated from marine invertebrates (Olafsen, 1988; Guyot, 1989; Raa, 1990; Seto, 1990). These compounds range from high molecular mass lectins in shellfish to low molecular mass peptides and quinones in sponges and tunicates, and they all seem to be part of the very efficient non-specific disease defence system of the animals.

Many of these compounds have an obvious potential in medical research and therapy. Some of the low molecular mass compounds have simple chemical structures and can easily be synthetized (Rinehart *et al.*, 1988).

Generally the marine environment is an immense pool of genetic material, and the search for interesting chemical compounds is only beginning.

References

Aas, T. (1982) Fiskeslo som fór til svin (Fish viscera as pig feed). In: *Husdyrforsøksmøtet* 1982. Norske Landbrukshøgskole, Ås, Norway, pp. 506–511.

Aguilera, J.M. and Figueroa, G. (1992) Combined methods technology in the preservation of pelagic fish muscle. In: Bligh, E.G. (ed.) *Seafood Science and Technology*, Proceedings International Conference on Seafood 2000, Halifax, Canada, 1990. Fishing News Books, Oxford, pp. 127–136.

Almås, K.A. (1990) Utilization of marine biomass for production of microbial growth media and biochemicals. In: Voigt, M.N. and Botta, P.B. (eds) *Advances in Fisheries Technology and Biotechnology for Increased Profitability*. Technomic Publishing Co., Lancaster, Pennsylvania, pp. 361–372.

Amano, K. (1962) The influence of fermentation on the nutritive value of fish with special reference to fermented fish products in South-East Asia. In: Heen, E. and Kreuzer, R. (eds) *Fish in Nutrition*. Fishing News Books, London, pp. 180–197.

Asgeirsson, B., Fox, J.W. and Bjarnason, J.B. (1989) Purification and characterization of trypsin from the poikilotherm *Gadus morhua*. *European Journal of Biochemistry* 180, 85–94.

Beddows, C.G. and Ardeshir, A.G. (1979a) The production of soluble fish protein solution for use in fish sauce manufacture. I. The use of added enzymes. *Journal of Food Technology* 14, 603–612.

Beddows, C.G. and Ardeshir, A.G. (1979b) The production of soluble fish protein solution for use in fish sauce manufacture. II. The use of acids at ambient temperature. *Journal of Food Technology* 14, 613–623.

Beltran, A. and Moral, A. (1989) Effect of smoking on lipid stability in sardine (*Sardina pilchardus* W.). *Zeitschrift für Lebensmittel-Untersuchung und Forschung* 189, 132–136.

Beuchat, L.R. (1974) Preparation and evaluation of a microbial growth medium formulated from catfish waste peptone. *Journal of Milk and Food Technology* 37, 277–281.

Bhuiyan, A.K.M.A., Ackman, R.G. and Lall, S.P. (1986) Effects of smoking on protein quality of Atlantic mackerel. *Journal of Food Processing and Preservation* 10, 115–126.

Biegler, P. (1960) Fischwaren-Technologie. Theorie und Praxis der Fabrikationsmethoden zur Konservierung von Fischen. In: Baader, C. (ed.) *Der Fisch*, Vol. 5. Verlag 'Der Fisch', Clara Baader, Lübeck.

Bligh, E.G., Shaw, S.J. and Woyewoda, A.D. (1988) Effects of drying and smoking on lipids of fish. In: Burt, J.R. (ed.) *Fish Smoking and Drying. The Effect of Smoking and Drying on the Nutritional Properties of Fish*. Elsevier Applied Science, London, pp. 40–52.

Brewer, P., Helbig, N. and Haard, N.F. (1984) Atlantic cod pepsin. Characterization and use as a rennet substitute. *Canadian Institute of Food Science and Technology Journal* 17, 38–43.

Campbell, D., Hellgren, L., Karlstam, B. and Vincent, J. (1987) Debriding ability of a novel multi-enzyme preparation isolated from Antarctic krill (*Euphausia superba*). *Experientia* 43, 578–579.

Cano-Lopez, A., Simpson, B.K. and Haard, N.F. (1987) Extraction of carotenoprotein from shrimp process wastes with the aid of trypsin from Atlantic cod. *Journal of Food Science* 52, 503–506.

Chen, L.B. and Issenberg, P. (1972) Interactions of wood smoke components with *epsilon*-amino groups in proteins. *Journal of Agricultural and Food Chemistry* 20, 1113–1115.

Corcoran, T.H. (1963) Roman fish sauces. *Classical Journal* 58, 204–210.

Cutting, C.L. (1955) *Fish Saving: A History of Fish Processing from Ancient to Modern Times*. Leonard Hill, London, 372 pp.

De Smedt, J. and De Ley, J. (1979) Identification of Ruiter's strains, isolated from browned marinated herring, as members of *Erwinia herbicola*. *International Journal of Systematic Bacteriology* 29, 183–187.

Doe, P. and Olley, J. (1990) Drying and dried fish products. In: Sikorski, Z.E. (ed.) *Seafood: Resources, Nutritional Composition, and Preservation*. CRC Press, Boca Raton, Florida, pp. 125–145.

Doe, P.E., Ahmed, M., Muslemuddin, M. and Sachithananthan, K. (1977) A polythene tent drier for improved sun drying of fish. *Food Technology in Australia* 29, 437–441.

Dvořák, Z. and Vognarová, I. (1965) Available lysine in meat and meat products. *Journal of the Science of Food and Agriculture* 16, 305–312.

Edin, H. (1940) Undersökningar angåande importavstegningens äggviteproblem (Investigations concerning the lack of protein caused by the import closure). *Nordisk Jordbruksforskning* 22, 142–158.

Eklund, M.W., Peterson, M.E., Paranjpye, R. and Pelroy, G.A. (1988) Feasibility of a heat-pasteurization process for the inactivation of nonproteolytic *Clostridium botulinum* types B and E in vacuum-packaged, hot-process (smoked) fish. *Journal of Food Protection* 51, 720–723.

Ellingsen, T. and Mohr, V. (1979) A new process for utilization of Antarctic krill. *Process Biochemistry* 14(10), 14–19.

Fichtl, P. and Schreiber, W. (1987) Umrötung von Fischfleisch bei Einsatz einer Starterkultur. *Informationen für die Fischwirtschaft* 34, 135–136.

Filsinger, B., Sisti, E. and Bergamaschi, N.J. (1987) Technical note: Chemical and sensory assessments in ripened anchovies. *International Journal of Food Science and Technology* 22, 73–76.

Gardiner, M.A. (1990) Survival of *Anisakis* in cold smoked salmon. *Canadian Institute of Food Science and Technology Journal* 23, 143–144.

Gildberg, A. (1989) Accelerated fish sauce fermentation by initial alkalification at low salt concentration. In: Miyachi, S., Karube, I. and Ishida, Y. (eds) *Current Topics in Marine Biotechnology*. Fuji Technology Press, Tokyo, pp. 101–104.

Gildberg, A., Hermes, J.E. and Orejana, F.M. (1984) Acceleration of autolysis during fish sauce fermentation by adding acid and reducing the salt content. *Journal of the Science of Food and Agriculture* 35, 1363–1369.

Gildberg, A., Batista, I. and Strøm, E. (1989) Preparation and characterization of peptones obtained by a two-step enzymatic hydrolysis of whole fish. *Biotechnology and Applied Biochemistry* 11, 413–423.

Góra, A. (1972) Opracowanie technologii słabego solenia śledzi i makreli oraz enzymatycznego solenia filetów śledziowych (The technology of light salting of herring and mackerel and of enzymatically salted herring fillets). *Research Report No. 203.3(T-95)*. Sea Fisheries Institute, Gdynia, 37 pp.

Grantham, G.J. (1981) Minced fish technology: a review. *FAO Fisheries Technical Paper* 216, 72 pp.

Guyot, M. (1989) Bioactive metabolites from marine invertebrates. In: Miyachi, S., Karube, I. and Ishida, Y. (eds) *Current Topics in Marine Biotechnology*. Fuji Technology Press, Tokyo, pp. 349–352.

Hamm, W.S. and Clague, J.A. (1950) Temperature and salt purity effects on the manufacture of fish paste and sauce. *Research Report* 24, Fish and Wildlife Service, US Department of the Interior. 11 pp.

Hardy, R.W., Shearer, K.D. and Spinelli, J. (1984) The nutritional properties of co-dried fish silage in rainbow trout (*Salmo gairdneri*) diets. *Aquaculture* 38, 35–44.

Hellgren, L. and Vincent, J. (1989) Debriding properties of krill enzymes in necrotic leg ulcers. *Archives of Dermatology* 125, 1006.

Hellgren, L., Karlstam, B., Mohr, V. and Vincent, J. (1991) Krill enzymes. A new concept for efficient debridement of necrotic ulcers. *International Journal of Dermatology* 30, 102–103.

Huss, H.H. (1981) The ecology of *Clostridium botulinum* type E. In: Roberts, T.A., Hobbs, G., Christian, J.H.B. and Skovgaard, N. (eds) *Psychrotrophic Microorganisms in Spoilage and Pathogenicity*. Academic Press, New York, pp. 463–473.

In, T. (1990) Seafood flavourants produced by enzymatic hydrolysis. In: Keller, S. (ed.) *Making Profits out of Seafood Waste*. Report No. 90-07, Alaska Sea Grant College Program, Fairbanks, pp. 197–201.

Indriati, N., Sudrajat, Y., Anggawati, A.M., Putro, S. and Madden, J. (1985) Insect infestation in dried-salted fish in Java. In: *Spoilage of Tropical Fish and Product Development*. FAO Fisheries Report No. 317, Supplement, Rome, pp. 297–304.

Jiang, S.T., Lan, C.C. and Tsao, Ch.Y. (1986) New approach to improve the quality of minced fish products from freeze–thawed cod and mackerel. *Journal of Food Science* 51, 310–312.

Johnsen, F. and Skrede, A. (1981) Evaluation of fish viscera silage as a feed resource. *Acta Agriculturae Scandinavica* 31, 21–27.

Karlstam, B., Vincent, J., Johansson, B. and Brynö, C. (1991) A simple purification method of squeezed krill for obtaining high levels of hydrolytic enzymes. *Preparative Biochemistry* 21, 237–256.

Keller, M. (1992) Fischindustrie und Küstenfischgroßhandel. In: *Jahresbericht über die Deutsche Fischwirtschaft 1990/91*, pp. 45–54.

Knøchel, S. and Huss, H.H. (1984a) Ripening and spoilage of sugar-salted herring with and without nitrate. I. Microbiological and related chemical changes. *Journal of Food Technology* 19, 203–213.

Knøchel, S. and Huss, H.H. (1984b) Ripening and spoilage of sugar-salted herring with and without nitrate. II. Effect of nitrate. *Journal of Food Technology* 19, 215–224.

Knorr, D. (1991) Recovery and utilization of chitin and chitosan in food processing waste management. *Food Technology* 45(1), 114–120, 122.

Kołodziejska, I., Sadowska, M. and Kołodziej, H. (1989) The effect of stabilizers on the functional properties of cod, blue whiting, and hake surimi. In: Spiess, W.E.L. and Schubert, H. (eds) *Engineering and Food. Vol. I: Physical Properties and Processing Control*. Elsevier Applied Science, London, pp. 315–324.

Kosak, P. and Toledo, R. (1980) Product quality and energy use in high temperature smoking of fish. *Journal of Food Science* 45, 1481–1486.

Kowalewski, W. (1989) Sterowanie procesami proteolitycznymi w dojrzewaniu ryb solonych – na przykładzie szprota bałtyckiego (Control of proteolytic enzymes in the ripening of salted Baltic sprats). PhD thesis, Technical University of Gdańsk, 53 pp.

Krogdahl, Å. (1985a) Fish viscera silage as a protein source for poultry. I. Experiments with layer-type chicks and hens. *Acta Agriculturae Scandinavica* 35, 3–23.

Krogdahl, Å. (1985b) Fish viscera silage as a protein source for poultry. II. Experiments with meat-type chickens and ducks. *Acta Agriculturae Scandinavica* 35, 24–32.

Krylova, N.N., Basarova, K.I. and Kuznetsova, V V. (1962) Interaction of smoke components with meat constituents. Paper presented at the *8th European Meeting of Meat Research Institutes*, Moscow, 9 pp.

Lanier, T.C. (1994) Functional food protein ingredients from fish. In: Sikorski, Z.E., Pan, B.S. and Shahidi, F. (eds) *Seafood Proteins*. Chapman & Hall, London, pp. 127–159.

Lea, C.H. (1933) Chemical changes in the fat of frozen and chilled meat. V. The effect of smoking and the influence of atmospheric humidity on the keeping properties of bacon. *Journal of the Society of Chemical Industry* 52, 57T–63T.

Lee, C.M. (1984) Surimi process technology. *Food Technology* 38(11), 69–80.

Lee, G.-H. (1990) Fish fermentation technology. A review. In: Reilly, P.J.A., Parry, R.W.H. and Barile, L.E. (eds) *Post-Harvest Technology, Preservation and Quality of Fish in Southeast Asia*. Echanis Press, Manila, pp. 1–13.

Levieva, L.C. (1956) Bufernost kak obektivnyi pokazatel stepeni zrelosti prezervov (Buffering capacity as an objective index of the degree of ripening of fish products). *Rybnoe Khozyaistvo* 32(5), 1–3.

Lewis, R.V. (1990) The aquatic leather industry: Opportunities for small manufactures in Alaska. In: Keller, S, (ed.) *Making Profits out of Seafood Waste*. Report No. 90-07, Alaska Sea Grant Program, Fairbanks, pp. 101–103.

Lower, E.S. (1984) Polymers from the sea. Chitin and chitosan. *Manufact. Chemist* Oct., 47, 49, 52.

Luijpen, A.F.M.G. (1959) De invloed van het kaken op de rijping van gezouten maatjesharing (The influence of gibbing on the ripening of salted matjes herring). PhD Thesis, Utrecht, 91 pp.

Mabesa, R.C., Carpio, E.V. and Mabesa, L.B. (1990) An accelerated process for fish sauce (patis) production. In: Reilly, P.J.A., Parry, R.W.H. and Barile, L.E. (eds) *Post-Harvest Technology and Quality of Fish in Southeast Asia*. Echanis Press, Manila, pp. 45–49.

Mackie, J.M. (1982) Fish protein hydrolysates. *Process Biochemistry* 17(1), 26–28, 31.

Martin, A.M. and Patel, T.R. (1991) Bioconversion of waste from marine organisms. In: Martin, A.M. (ed.) *Bioconversion of Waste Materials to Industrial Products*. Elsevier Applied Science, London, pp. 417–420.

Meyer, V. (1965) Über die Bedeutung proteolytischer Fermente bei der Herstellung nichtsterilisierter Fischwaren. *Archiv für Fischereiwissenschaft* 15, 245–253.

Miler, K.B.M. and Sikorski, Z.E. (1990) Smoking. In: Sikorski, Z.E. (ed.) *Seafood: Resources, Nutritional Composition, and Preservation*. CRC Press, Boca Raton, Florida, pp. 163–180.

Murata, Y., Hayashi, T., Watanabe, E. and Toyama, K. (1991) Preparation of skipjack spermary extract by enzymolysis. *Nippon Suisan Gakkaishi* 57, 1127–1132.

Naczk, M. and Artyukhova, A.S. (1990) Canning. In: Sikorski, Z.E. (ed.) *Seafood: Resources, Nutritional Composition, and Preservation*. CRC Press, Boca Raton, Florida, pp. 181–198.

Norland, R.E. (1990) Fish gelatin. In: Voigt, M.N. and Botta, J.R. (eds) *Advances in Fisheries Technology and Biotechnology for Increased Profitability*. Technomic Publishing Co., Lancaster, Pennsylvania, pp. 325–333.

Olafsen, J.A. (1988) Role of lectins in invertebrate humoral defense. *American Fisheries Society Spec. Publ.* 18, 189–205.

Olley, J. and Thrower, S.J. (1977) Abalone – an esoteric food. In: Chichester, C.O. *et al.* (eds) *Advances in Food Research*, Vol. 23. Academic Press, New York, pp. 143–186.

Olley, J., Doe, P.E. and Heruwati, E.S. (1988) The influence of drying and smoking on the nutritional properties of fish: an introductory overview. In: Burt, J.R. (ed.) *Fish Smoking and Drying: The Effect of Smoking and Drying on the Nutritional Properties of Fish*. Elsevier Applied Science, London, pp. 1–21.

Olsen, R.L., Øverbø, K. and Myrnes, B. (1991) Alkaline phosphatase from the hepatopancreas of shrimp (*Pandalus borealis*): A dimeric enzyme with catalytically active subunits. *Comparative Biochemistry and Physiology* 99B, 755–761.

Opshaug, K. (1983) Fremgangsmåte for enzymatisk hurtigmodning av sild (Procedure for accelerated enzymatic ripening of herring). *Norwegian Patent* No. 148,207.

Opstvedt, J. (1982) Fiskeslo som főr til kalver (Fish viscera as calf feed). In: *Husdyrforsøksmøtet 1982*. Norske Landbrukshøgskole, Ås, Norway, pp. 499–505.

Opstvedt, J. (1988) Influence of drying and smoking on protein quality. In: Burt, J.R. (ed.) *Fish Smoking and Drying – The Effect of Smoking and Drying on the Nutritional Properties of Fish*. Elsevier Applied Science, London, pp. 23–39.

Orejana, F.M. and Liston, J. (1982) Agents of proteolysis and its inhibition in patis (fish sauce) fermentation. *Journal of Food Science* 47, 198–203.

Osborne, R., Voigt, M.N. and Hall, D.E. (1990) Utilization of lumpfish (*Cyclopterus lumpus*) carcasses for the production of gelatin. In: Voigt, M.N. and Botta, J.R. (eds) *Advances in Fisheries Technology and Biotechnology for Increased Profitability*. Technomic Publishing Co., Lancaster, Pennsylvania, pp. 143–150.

Owens, J.D. and Mendoza, L.S. (1985) Enzymatically hydrolysed and bacterically fermented fishery products. *Journal of Food Technology* 20, 273–293.

Pan, B.S. (1988) Undesirable factors in dried fish products. In: Burt, J.R. (ed.) *Fish Smoking and Drying. The Effect of Smoking and Drying on the Nutritional Properties of Fish*. Elsevier Applied Science, London, pp. 61–71.

Pedersen, T. (1981a) Törrfisk (Stockfish). In: *Prosesser og produkter i norsk fiskeindustri*, Vol. 4. Universitetsforlaget, Oslo, pp. 26–61.

Pedersen, T. (1981b) Produksjon av klippfisk (Production of klipfish). In: *Prosesser og produkter i norsk fiskeindustri*, vol. 4. Universitetsforlaget, Oslo, pp. 95–124.

Pedersen, T. (1983) Hermetisering (Canning). In: *Prosesser og produkter i norsk fiskeindustri*, vol. 5. Universitetsforlaget, Oslo, pp. 9–31.

Pelroy, G.A., Eklund, M.W., Paranjpye, R.N., Suzuki, E.M. and Peterson, M.E. (1982) Inhibition of *Clostridium botulinum* types A and E toxin formation by sodium nitrite and sodium chloride in hot-process (smoked) salmon. *Journal of Food Protection* 45, 833–841.

Petersen, H. (1953) Acid preservation of fish and fish offal. *FAO Fisheries Bulletin* 6, 18–26.

Poulter, R.G. (1988) Processing and storage of traditional dried and smoked fish products. In: Burt, J.R. (ed.) *Fish Smoking and Drying. The Effect of Smoking and Drying on the Nutritional Properties of Fish*. Elsevier Applied Science, London, pp. 85–90.

Raa, J. (1986) Modern biotechnology: impact on aquaculture and the fish processing industry. Paper presented at the *5th World Productivity Congress*, Jakarta.

Raa, J. (1990) Biotechnology in aquaculture and the fish processing industry: A success story in Norway. In: Voigt, M.N. and Botta, J.R. (eds) *Advances in Fisheries Technology and Biotechnology for Increased Profitability*. Technomic Publishing Co., Lancaster, Pennsylvania, pp. 509–524.

Raa, J. and Gildberg, A. (1976) Autolysis and proteolytic activity of cod viscera. *Journal of Food Technology* 11, 619–628.

Raa, J. and Gildberg, A. (1982) Fish silage: A review. *CRC Critical Reviews of Food Science and Nutrition* 16, 383–419.

Reay, G.A. (1933) The influence of freezing temperatures on haddock's muscle. *Journal of the Society of Chemical Industry* 52(34), 265T–270T.

Richardsen, R. (1991) Utilisation of roe and milt from herring. In: Burt, J.R., Hardy, R. and Whittle, K.J. (eds) *Pelagic Fish: The Resource and its Exploitation*. Blackwell Scientific Publications, Oxford, pp. 299–306.

Rinehart, K.L., Kishore, V., Bible, K.C., Sakai, R., Sullings, D.W. and Li, K.-M. (1988) Didemins and tunichlorin: Novel natural products from the marine tunicate *Trididemnum solidum*. *Journal of Natural Production* 51, 1–21.

Rodger, G., Hastings, R., Cryne, C. and Bailey, J. (1984) Diffusion properties of salt and acetic acid into herring and their subsequent effect on the muscle tissue. *Journal of Food Science* 49, 714–720.

Rørvik, L.M., Yndestad, M. and Skjerve, E. (1991) Growth of *Listeria monocytogenes* in vacuum-packed, smoked salmon during storage at 4°C. *International Journal of Food Microbiology* 14, 111–118.

Ruiter A. (1972) Substitution of proteases in the enzymic ripening of herring. *Annales de Technologie Agricole* 21, 597–605.

Ruiter, A. (1979) Color of smoked foods. *Food Technology* 33(5), 54, 56, 58–60, 63.

Seto, A. (1990) Overview of marine biotechnology in Japan and its commercial application for aquaculture. In: Voigt, M.N. and Botta, J.R. (eds) *Advances in Fisheries Technology and Biotechnology for Increased Profitability*. Technomic Publishing Co., Lancaster, Pennsylvania, pp. 525–541.

Shenderyuk, V.I. and Bykowski, P.J. (1990) Salting and marinating of fish. In: Sikorski, Z.E. (ed.) *Seafood: Resources, Nutritional Composition, and Preservation*. CRC Press, Boca Raton, Florida, pp. 147–162.

Sikorski, Z.E. (1988) Smoking of fish and carcinogens. In: Burt, J.R. (ed.) *Fish Smoking and Drying. The Effect of Smoking and Drying on the Nutritional Properties of Fish*. Elsevier Applied Science, London, pp. 73–83.

Sikorski, Z.E. and Bykowski, P.J. (1994) Research on chitin and chitosan in Poland. Paper presented at the *6th International Conference of Chitin and Chitosan*, Gdynia. Abstracts, p. 3.

Sikorski, Z.E. and Pan, B.S. (1991) Preservation of seafood quality. Paper presented at the *8th World Congress of Food Science and Technology*, Toronto. Abstracts of Papers, p. 12.

Sikorski, Z.E. and Ruiter, A. (1994) Changes in protein and nonprotein compounds in cured, fermented, and dried seafoods. In: Sikorski, Z.E., Pan, B.S. and Shahidi, F. (eds) *Seafood Proteins*. Chapman & Hall, London, pp. 113–126.

Simko, P. (1991) Changes in benzo(a)pyrene contents in smoked fish during storage. *Food Chemistry* 40, 293–300.

Simpson, B.K., Simpson, M.V. and Haard, N.F. (1989) On the mechanism of enzyme action: Digestive proteases from selected marine organisms. *Biotechnology and Applied Biochemistry* 11, 226–234.

Skaugrud, Ø. and Sargent, G. (1990) Chitin and chitosan: Crustacean biopolymers with potential. In: Keller, S. (ed.) *Making Profits out of Seafood Wastes*. Report No. 90-07, Alaska Sea Grant College Program, Fairbanks, pp. 61–69.

Solberg, S., Ofstad, R. and Solberg, T. (1992) The influence of $CaCl_2$ and $MgCl_2$ in washing water on functional properties of cod (*Gadus morhua*) surimi: a multivariate data analysis. In: Bligh, E.G. (ed.) *Seafood and Science Technology*, Proceedings International Conference Seafood 2000, Halifax, Canada, 1990. Fishing News Books, Oxford, pp. 317–324.

Sripathy, N.V., Kadkol, S.B., Sen, D.P., Swaminathan, M. and Lahiry, N.L. (1963) Fish hydrolysates. III. Influence of degree of hydrolysis on nutritive value. *Journal of Food Science* 25, 365–369.

Stefansson, G. and Steingrimsdottir, U. (1990) Application of enzymes for fish processing in Iceland – Present and future aspects. In: Voigt, M.N. and Botta, J.R. (eds) *Advances in Fisheries Technology and Biotechnology for Increased Profitability*. Technomic Publishing Co., Lancaster, Pennsylvania, pp. 237–250.

Sugiyama, K., Egawa, M., Onzuka, H. and Oba, K. (1991) Characteristics of sardine muscle hydrolysates prepared by various enzymatic treatment. *Nippon Suisan Gakkaishi* 57, 475–479.

Tarr, H.L.A. (1948) Possibilities in developing fisheries by-products. *Food Technology* 2, 268–277.

Tarr, H.L.A. and Deas, C. (1949) Bacteriological peptones from fish flesh. *Journal of the Fisheries Research Board of Canada* 7, 552–560.

Tatterson, I.N. and Windsor, M.L. (1974) Fish silage. *Journal of the Science of Food and Agriculture* 25, 369–379.

Tressler, D.K. (1920) Some considerations concerning the salting of fish. Doc. 884, US Bureau of Fisheries, 55 pp.

Tülsner, M. (1978) Die bestimmten stofflichen Veränderungen in Fisch beim Salzen und Marinieren. *Lebensmittelindustrie* 25, 169–173.

Uchida, Y., Hukuhara, H., Shirakawa, Y. and Shoji, Y. (1990) Bio-fish flour. In: Keller, S. (ed.) *Making Profits out of Seafood Waste*. Report No. 90-07, Alaska Sea Grant Program, Fairbanks, pp. 95–99.

Vecht-Lifshitz, S.E., Almås, K.A. and Zomer, E. (1990) Microbial growth on peptones from fish industrial wastes. *Letters to Applied Microbiology* 10, 183–186.

Vinot, C., Bouchez, P. and Durand, P. (1989) Extraction and purification of peptides from fish protein hydrolysates. In: Miyachi, S., Karube, I. and Ishida, Y. (eds) *Current Topics in Marine Biotechnology*. Fuji Technology Press, Tokyo, pp. 361–364.

Whittle, K.J. and Hardy, R. (1992) Under-used resources – recent process innovations. In: Bligh, E.G. (ed.) *Seafood Science and Technology*, Proceedings International Conference Seafood 2000, Halifax, Canada, 1990. Fishing News Books, Oxford, pp. 101–115.

Wignall, J. and Tatterson, I. (1976) Fish silage. *Process Biochemistry* 11(10), 17–19.

Willard, T. (1990) Utilization of marine by-products in pet foods. In: Keller, S. (ed.) *Making Profits out of Seafood Waste*. Report No. 90-07, Alaska Sea Grant Program, Fairbanks, pp. 121–125.

Wood, A.L. (1963) Accelerated cooling of wet, heavily salted fish. *Journal of the Fisheries Research Board of Canada* 20, 997–1000.

Wood, C.D. (1981) The prevention of losses in cured fish. *FAO Fisheries Technical Paper* 219, 87 pp.

Wyatt, B. and McGourty, G. (1990) Use of marine by-products on agricultural crops. In: Keller, S. (ed.) *Making Profits out of Seafood Wastes*. Report No. 90-07, Alaska Sea Grant Program, Fairbanks, pp. 187–195.

Ya, T., Simpson, B.K., Ramaswamy, H., Yaylayan, V., Smith, J.P. and Hudon, C. (1991) Carotenoproteins from lobster waste as a potential feed supplement for cultured salmonids. *Food Biotechnology* 5, 87–93.

Zaitsev, V., Kizevetter, L., Makarova, T., Minder, L. and Podsevalov, V. (1969) *Fish Curing and Processing*. MIR Publishers, Moscow.

12 Fish Meal and Fish Oil – Not Only By-products

WALTHER SCHMIDTSDORFF
Technological Laboratory, Ministry of Fisheries, Technical University of Denmark, Lyngby, Denmark.

Introduction

The annual world fish catch has been over 90 Mt since 1986 and reached a total of 100 Mt in 1989 (see Chapter 1), but only about 1% of people's food is fish, although nearly 10% of the animal protein intake is fish protein. The pelagic-type species represent between 27 and 30% of the catch, and of the pelagic species some are specifically landed for reduction to meal and oil. This part of the catch has not increased at the same rate as the total world catch, but has increased slowly from about 20 to 25 Mt in the last 20 years (Bimbo, 1987). One must remember that a variable but significant portion of the food fish is converted to fish meal and oil.

The fish meal and oil industry started in northern Europe and North America at the beginning of the 19th century and was solely based on surplus catches of herring from seasonal coastal fisheries. It was essentially an oil production activity, the oil finding industrial use in leather tanning and in the production of soap, glycerol and other non-food products. The residue was originally used as a fertilizer, but since the beginning of the 20th century it has been dried and ground into fish meal for animal feeding. In fact one definition of fish meal is that it is a ground solid product that has been obtained by removing most of the water and some or all of the oil from fish or fish waste.

From time to time voices have been raised in favour of reserving the total world catch, which probably has reached its highest economically feasible size at about 100 Mt, for feeding humans. Only where it is uneconomic or impracticable should the catch be reduced to fish meal and oil. However, 90% of the fish that are currently reduced to fish meal and oil, the so-called industrial fish (menhaden, capelin, sand eel, anchovy, sardine, pilchard, etc.) are presently unmarketable in large quantities as human food. The reasons why this portion of the catch cannot be used for direct consumption include that the fish are too small, break down or turn rancid too quickly for economic storage and subsequent heading, gutting, cleaning and processing. Turning high quality fish into fish meal and oil should not be encouraged. It is obviously more efficient, however, in a hungry world to harvest those species for feeding to animals and for production of a valuable oil, subsequently consumed by man, than to leave them unharvested in the sea.

Small oily fish are the foundation of the fish meal and oil industry. Even in frozen storage these fish turn rancid rapidly unless special and expensive precautions are taken. With present knowledge they can be used best by reducing them to fish meal for animal feeding and for fish farming, which is a rapidly growing industry. No economically feasible substitute for fish meal as the main ingredient of fish feed production has so far been found. The fish oil still finds its main market in the edible oil industry as an ingredient used for edible purposes after being hydrogenated, e.g. for margarine production. But, as will be explained later, new evidence about the unique qualities of the polyunsaturated n-3 fatty acids in fish oils has opened up a new development in usage of unhydrogenated fish oils in human nutrition.

There is a good world market demand for high quality fish meal and oil, and production can be quite profitable if suitable raw material is available. The industry can also take care of the offal, preferably in a fresh state, from filleting, gutting and other fish processing operations, which often pose disposal problems.

Fish Protein Concentrate (FPC)

The most efficient course would be to feed the powdered products resulting from reduction directly to humans. This introduces the technology of fish protein concentrates (FPC). Direct feeding reduces the losses of cycling through poultry or pigs, which may take 3 kg of fish to produce 1 kg of edible chicken or pork, and it will also reduce the losses involved in preparing fish for human consumption. A fillet may contain only half the protein of the original fish after heading, gutting, filleting and skinning.

The production of FPC is within the capacity of the fish meal and oil industry; indeed the industry is in a phase of its development where the production and marketing of special quality products with higher profitability pose very similar challenges to the industry. The fish must be fresh and sound and the plant must be easy to clean and to sterilize after use. These are technological problems and, although the solutions are expensive, the industry already possesses the design and organizational expertise. Two types of FPC have been produced, i.e. a type A, with a fat content of less than 0.5%, and a type B, with a fat content less than 10%. The major problems of use are social rather than technological; these are the difficulty and expense of incorporating new powdered products in the diet. FPC type B tastes fishy after production and 'fishmealy' or rancid after storage, because most of the flavour is in the oil fraction. In some areas the rancid flavours are accepted as they are similar to those of traditional dried fish products. FPC type A is flavourless, but the removal of the oil by solvent extraction is expensive. It must also be milled finely as it has little water-holding capacity and would taste gritty when incorporated in food; this process is expensive. It is hoped that technological solutions to these problems will be developed and that FPC for human feeding can be made available at competitive prices. Great efforts have been made in the search for socially acceptable methods to incorporate FPC in food. For example, type A can be chemically refined to impart functional capacities (water holding, gelling) to the powder. This raises the cost, but the product is nevertheless cheaper than meat, and can be used profitably as a meat extender, for instance in sausages. From the above it can be seen that the production of FPC is an enterprise which will be embarked upon only when a ready market for the products has been established.

Great efforts have been made by governmental bodies (USA), private aid organizations and the fish meal industry (Norway) and by pharmaceutical firms (Astra Nutrition) both in product development of FPC type A and in the acceptability testing for large groups of populations in so-called Third World countries as reported by the Food and Agriculture Organization of the United Nations (Tagle, 1976) and elsewhere. Norway led the way by FPC type B production and use in feeding programmes as late as in the mid-1980s (documented by a series of publications: International FPC News). The records show, however, that FPC is no longer used and regarded as a panacea for protein malnutrition as suggested by Pariser *et al.* (1978). FPC type B is still produced and marketed in Scandinavia. It can be bought in certain health food shops, but is not accepted by the majority of the population as a foodstuff. One might speculate that the failure of marketing FPC in the Third World in spite of its undisputed nutritional qualities has a close connection to this fact.

Raw Material

Resources

Fish used for reduction to meal and oil are divided into three categories of decreasing importance: (i) fish caught solely for the purpose of fish meal production as by most of the big producer nations (Chile, Peru, Norway, Denmark, South Africa and the USA); (ii) by-catches from other fisheries by most fish-producing countries; and (iii) fish offals and offals from the consumption industry. In many cases, however, especially if the geographical distance to animal consumers such as pigs is not too great, the production of liquid or semiliquid acid fish silage will be more economically attractive because of the very low investment costs and low infrastructure demanded in comparison to fish meal plant construction.

A fish meal industry requires a regular supply of raw material. Before planning the construction of fish meal factories it is necessary to have a good knowledge of the species available, the length of the fishing season, the location of the fish, the catchability of the fish by different fishing gear and the attainable catches per year for a continuous period. This information forms the basis for any economic calculation of the feasibility of a fish meal industry.

Practically all fish species may be converted into fish meal and oil. The composition and, primarily, the quality of the raw material are predominant factors in determining the properties and yield of the products. A separation of oil from the other constituents is one of the major operations in the manufacture of fish meal and oil. As previously mentioned, the pelagic small fatty fish species, in which the clupeids (herring-type fishes) are dominant, provide the largest single source for the production of fish meal and oil. Such species are anchoveta, anchovy, herring, pilchard and sprat. Sand eel and capelin are very similar in size and composition, though of other species. Mackerel and horse mackerel are bigger fish but also fatty pelagic species. The main constituents of the fish vary little as regards protein and inorganic matter. Fat (oil) and water contents, which make up the remaining *c.* 72–78% of the fish, may, on the other hand, vary very much. In fatty species, fat may vary from below 2% to more than 30%, depending on season. A few lean pelagic species are used to a small extent, but total fat content may for these species be as high as 5% and located in the liver.

Economic evaluation of the raw material

From reliable analysis of the raw material one may estimate the amount of fish meal and oil which may be produced, and thus the value of the raw material. The water content determines the figures for the cost of drying, and increasing fat content will lead to improved oil yield, reduced demand on drying energy and increased processing capacity of the plant. The extent of deterioration of the fish will, however, also play an important part in the functioning of the processing plant apart from its decisive influence on the quality of the finished goods, meal and oil.

Evaluation requires that representative samples from each catch be submitted for analyses at a control laboratory. Sampling is not straightforward because considerable variations may exist within the same catch. Sampling is done most conveniently during unloading of the fishing vessel. An automatic sampling device, installed directly after the weighing equipment, is recommended. Stratified spot sampling of 4–5 kg samples of fish should be made several times during unloading of a vessel containing industrial fish. The sampling container may contain more than 200 kg of raw fish depending on the size of the catch. The fish has to be ground and thoroughly mixed before subdivision of the sample can take place. The final size of the fish samples for laboratory examination will be approximately 500 g.

The evaluation of the raw material requires the analysis for protein, fat, water and 'ash' content, and the determination of volatile basic nitrogen. The rapid deterioration of the raw fish is due to the action of spoilage bacteria from the surface and particularly from the digestive tract and to autolytic breakdown caused by enzymatic action in the tract. Bacterial and autolytic deterioration result in breakdown of both lipids and proteins. Particularly in small fatty species used for fish meal production, the digestive enzymes may cause extensive autolysis leading to softening of the meat, rupture of the belly wall and formation of considerable amounts of blood water containing both protein and oil. Such solubilization causes difficulties in handling and processing and may lead to serious losses. This will again cause environmental pollution problems by both liquid and gaseous effluents.

Some important end products of the bacterial spoilage are the volatile basic nitrogenous compounds, mainly trimethylamine and ammonia (TVN, total volatile nitrogen). The trimethylamine is formed first by bacterial reduction of trimethylamine oxide. But the extensive formation of ammonia in a load of deteriorating fish is produced by bacterial metabolism of amino acids, in turn breakdown products of protein, so that ammonia formation is directly correlated to protein losses.

Preservation of the raw material

The production of fish meal and oil from fresh raw material provides the highest yield and the best quality final products. In many cases, however, it is difficult to avoid partial spoilage of the fish as it has to be collected in remote areas.

The search for economic means of preserving the catch during periods of transport and storage, exceeding a period of 30 h, has been a continuing challenge to the industry. It should be mentioned here that the breakdown of fish protein and lipids, due to both autolytic and bacterial spoilage, is highly temperature dependent.

The storage life of fish can be extended either by physical or chemical means. Chemical preservation of raw fish immediately acts on the bacteria on the surface of the fish, but due to long delays in action on the interior this method is only used when the fish has to be stored for long periods at temperatures close to 0°C. A number of chemical preservatives have been evaluated, but only sodium nitrite has demonstrated favourable properties for preserving certain species at low temperatures. Unless nitrites are added in strictly controlled small amounts, they may react with other components of the raw material to form N-nitroso compounds, which show carcinogenic properties. Formerly formaldehyde was in wide use as both a preservative and a processing remedy because, in addition to its bacteriocidal activity, it has a firming effect on the raw material so that it becomes more amenable to pressing after cooking. Formaldehyde, however, combines with protein by a mechanism in which the essential amino acid, lysine, is involved. This may result in lower protein value of the fish meal and, in combination with regulations for protection of workers, this led to almost total abandonment of its use. The only reliable preservation method for storing of ungutted fish in bulk is, therefore, lowering the temperature or chilling.

During storage the rate of breakdown caused by bacteria and digestive and tissue enzymes doubles with an increase in temperature of approximately 4°C. At temperatures higher than 5°C hydrogen sulfide is produced by bacteria and its formation rapidly increases with temperature. At 0°C hydrogen sulfide is not formed until storage exceeds 9–10 days (Technological Laboratory, Danish Ministry of Fisheries, 1971).

In principle, two methods of chilling may be considered, i.e. refrigerated or chilled water systems and mixing with ice.

In refrigerated water systems prolonged storage of fish in sea water is limited by absorption of salt; high salt levels are not desired in fish meal. It is therefore preferable to use diluted sea water or fresh water. The method comprises circulating refrigerated water through the mass of fish. Merely pumping the water over the fish does not effectively chill the whole mass. Advanced systems are based on pumping the chilled water upwards through the fish. The method is expensive, however, and mostly used for preservation of fish catches for consumption.

Mixing fish and ice in the right proportions for chilling the fish at 0°C is an effective method for preserving whole fish.

In Scandinavia, particularly in Denmark, an ice mixing and fish grading system has been installed in a large section of the industrial fishing fleet. The system has basically been developed by the Technological Laboratory of the Danish Ministry of Fisheries and is still being improved (Olsen and Hansen, 1982; Olsen, 1992a; 1992b).

The mechanism of the icing system works in the following way: A receiving box for the emptying trawl's cod end is installed on deck. The box is equipped to remove the few large fish and other large objects before the fish enter a conveyer leading to a grader. The box may receive loads of up to 10 t of fish, which are converted to a continuous flow of up to 2000 kg min^{-1}. The larger food fish pass through the length of the grading cylinder, while the smaller industrial fish fall through the apertures in the cylinder into a trough equipped with a continuous supply of small pieces of ice to chill the fish to 0 °C and maintain this temperature until the fish are landed.

The continuous supply of ice to the trough under the grader is taken from the store of bulk ice in the hold. At the bottom of the store a horizontal conveyer feeds the ice into a vertical conveyer to the trough. Fish at 15 °C when caught require about 23% of their weight in ice to be chilled and maintained at 0 °C for 4 days in an insulated hold.

The Process of Fish Meal and Oil Manufacturing

The universal process of fish meal and oil manufacture is the so-called wet rendering process. As the name implies, the process separates liquid from solids by use of hot water or steam. The purpose of the process is to separate the three fractions of the raw material, i.e. solids (fat-free dry matter), oil and water. The following processing steps are included: (i) heating, which coagulates the protein, ruptures the fat depots and liberates oil and water; (ii) pressing (or centrifugation), which removes a large fraction of the liquids from the mass; (iii) separation of the liquid into oil and water (stickwater); (iv) evaporation of the stickwater into a concentrate (fish solubles); (v) drying of the solid material (presscake) plus added solubles, which removes sufficient water from the wet material to produce a stable meal.

The composition of the raw material, expressed in terms of dry matter, protein and fat, will determine what yield of products may be expected.

Evaporators for the stickwater, however expensive, are today considered standard items of equipment for fish meal plants, because they recover dry matter that will increase the yield of meal by 20% or more, depending upon the freshness and nature of the raw material. Furthermore they eliminate a serious pollution problem.

Location of the plant relative to settlements, closed harbours or natural resorts is another feature for trouble-free operation. Public regulations protecting the environment against pollution of air and water should be carefully studied, as these will affect the selection of equipment and factory premises and determine the amount of investment necessary in financially unproductive odour abatement and cleaning of wastewater emission.

Figure 12.1 is a diagram of a typical fish meal plant. The raw material is first unloaded from the fishing vessel by wet fish pump, pneumatic elevator or some sort of mechanical conveyer. The fish is weighed or

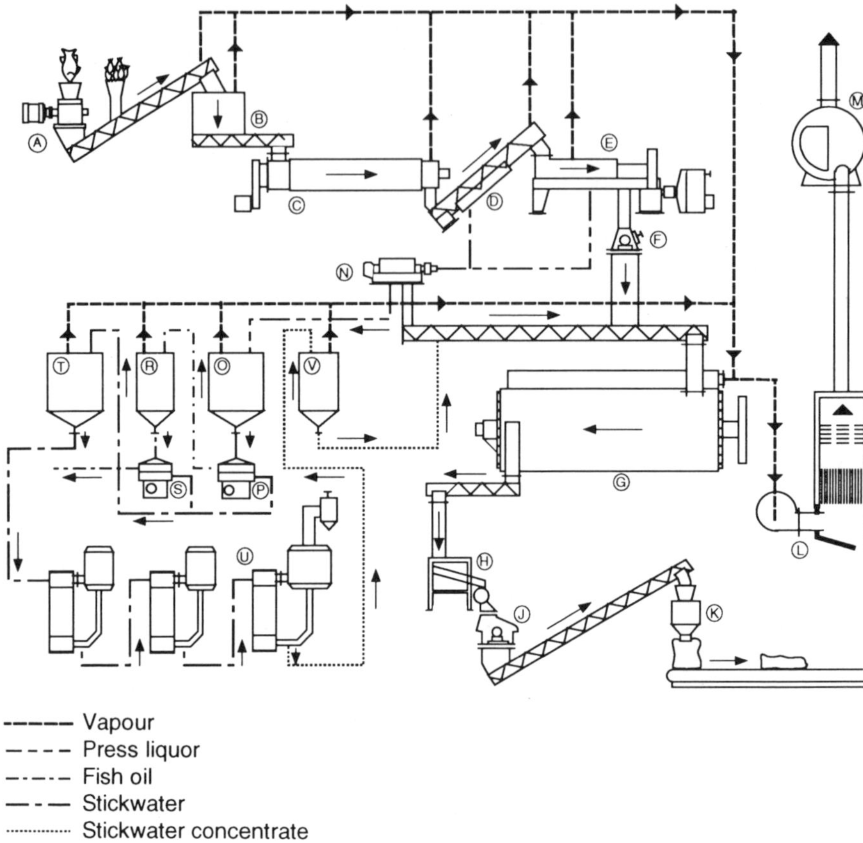

Fig. 12.1. Diagram of a fish meal plant. See text for more details.

measured by volume before it is transported to the pits or tanks for storage of the raw material.

Large fish (if present, only sharks or other low value fish) are hashed (A), while smaller (normal) fish are fed directly at a constant rate by the feeding machine (B) to the indirect steam cooker (C). The coagulated mass is pre-strained in a strainer conveyer (D) before entering the twin screw press (E).

The products from the press (presscake and press liquor) are treated as follows: The presscake is disintegrated in the tearing machine (wet mill) (F) to facilitate mixing with stickwater concentrate and drying in an indirect steam dryer or in a direct fired flame dryer (G). The meal passes through a vibrating screen (H) furnished with a magnet to remove unwanted matter such as pieces of wood and metal (e.g. fish hooks) before entering the hammer mill (J). Before milling, cooling of the meal in a special meal cooler may have taken place. The ground meal is automatically weighed out in bags by the scales (K), the bags are closed (e.g. by sewing) and conveyed to the store. Alternatively the meal is stored in holding and blending silos, before bagging, pelleting and storing in bulk.

To remove most of the sludge, the press liquor passes through a decanter centrifuge (N). The press liquor then passes through a buffer tank (O) before separation into oil, stickwater and fine sludge in the stickwater centrifuge (P). The sludge is added to the presscake. The fish oil passes through a buffer tank (R) before water and impurities are removed in the oil separator (polishing) (S). After polishing the fish oil passes through an inspection before storage in the oil tank. The stickwater passes through a buffer tank (T) before concentration in a multi-effect evaporator (U). After the buffer tank (V), the concentrate is mixed thoroughly with decanter sludge and presscake in the conveyer leading to the dryer. In some cases the stickwater concentrate, called condensed fish solubles, is sold separately.

As indicated in Fig. 12.1 by the heavy dashed line, the factory can be deodorized by air suction from all machinery and tanks. The air passes through a scrubber (L) and, after passing through a demister and a heater to avoid condensation, it is used as combustion air in the steam boiler. Other methods of oxidation of the (small) amounts of obnoxious compounds in the effluent air are available, but mostly not effective or very expensive.

Figure 12.2 gives an example of a mass balance. Here the streams of the three fractions of the raw material can be followed, i.e. solids (fat-free dry matter), oil and water, through the factory. The actual figures will vary with the composition of the raw material, particularly with the water content, but the diagram is sufficient to illustrate the general trend. Prospective manufacturers may estimate their expected yield of meal on the basis of the dry matter content of the raw material plus residual moisture and oil in the meal. Likewise the oil yield will be the fat content of the raw material less the small amount remaining in the meal.

In the following, a closer look is taken at some of the unit operations most important in the fish meal process.

Heating

Heating or cooking has the purpose of liberating the oil from the fat depots of the fish, and to condition the material for the subsequent treatment in the various processing units of the plant. It is therefore a key process of great importance for the whole functioning of the factory. Until fairly recently, the general view has been that the best results would be obtained at the highest possible temperatures, which, at atmospheric pressure, would

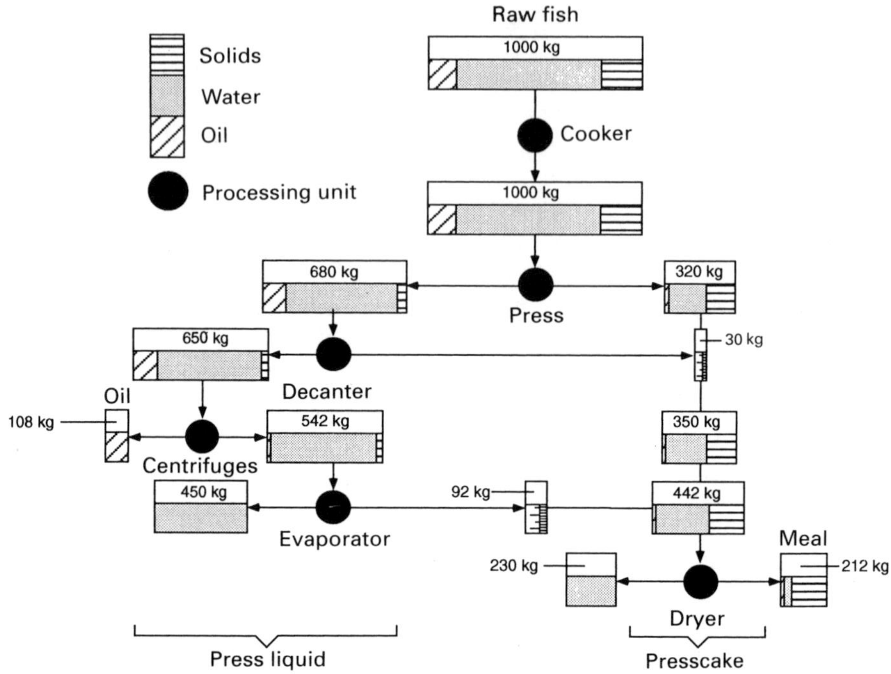

Fig. 12.2. Mass balance in meal production.

be 100 °C. New experiments have shown that the walls of the fat cells are broken down before the temperature reaches 50 °C. The oil is then free, and theoretically it should be possible to separate it from the solid material.

Another important recent observation is that coagulation of the fish material is completed at much lower temperatures than 100 °C depending on pH, ionic concentrations, etc., and, furthermore, that the process may be very rapid. There might be cases where optimal cooking temperatures are 60–75 °C, depending on fish species and other circumstances. The problem is primarily a matter of heat transfer and temperature control to ensure a uniform, optimum temperature through the whole mass. As reduction of heat load on the material, i.e. the combined effect of temperature and time, tends to improve the quality of the products, the cooking conditions for each separate type of raw material heated in a particular cooker should be investigated, if possible in pilot-scale experiments, in order to find optimum conditions.

The most common practice is to cook the fish in a steam cooker, through which it is conveyed continuously. Heat is generally transferred indirectly from a steam-heated surrounding jacket and a heated screw conveyer, or the heat may be conveyed through stationary heating plates and a rotating shaft equipped with scrapers and shovels to transport the fish material through the cooker, in principle in a piece of equipment very similar to the dryer shown in Fig. 12.5 (p. 363). These methods are improvements over the direct steam injection cooker, in which water is condensed in the mass during the process, and has to be removed in the following stages. Provision for injection of live steam directly into the mass is made, however, even in indirect cookers, to facilitate heat transfer by breaking up the mass.

Production of a cooked material which can be readily pressed is dependent on the raw material quality and on the process conditions. A precise time–temperature programme for this process can therefore not be set up and, as mentioned above, a process of trial and error is generally required when fish of unknown history are cooked. The most common practice of cooking good raw material, however, is to heat to 95–100 °C within 15 to 20 min. The proof of good cooking is good pressability of the mass, which leads to proper removal of press liquor and, in particular for fatty fish species, efficient recovery of oil, giving a meal with a low fat content which is a criterion of quality.

Cookers are available in sizes which can process from below 10 t to above 600 t of raw material per 24 h. The capacity of a heat exchanger, such as an indirect steam cooker, is proportional to the area of the heating surfaces and to the temperature difference between the two sides of the wall.

In order to utilize waste heat, either from the evaporators or from the dryers, a preheater is often installed. A type consisting of a set of tubes coupled together and surrounded by a cylindrical jacket is mostly used. The raw material is transported through the tubes by pumping and the waste vapour condenses on the outside of the tubes.

Pre-straining

Pre-straining is a process in which the oil and water released in the cooker are removed by a simple draining process in order to facilitate the functioning of the press. The liquid liberated in the cooker is drained from the coagulated fish pulp in a strainer conveyer set on an incline between the cooker and the press. The mass is transported in a trough by a screw conveyer. Strainers with different sizes of perforations may be required for various types of fish.

Pressing

Pressing has the purpose of squeezing out as much liquid as possible from the solid fish pulp. This is important not only to improve the oil yield and the quality of the meal, but also to reduce the moisture content of the presscake as far as possible, thereby reducing the fuel consumption of the dryers and increasing their capacity.

The dominant type of press used in the fish meal industry is the double or twin screw press. The performance of the press is largely determined by the profile and the compression ratio of the screws, i.e. the ratio between the flow volumes of the inlet and outlet flows. The twin screw press should be able to produce presscakes with moisture contents somewhat below 50%.

Processing problems may be encountered under two entirely different conditions. One relates to completely fresh fish, which tend to retain more oil and water than desirable. For the time being there is no solution to this problem except by resorting to one of two equally deplorable measures: either by reducing the speed of the press and thereby the capacity of the whole plant, or by storing the fish for a day or two before processing, thus leading to deterioration of quality.

The other problem occurs with soft and autolysed fish. The solution to this problem is to reduce the speed of the pre-strainer and bleed off more liquid and fines to be handled by the decanters. Some processors will often resort to the use of formaldehyde as a coagulating agent, which helps to solidify the material and improve the performance of the press. This practice, however, should be restricted as far as possible.

To separate solids from liquid by centrifugation is a standard operation in many industries including the fish meal industry. With the development of centrifuges that can handle materials with high contents of solids and high rates of throughput, it is now possible to use so-called decanters (described in the next section) instead of presses for separation of cooked fish. The advantages are several. First, it presents a simplification of the process. Secondly, centrifugation is a more controllable unit operation than pressing and filtration. Thirdly, centrifugation is a much quicker process than pressing and significantly reduces the heat load on the material, a factor of importance for the manufacture of special high quality products. Perhaps the most important advantage is the ability to process soft and very fluid material where the press has difficulties as described in the above paragraph. Better hygiene and simpler procedures for cleaning operations are further features on the plus side.

On the negative side it should be noted that the decanter centrifuge will discharge the solids with a higher moisture content than the press. This means increased fuel consumption, often considerable, in the drying process. Furthermore the centrifuge tends to produce more emulsions and fines, causing problems in the subsequent separation of oil, water and sludge in the liquid phase.

Plants with decanters instead of presses are in operation in many parts of the world, generally with small or medium production capacity.

Separation

The separation of the three fractions of press liquor coming from the press and pre-strainer, i.e. sludge, oil and water, is based on their different specific gravities.

For efficient separation centrifugation is the method of choice. An important prerequisite is high temperature, implying that the press liquor should be reheated to 90–95 °C before entering the centrifuges. This applies to sludge removal as well as to separation of oil and water.

The suspended solids are first to be removed. This is done in a horizontal centrifuge, of similar construction to the centrifuge described under Pressing, a so-called decanter or desludger. It consists of a partly cylindrical and partly conical rotor drum (bowl) and, inside this, a screw conveyor of the same shape. The press liquor is fed into the rotor where, by centrifugal force, it is thrown toward the bowl's periphery. The denser solids are rapidly precipitated along the inside rotor surface. The screw conveyer rotates with the bowl, but at a rate some 30 to 50 rpm faster than the speed of rotation of the drum; the deposited solids are thus scraped off continuously.

Separation of stickwater takes place in vertical disc centrifuges, either of the nozzle type, which discharge the stickwater and remaining sludge continuously, or of the self-cleaning type, which are often preferred. In the latter the stickwater is continuously discharged, whereas the sludge is collected in the bowl and periodically ejected. Stickwater with a dry matter content of 6–9% is concentrated in the evaporators. The sludge is pumped to the presscake before drying.

The final refining step of the oil is carried out in special separators. This step, called polishing, is facilitated by using hot water, which extracts impurities from the oil and ensures a better stability during storage.

Evaporation

Evaporation of stickwater is a relatively new process in the fish meal industry. Evaporators did not become standard equipment all over the world until the late 1970s. The evaporation plant, however, has proved to be an indispensable part of a fish meal factory for the following reasons: First, it adds an extra 20–25% to the fish meal output, mainly as protein. Secondly, this stickwater loss used to cause great pollution problems both in the receiving sea water and on shore along beaches. And thirdly, evaporators are energy-saving in comparison to the direct or indirect dryers. The energy demand for evaporation of 1 kg of water in the evaporator system can be less than a quarter of the necessary energy needed in the dryers.

Stickwater coming from separators after decanter treatment of press

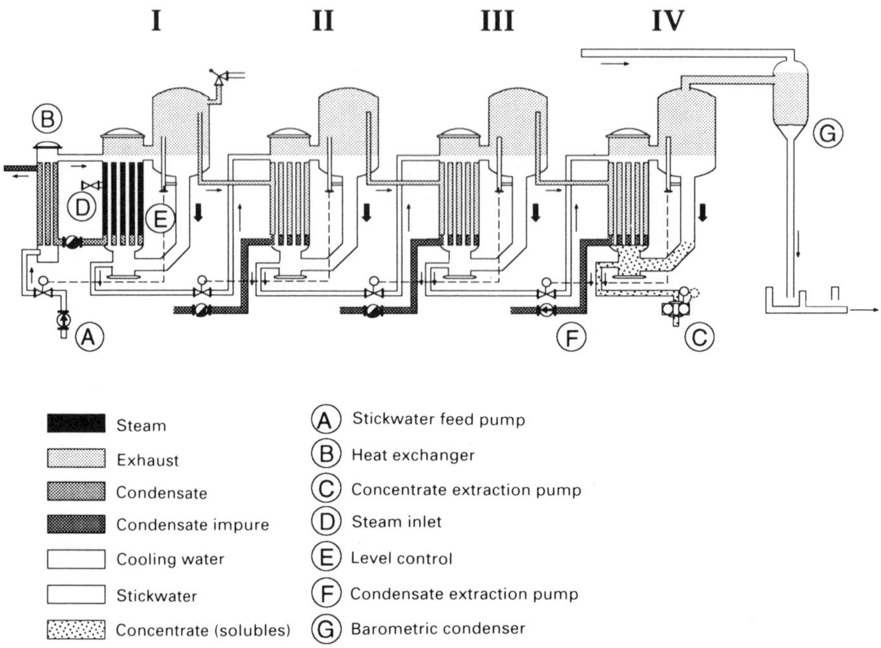

Fig. 12.3. Evaporator plant.

liquids may for practical purposes be estimated to amount to about 65% of the raw fish material. Besides water, stickwater will contain dissolved protein, suspended protein, residual oil, inorganic matter, amines plus ammonia and vitamins, ranked according to decreasing concentration. The content of residual oil will depend upon the efficiency of the separation process and should be as low as possible, certainly well below 1%. The other components put together, usually called dry matter, amount to 5–6% for fresh fish and correspond to about 20% of the yield of meal. Dry matter percentages in stickwater may become very high, up to 10–15%, depending on the degradation of the raw material.

Figure 12.3 shows a quadruple effect evaporator plant. This system works with the I–II–II–IV mode of operation, i.e. the stickwater flows in parallel with the flow of steam. It is fed continuously to stage I and progressively concentrated during the subsequent passage through stages II, III and IV. Live steam from the boiler is supplied to the preheater and to the heat exchanger of the first stage, while the vapour emanating from this stage is used for heating in the second. Vapour from the second stage is used in stage III, and so on to stage IV. Vapour from the last stage is normally condensed in a condensing tower, but it is increasingly being used to preheat the raw material.

Most factories prefer to feed dilute solution to the second stage and withdraw concentrate from the first. This subjects the concentrate to the highest available temperature in the system (generally above 100 °C), which has the following advantages: (i) reduction of viscosity of the concentrate; (ii) effective decomposition and removal of nitrogenous contaminants, e.g. nitrites and N-nitroso dimethylamine; and (iii) destruction of any pathogenic bacteria, including *Salmonella*, that might have infected the material in earlier stages.

Concentrated stickwater may be sold separately under the name of 'fish solubles'. The most common way of utilizing stickwater concentrate, however, is to mix it with the presscake and to dry the mixture to the so-called whole meal or whole fish meal. The dry matter content of the concentrate may vary between 30 and 50%.

The commonly used evaporators operate with vertical tubes partly filled with boiling liquid, gradually giving off water vapour while the liquid moves upward in the tubes and into the chamber, separating liquor and vapours. The volume of liquid is quite large and it therefore takes a long time before the desired concentration is reached, and concentrate starts flowing to the dryers. This time lag between pressing and concentrating is particularly undesirable and causes problems in factories with intermittent supplies of raw material. In such cases the falling film evaporator becomes of special interest because it operates with a short holding time and a small liquor volume. Here the stickwater enters at the top of the tubes, which are heated from the outside by steam or hot gases. On its way down, water evaporates and the liquid becomes more and more concentrated and finally ends at the bottom as concentrate. The falling film evaporator also offers advantages in connection with energy-saving systems using vapour compression.

Drying

The purpose of the drying process is to convert the wet and unstable mixture of presscake, decanter sludge and concentrate into a dry and stable fish meal. In practice this means drying to a moisture content below 10%, which generally may be considered low enough to control microbial activity. This drying is done by heating the material to a temperature where the rate of evaporation of the water is considered satisfactory. With the equipment and conditions normally used in the fish meal industry, the temperature of the drying material should not exceed 90 °C in order not to impair nutritional quality. Indeed, vacuum dryers are now in practical use for production of special high quality fish meals which will not exceed drying temperatures of 70 °C or lower if wanted.

Vacuum dryers or certain types of indirect dryers will also minimize access of oxygen (air) to the drying mass, which may be considered an advantage, because oxygen easily attacks the highly unsaturated residual fat and forms reactive oxidation products which in turn may decompose vitamins and amino acids of the fish meal. Though antioxidants will partly be removed (evaporated, destroyed) in the drying process, they are sometimes added even before the drying.

A prerequisite for optimum drying conditions is that the material should be divided into relatively small particles so as to facilitate the escape of water vapour from below the surface. The presscake is, therefore, passed through a wet mill where it is disintegrated by fast-moving hammer heads mounted on a rotor. To avoid sticking and lump formation during drying of presscake to which stickwater concentrate is added in one stage, thorough mixing is necessary. This may be achieved by heating the concentrate to about 100°C before mixing. Such heating also serves to destroy bacteria if present.

The two main types of drying are direct heat drying and indirect steam drying.

The direct rotary dryer, also called flame dryer or direct hot air dryer, is likely to lose its dominant position in drying most of the world production of fish meal if the trend towards the production of special quality fish meal continues. Furthermore it is almost impossible to abate the fish odour in the large volumes of gases emitted from this type of dryer. It may be estimated, however, that over 50% of the world production of fish meal is still produced in direct rotary dryers.

A direct-fired rotary dryer is illustrated in Fig. 12.4. Rotating at a peripheral speed of about $1\,m\,s^{-1}$, the cylindrical compartment is equipped with horizontal and helical flights, which provide cascading and good agitation of the fish material, a large area of contact between fish particles and air and, accordingly, means for efficient dehydration. Because large seasonal catches of fish must be processed within relatively short periods of time, there has been a demand for large sizes of rotary dryers capable of processing 1000 t or more of raw material per 24 hours.

The direct action of the drying medium is both the strength and the weakness of the system. It represents the most efficient mode of heat and mass transfer but, unless operated properly, it may constitute a source of contamination of the product. Flue gas contaminants may range from products of incomplete combustion, e.g. particles of soot, to oxides of metals, sulfur and nitrogen, which are capable of reacting with nutritional components (such as oil and protein) of the meal. The temperatures of combustion must be high enough to ensure complete fuel combustion. Inlet temperatures on the other hand must not be too high; temperatures of 500–600°C are considered safe levels.

In indirect dryers the fish meal is at no stage in direct contact with flue gases. Figure 12.5 shows a modern so-called ring channel dryer. It consists of a stationary steam-heated cylindrical jacket and a steam-heated rotor

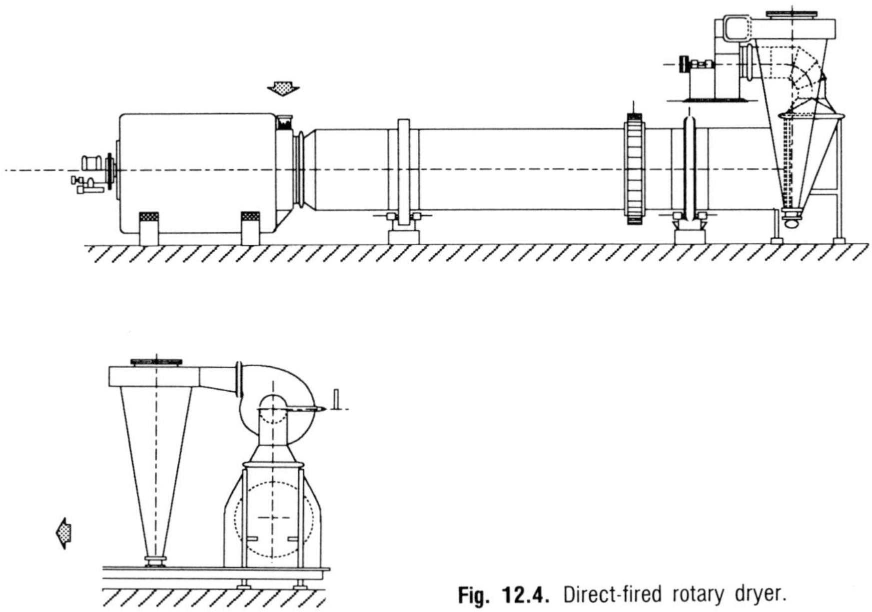

Fig. 12.4. Direct-fired rotary dryer.

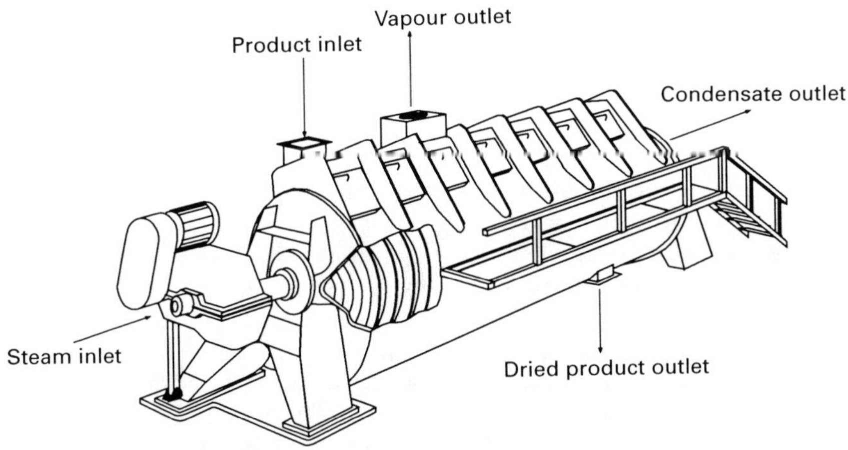

Fig. 12.5. Ring channel dryer.

equipped with steam-heated double-walled plates with concentric steam channels perpendicular to the rotor, which provide good agitation and heat transfer to the drying fish particles. These dryers are normally designed to remove up to 3000 kg of water per hour, corresponding to 350 t of raw fish per 24 hours. One of the advantages of these dryers is their ability to dry stickwater concentrate together with presscake in one stage without deposits on the heating surfaces. The gentle transport of material through the stationary cylinder of the ring channel dryer appears to yield a product with a physical structure different from that of the rotary dryer. The particles are longer and less prone to spillage and thus more acceptable for certain feeding uses.

Other types of indirect steam dryers, basically similar to the ring channel dryer, are the rotary disc dryer and the steam-heated coil dryer. The main difference between the dryers mentioned is the design of the rotating heating elements and rotating transport shovels.

From a nutritional point of view, direct and indirect drying methods can yield products of equivalent values, provided that direct drying in particular is properly controlled. But special high quality fish meal, e.g. for feeding carnivorous fish in aquaculture, can only be produced by using fresh raw fish and gentle or very fast drying. The indirect steam dryer also has an important advantage with regard to air pollution.

Milling and Storage of Fish Meal

The ideal in milling is to produce small particles averaging around 40 mesh Tyler screen and of as even a size as possible. In practice, however, there is a great variation in particle sizes, ranging from 10 mesh to over 100 mesh. Most purchasing specifications require the fish meal to pass through a 10 mesh screen, otherwise it is too coarse for uniform incorporation into feed mixes. Production of excessive fines must be avoided for several reasons. They cause dusting when handled, sift through woven bags resulting in loss and in pollution, cause compacting of bulk meal and tend to clog the nostrils of chickens eating the feed.

There are numerous types of mills on the market. In view of the need for high rates of throughput and ready access for cleaning, hammer mills have proved particularly successful. The coarse meal is disintegrated by the impact of rapidly rotating hammers, pivoted on horizontal or vertical shafts. A grating is usually attached around the rotor which retains the meal until it is fine enough to pass through the perforations. The ground meal is cooled by air which is also used for transportation of the meal.

Reactive fish meals are stabilized by means of antioxidant applied immediately after manufacture, and may be stored in bulk or bags or shipped as soon as they are properly cooled. The amount of antioxidant required for avoiding undue heating depends on the degree of unsaturation (characterized by the iodine value) of the oil and varies much with fish species. Considerable amounts of ethoxyquin, i.e. the optimal antioxidant for fish meal, are added, however. Thus South African pilchard meal (iodine value about 180) is normally stabilized with 400 mg of ethoxyquin per kg (though 200 mg would suffice), and herring meal (iodine value about 120) by 200 mg kg^{-1}. But certain customers may demand 400 mg kg^{-1} added. Direct-dried anchoveta and similar meals (iodine value in residual oil up to above 200) are in special circumstances protected with more than 1000 mg of ethoxyquin per kg.

The antioxidant is added immediately after drying, often by spraying over the meal in the screw conveyer leading from the dryer to the mill so that mixing can occur *en route*, or in specially designed equipment after the mill. In any case, careful control is necessary.

Fish meal is frequently stored and transported in bags. These usually hold 50 kg each.

The freshly prepared fish meal must be cooled from around 60–70 °C to room temperature, whether treated with antioxidant or not, and whether bagged or stored in bulk (silos). Bags of stabilized fish meal are left standing in single or double rows or singly spaced pallets for a few days. Unbagged fish meal should be continually turned and cooled by air as done in properly designed silos. A so-called curing of unstabilized reactive fish meal has been used by applying the above-described cooling procedures for much longer periods of time, e.g. 28 days.

Like other feedstuffs fish meal may be pelletized. This is mainly done by large producers and facilitates storage in bulk and transport. The flow properties are improved and dusting is reduced. The bulk density of pellets is the same as that of fish meal (generally 600 to 700 kg m^{-3}). During handling, however, some of the pellets break. The broken pieces and meal formed in this way occupy the spaces between the pellets and thus increase the bulk density.

Storage methods for fish meal vary, depending on many factors, including climatic conditions, use of antioxidant, and transport and marketing arrangements. Factories should have storage capacities for a reasonable buffer stock. In case of difficult marketing and shipping conditions, for instance, storage sufficient to hold 30 days' production may be required. Fish meal must be protected from contamination with pathogenic bacteria, of which *Salmonella* is a special nuisance, mostly as an infection carried by rodents, seabirds and humans (by imperfect cleaning of boots). It must also be protected from moisture and from undue self-heating. For this purpose well-known methods from storage of feedstuffs are being used, such as storage of bags in covered store houses or storage of meal or pellets in bulk in silos.

Pollution Abatement

In most countries, legislation on pollution requires the abatement of malodours and of liquid effluent from the fish factories.

Pollution by malodours from fish meal factories is not considered hazardous to health, but is disagreeable to such an extent that it may be considered the most serious problem fish meal factories may cause to their environment. The human nose is very sensitive to bad smells and, roughly speaking, the smell impression is only reduced to a half if the concentration of fish odours is reduced by 90%; in other words, a 99.9% reduction of concentration reduces the smell impression to only one eighth.

The odour coming from fresh fish of human consumption quality, during processing into fish meal, is generally acceptable; but when stale fish is processed highly obnoxious substances such as trimethylamine, methyl mercaptan and hydrogen sulfide are released, which have to be eliminated from the exhausts from the factory by special techniques. It is therefore important that the fish raw material is as fresh as possible before entering the process.

During processing, malodours may be formed, especially by oxidation and pyrolyses in direct-fired dryers.

Apart from the above-mentioned freshness of raw material, attention should be paid to (i) volumes and odour load of gases handled; (ii) centralization of exhausts; (iii) method of drying; and (iv) location of the factory and meteorological conditions. As explained earlier, effluent gases from direct rotary dryers may be difficult to handle, partly due to great volumes, and partly due to problems with scorching and aerosol formation.

The first step in cleaning of effluent gases is always the scrubbing of the gases. The second step is, in most cases, high temperature combustion. Heating the gases to above 750°C for about one second effectively destroys the content of malodorous components. This observation is the basis for the most widely used method of deodorization today.

All equipment from which heavily loaded gases may evolve, such as cookers, presses, dryers, tanks and centrifuges, are completely encapsulated and kept at a small underpressure by means of a ducting system leading to the water scrubber, followed by a fan which in turn delivers the uncondensed gases to the boiler plant, where they are exposed to high temperatures. Especially during stoppages in the production when there is no need of steam production and the furnace accordingly is closed down or, for any other reason, the exhaust exceeds the need of the furnace, the odorous gases then need to be treated by some other method such as catalytic incineration in special equipment or by inactivation in special chemical scrubber installations. The latter method is still being discussed with regard to its efficiency. In certain cases it has been reported that improperly designed or managed chemical scrubbers may aggravate the odour problems by only partly oxidizing or adding chlorine to the odorous compounds by insufficient chemical reactions.

Work on developing more effective deodorization systems besides the seawater scrubbing followed by boiler incineration is still going on; possibilities include using bio-scrubbers, activated carbon filters, etc.

Pollution from liquid effluents appears to be coming into focus. Pollution of fresh water by factory operations occurs from the centrifuges, pumps and especially from cleaning operations. The organic effluent must be kept to a minimum in properly managed factories. Unloading by pumping by addition of seawater and fluming appears to cause large problems and will normally not be allowed inside harbours, for which reason dry offloading by elevators or pneumatic unloaders is preferred in such places. In cases where fresh water is used for cleaning, the effluent can be pumped to the bloodwater tank of the factory and thus enter into the normal production.

Sea water is used in large quantities in the condensers of the stickwater evaporators and in the scrubbing towers for gaseous effluents. Contaminants are volatile components from the raw fish, dust particles from the dryers and meal equipment and possible carryover from the evaporators. It appears that minimizing the pollution is directly connected with strict control of raw fish quality and with the efficiency of, for example, cyclone and evaporator operations. It is often stated that the components of the seawater effluent consist of organic matter that is easily decomposed. But, even so, high concentrations of pollutants are not normally acceptable.

The Products

In 1991 the world production of fish meal (Table 12.1) and oil was around 6.4 Mt and 1.4 Mt, respectively. World imports/exports of fish meal were 3.3 Mt with a total purchase value of 1770 million US$ ($536 per t). World exports of fish oil were 0.65 Mt with a total value of 230 million US$ ($354 per t). Fish meal protein is used to supplement other proteins, particularly vegetable proteins, in diets for farmed animals.

A growing market has appeared within the last 10 to 15 years for fish meal (and fish oil) as the main ingredients in diets for farmed fish and shrimps. Fish meal and oil may, especially in diets for salmonids, only be supplemented with other feed ingredients at a less than 10% level. An estimate of the outlet of fish meal for fish farming is more than 1 Mt in 1993 at of a value well above normal the ordinary fish meal sales value.

About 90% or more of world production of fish meal is from oily fish species. A minor part is from fish offal, and some of this may be produced from lean fish such as cod and haddock and is called white fish meal. Less than 1% is produced from other species such as shellfish and mammals. The majority of fish meal is so-called whole meal, i.e. only water and some oil is extracted from the fish.

The quality of the protein in the fish, i.e. the composition and availability of the amino acids in relation to the animal's requirement, is high, particularly in whole fish. The value of the protein in skin, connective

Table 12.1. Fish meal[a]: World production (thousands of tonnes). Source: FAO and OWW (Bowman, 1993).

	1988	1989	1990	1991	1992
Argentina	15	17	17	17[g]	18[b]
Brazil	30	27	27	27	27[b]
Canada	69	71	68	62	63[b]
Chile	1122	1381	1073	1228	1301
China	59[g]	63[g]	110	100	82[b]
Denmark	366	346	275	367	358[b]
Ecuador	166	105	42	45	42[b]
Faeroe Islands	29	23	17	13	24[b]
France	21	22	21	20	22[b]
Germany	25	25	26	30	26[b]
Iceland	192	144	150	83	177[b]
India	33[g]	55	58	63	52[b]
Ireland	14[f]	13[f]	15[f]	20[f]	22[f]
Japan[c]	1096	1049	977	788	430[b]
Korea (South)	72	36	87	60	52[b]
Malaysia	44	41	43	51	39[b]
Mexico	79	90	83[g]	80	87[b]
Morocco	30[g]	35[g]	32[g]	32[g]	34[b]
Namibia	na	na	49	32	na
Norway	184	200	167	208	267[b]
Pakistan	33	33	35	36	36[b]
Panama	20	33	23	25	20[b]
Peru	1126	1169	1204	1312	1367
Poland	69	62	52	51	49[b]
Romania	12	9	4	4	4[b]
South Africa[d]	240	132	54	55	152[b]
Spain	136	128	135	75	78[b]
Sweden	14	13	11	12[g]	12[b]
Taiwan	5	8	7	14	14[e]
Thailand	237	269	285	285	257[b]
Turkey	20[g]	11[g]	3[g]	3[g]	3[e]
UK	50	49	51	50	48[b]
Uruguay	8	8	4	10	10[e]
USA[c]	388	386	403	398	316[b]
USSR (ex)	769	752	695	632	530[b]
Venezuela	5	4	10	8	8[e]
Total (36)	6768	6809	6313	6296	6023
Others	69	66	59	71	70
Total, world	6837	6875	6372	6367	6093

[a] Meals, solubles and similar animal feedingstuffs of aquatic animal origin. Including fish meals from 'white fish', oily fish and fish solubles. (Excluding whale meal and solubles, crustacean and seal meal and fish silages.)
[b] Estimated data from OWW (Oil World. The Weekly Forecasting and Information Service for Oilseeds, Oils, Fats and Oilmeals, Hamburg).
[c] Solubles, data reported by FAO as wet weight, OWW as 50% weight.
[d] For 1988, 1989, 1992 includes Namibia.
[e] Data estimated by IFOMA.
[f] Data supplied by Fish Industries, Dublin, Ireland.
[g] Estimated, provisional or unofficial data.

tissue and bone is lower; consequently, the quality of the protein in offal is lower than in whole fish. For example the proportion of methionine and lysine in white fish offal is about 10% lower than in whole oily fish.

Freshness of the raw material is important in its effect on the quality of the end product. As a guide to freshness of raw material the content of total volatile nitrogen (TVN) in the fish is measured in the factory laboratories. It should be less than 80 mg N per 100 g of raw material for production of ordinary fish meal for pig or poultry feeding. For special types or brands of fish meal, e.g. 'eel meal', 'salmon meal' or LT meal (explained below), the manufacturer often guarantees lower TVN values, e.g. maximum 50 or 60. The lowest limit is related to the natural content of trimethylamine oxide of the fish and is between 15 and 50 depending on fish species.

Process control in the factory is necessary for the manufacture of high quality fish meal. Excessive temperatures for prolonged periods should be avoided. It has been shown that the rate of loss of available lysine is less than 1% per hour at temperatures below 120°C; above this temperature the rate increases very rapidly (10% per hour at 140°C; Carpenter and Booth, 1973). But the lucrative market possibilities for high quality fish meals for aquaculture have required decreasing heat load in the process in order to preserve the nutrients of the raw fish as well as possible. Fish meals produced from absolutely fresh raw material at gentle heating (fast cooking, low drying temperatures, e.g. by vacuum drying) are now marketed as LT fish meals (LT means low temperature). However, as yet no universally accepted requirements have been agreed upon for LT meals which are widely used in the more profitable end of the market, e.g. as milk replacers for early weaned pigs or for salmonid feed.

Good hygiene is necessary at all stages in handling and storage. Precautions must particularly be taken against *Salmonella* contamination. This organism is destroyed in the cooking plus pressing process provided the temperature is raised above 80°C, but poor hygiene beyond this point could result in reinfection of the meal. *Salmonella* is not considered a natural part of the bacterial flora of the raw fish. It must be stressed that the fish material in the dryer may not reach sufficiently high temperatures for sufficient time to destroy *Salmonella*. To minimize the risk the manufacturers should observe the precautions, which are common to all protein feed manufacturers.

A specific factory cleaning scheme is necessarily combined with regular bacteriological inspection; should *Salmonella* nevertheless be detected in fish meal, the contaminated lots must be disinfected either by a prescribed heating and re-drying procedure or, alternatively, biocides such as formic or propionic acids can be used to treat the infected fish meal. These are expensive processes, however, and it is far more economical to comply with the outlined precautions.

Table 12.2. Proximate analyses for herring-type meal and anchoveta-type meal.

	Moisture (%)	Crude protein (%)	Fat (%)	Ash (%)
Herring-type fish meal	8	72	9	10
Anchoveta-type fish meal	10	65	9	16

The Role of Fish Meal in Animal Diets

As well as being a rich source of high quality protein, fish meals have a relatively high energy content and are rich in important minerals such as calcium, phophorus, selenium, B vitamins and in long-chain n-3 polyunsaturated fatty acids. The nutrient analyses for fish meals for the east Atlantic or herring-type meal and South American or anchoveta-type meal are given in Table 12.2.

Amino acid composition and availabilities of amino acids to the animals are especially important in animal feeding. Here reference should be made to a number of *Technical Bulletins* and tables published by the International Fishmeal and Oil Manufacturers Association (IFOMA).

Fish meal is used in feeds for poultry, pigs, ruminants, farmed fish and fur-producing animals. It increases productivity and improves the efficiency with which feed is converted to animal products (feed conversion). It is of special value in diets for young animals, for example in starter diets for broilers, diets for early weaned pigs, and for farmed fish and carnivorous fur-producing animals.

Fish meal is particularly beneficial in situations which are less than ideal, i.e. where feed mixing and quality is poor, where husbandry standards are low and where disease problems are prevalent. A detailed assessment of fish meal in diets for poultry, pigs, ruminants and aquaculture is given in *IFOMA Technical Bulletins*. In diets for very young animals such as early weaned pigs, ruminants and for aquaculture, it is recommended that special quality fish meal is used. It should be made from very fresh raw material and, for ruminants, have a low content of soluble nitrogen.

The composition of a special fish meal for ruminants produced from presscake without added stickwater is shown in Table 12.3.

The essential amino acid composition of LT meals (whole meals) is equal to the body protein of, for example, rainbow trout, Atlantic salmon and Pacific salmon species.

Other potential animal protein sources for dry feeds are blood meal, hydrolysed feather meal, meat and bone meal, skimmed milk powder and

Table 12.3. Composition of a special fish meal for ruminants.

Crude protein	Min. 71%
Crude fat	Max. 12%
Water	Max. 10%
Digestibility	Min. 83%
Volumetric titration value	Max. 80 ml 10 g^{-1} dry matter
Buffer solubility	Max. 20%

casein. Except for skimmed milk powder and casein, they all have a poorer balance of essential amino acids than fish meal, though their protein content is high as compared to most vegetable protein sources. Their use in diets for carnivorous fish is generally limited to less than 10% of the diet. Vegetable protein sources relatively low in protein and showing varying contents of fibre, and which can contain toxins such as trypsin inhibitors in improperly treated soyabean meals, are used at very low levels, generally less than 5%, if at all. Attempts are being made to produce soya protein concentrates with little if any fibre and supplement with amino acids to replace some of the fish meal in feeds for salmonids and other carnivores.

Lipids from wild fish, particularly marine species, contain comparatively high levels of n-3 polyunsaturated fatty acids (PUFAs). For many farmed fish, and particular marine fish, long-chain n-3 PUFAs must be supplied in the diet. The marine fish can only meet their requirement for n-3 fatty acids from eicosapentaenoic acid (EPA) and docosahexaenoic acid (DHA), C20:5 n-3 and C22:6 n-3, both of which occur in fish oils and residual lipids in fish meal, if the meal is a special (LT) meal. For the consumers, farmed fish can thus provide a rich and palatable source of these fatty acids.

When producing LT meal high demands are placed on the freshness/quality of the raw material, i.e. low TVN values (max. 50 at the cooker inlet). The production of LT meal implies reduced heating (70°C or lower in the dryer instead of 90°). A specification for an LT meal is shown in Table 12.4.

In addition to the above-mentioned meal types, other special fish meals are produced, particularly in Denmark and Norway. These types are typically produced by adding the concentrated stickwater to the presscake in various proportions or by changing other process parameters. Table 12.5 illustrates the primary use of the different product types.

Though standard meal for poultry and pig feeds is still the major product type and these two applications are also the major ones, the table illustrates that the potentials of the aquaculture sector have been recognized by the advanced fish meal and oil industry.

Table 12.4. Specification for an LT fish meal.

Crude protein	Min. 72%
Crude fat	Max. 12%
Water	Max. 10%, min. 6%
Salt (NaCl)	Max. 2.5%
Ash (salt-free)	Max. 14%
Ammonia (NH_3—N)	Max. 0.18%
Water-soluble protein	Max. 32%, min. 20%
Digestibility	Min. 90%
TVN in raw material	Max. 50 mg N 100 g^{-1} fish
Cadaverine	Max. 1000 mg kg^{-1}
Antioxidant	Ethoxyquin, 400 ppm added

Table 12.5. Primary use of some types of fish meal.

Fish meal type/primary use	Poultry	Pigs	Piglets	Ruminants	Aquaculture
Standard meal	X	X			
Special meal for ruminants				X	
LT meal			X		X
Other special meal types					X

Fish Oil

Fish oils contain mainly triglycerides of fatty acids with variable amounts of phospholipids, glycerol ethers and wax esters. They contain a wide range of long-chain fatty acids with the number of carbon atoms ranging from 14 to 22 and with high degrees of unsaturation.

The complex nature of fish oils depends on a number of factors. The fatty acid patterns of the oils vary with fish species, with the composition of the plankton and the time of the year. These influence the properties of oils with regard to both edible and technical applications. The oils contain variable but small amounts of unsaponifiable components such as hydrocarbons, fatty alcohols, waxes and ethers.

In order to manufacture oil of desirable properties, the following should be observed: the fish should be as fresh as possible; the oil should be properly separated and polished with hot water (100°C); it should be cooled before delivery to the storage tank which must be regularly cleaned and dried. Any water should be tapped off when settled, and the tanks should be filled as fast as possible to avoid oxidation of the oil.

Fish oil is still mainly used for the manufacture, by hardening (hydrogenation), of edible products such as margarines, shortenings, and, especially in South America, salad oils. Their physical properties have made fish oils attractive constituents in margarines because margarines prepared from hardened vegetable oils sometimes recrystallize on storage. This makes these margarines hard and crumbly. As fish oils have a widely varied chain length, margarines prepared from these have an excellent plastic consistency. The value of hardened fish oil lies in its creaming power, particularly in cake making.

The highly unsaturated properties of unhydrogenated fish oils make the oils (particularly the highly unsaturated fractions, e.g. from menhaden oil) suitable for a number of technical applications such as drying oils and varnishes. The saturated fat fraction is a disadvantage for these purposes and some fish oils are therefore not suitable for fractionation and technical uses. Fish oils may also be used for production of fatty acids with a spectrum of chain lengths for a number of technical applications.

Medical and nutritional investigations in recent years have shown that the long-chain polyunsaturated n-3 fatty acids (n-3 PUFAs) particular to marine oils (lipids from fish or from marine mammals) have various beneficial effects on human health (Barlow *et al.*, 1990).

The major factors in the refined, but unhydrogenated, marine oils which had very positive effects on cardiovascular diseases were discovered to be the long-chain PUFAs, EPA and DHA, by the Danish physicians Bang and Dyerberg in the 1970s by clinical examinations of Greenland seal hunters. (See Chapter 4 for more details.)

Further medical research has led to the conclusion that the beneficial role of fish oil in human diets can be divided into two main areas: (i) to sustain normal healthy life; and (ii) to alleviate certain diseases.

Research results conclude that the fetus and the young child have a requirement for DHA, particularly for normal brain and nervous tissue development. Animal experiments have shown that the brain and nervous tissue in animals, as in humans, are extremely high in DHA and that even maintaining these animals on diets deficient in n-3 fatty acids over generations results in only a modest decrease in DHA in the brain and nervous tissue. This suggests that these organs are maintaining the DHA, probably because it is essential to their function. Animal experiments have suggested that a deficiency of n-3 fatty acids decreases the animal's ability to learn and its ability to see properly. It can be concluded from these and similar experiments that in properly constituted milk fats for infants a ratio of n-3 : n-6 PUFAs of 1 : 4 appears appropriate. This is actually not the case in most commercial infant formulas, so these have to be corrected.

From the considerable evidence available it is apparent that n-3 fatty acids have a place in the dietary requirements of the normal human with intakes of about 1 g per day being sufficient (Bjerve *et al.*, 1992). Patients suffering from a number of blood vessel disorders and inflammatory diseases can positively benefit from including fish oil in their diet with effects being observed when the intake of EPA and DHA is 3 g per day or more.

Barlow *et al.* describe in the above-cited publication (1990) how, apparently for the first time, fish oils of different origin have successfully been refined (i.e. deacidified, bleached and deodorized), effectively protected by addition of proper antioxidants and incorporated in a number of foods as substitutes for less unsaturated fats.

The above work was undertaken under the auspices of the IFOMA, and is now being continued in a number of institutes in many countries including the Center for Food Research at the Technical University of Denmark to which the author of this chapter is attached.

The FAO's Food Balance Sheet (1984–1986) indicates an average global consumption of only two thirds of the minimum requirement of 1 g n-3 PUFAs per day. Currently fish and fish dishes provide an average intake of 0.2 g n-3 fatty acids per day, mainly as EPA and DHA. To reach the proposed target requires a much greater consumption of fish or the use of fish oil in foods (apart from the possibility of using pharmaceutical concentrations). The provision of a wide range of food with an increase in the content of n-3 fatty acids and a more desirable n-3/n-6 ratio is the best approach. This can be achieved by inclusion of antioxidant or otherwise protected fish oil in the diets of poultry and pigs, thereby increasing these fatty acids in the meat and eggs, or it can be achieved by incorporation of protected fish oils in foods.

A number of problems, especially concerning the refining and protecting of fish oil as a bland oil with no flavour or flavour reversion, problems are still being investigated and have been solved in so far as edible, refined, fish oils for incorporation in foods, e.g. mayonnaise, margarine and salad oil, are actually in the market. It is estimated that an increasing part of fish oil in the future may be sold as a refined premium quality oil.

Acknowledgements

The author of this chapter acknowledges his thanks to Dr Stuart M. Barlow, Director General of the International Fishmeal and Oil Manufacturers Association (IFOMA) for permission to use illustrations and tables belonging to the association or to members of its Scientific Committee, who also gave their kind permission to use their contributions to *FAO Fisheries Technical Paper* 142, rev. 1.

References and Further Reading

Anon. (1981) Nutrient analysis tables for UK. *IAFMM Nutrient Analysis*. IFOMA, St Albans, UK, 7 pp.

Barlow, S.M. and Pike, I.H. (1977) The role of fat in fish meal in pig and poultry nutrition. *IFOMA Technical Bulletin No. 4*. IFOMA, St Albans, UK, 38 pp.

Barlow, S.M., Collier, G.S., Juritz, J.M., Burt, J.R., Opstvedt, J. and Miller, E.L. (1984) Chemical and biological assay procedures for lysine in fish meals. *IFOMA Technical Bulletin No. 20*. IFOMA, St Albans (UK), 12 pp.

Barlow, S.M., Young, F.V.K. and Duthie, I.F. (1990) Nutritional recommendations of n-3 polyunsaturated fatty acids and the challenge to the food industry. *Proceedings of the Nutritional Society* 49, 13–21.

Bimbo, A.P. (1987) The emerging marine oil industry. *Journal of the American Oil Chemist's Society* 64, 706, 708–715.

Bjerve, K.S., Thoresen, L., Bønaa, K., Vik, T., Johnsen, H. and Brubakk, A.M. (1992) Clinical studies with alpha-linolenic and long chain n-3 fatty acids. *Nutrition* 8, 130–135.

Bowman, S.J. (1993) *Digest of Selected Statistics*. IFOMA, St Albans, UK, p. 3. 48 pp.

Carpenter, K.J. and Booth, V.H. (1973) Damage to lysine in food processing: its measurement and significance. *Nutrition Abstracts Review* 43, 423–451.

FAO Fishery Industries Division (1986) The production of fish meal and oil. *FAO Fisheries Technical Paper* 142, rev 1. FAO, Rome, 63 pp.

Gulbrandsen, K.E. (1984) Fish meal in diets for young pigs. *IFOMA Technical Bulletin No. 17*. IFOMA, St Albans, UK, 20 pp.

Miller, E.L. (1970) Available amino content of fish meals. *IFOMA Technical Bulletin No. 1*. IFOMA, St Albans, UK, 11 pp.

Olsen, K.B. (1992a) Shipboard handling of pelagic fish with special emphasis on fast handling, rapid chilling and working environment. In: Burt, J.R., Hardy, R. and Whittle, K.J. (eds) *Pelagic Fish, the Resource and its Exploitation*. Fishing News Books, London, pp. 55–69.

Olsen, K.B. (1992b) Handling and holding fish on fishing vessels in Denmark. In: Huss, H.H., Jakobsen, M. and Liston, J. (eds) *Quality Assurance in the Fish Industry*. Developments in Food Science 30. Elsevier, Amsterdam, pp. 185–195.

Olsen, K.B. and Hansen, P. (1982) Handling mixed catches. In: *Fish By-Catch . . . Bonus from the Sea*. IDRC, Ottawa, pp. 59–60.

Opstvedt, J., Miller, E.L. and Pike, I.H. (1991) Complementary effects of fish meal with soybean meal replacers in broiler diets. *IFOMA Technical Bulletin No. 26*. IFOMA, St Albans, UK, 16 pp.

Pariser, E.R., Corkery, C.J., Wallerstein, M.B. and Brown, N.B. (1978) *Fish Protein Concentrate: Panacea for Protein Malnutrition*. International Nutrition Policy Series No. 3, MIT Press, Cambridge, Massachusetts, 296 pp.

Pike, I.H. (1979) The role of fish meal in diets for pigs. *IFOMA Technical Bulletin No. 6*, IFOMA, St Albans, UK, 27 pp.

Pike, I.H. (1990) The role of fish oil in feeds for farmed fish. *IFOMA Technical Bulletin No. 25*. IFOMA, St Albans, UK, 12 pp.

Pike, I.H., Curran, M.K., Edge, M. and Harvey, A. (1984) Effect of nutrient density, presence of fish meal and method of feeding of unmedicated diets on early weaned pigs. *IFOMA Technical Bulletin No. 21*. IFOMA, St Albans, UK, 9 pp.

Pike, I.H., Andorsdóttir, G. and Mundheim, H. (1990) The role of fish meal in diets for salmonids. *IFOMA Technical Bulletin No. 24*, IFOMA, St Albans, UK, 35 pp.

Tagle, M.A. (1976) *Report on the Acceptability Testing of FPC Type B*. Norwegian Funds-in-Trust FAO/TF/INT 120 (NOR)-PHASE 1, FAO, Rome, 154 pp.

Technological Laboratory, Danish Ministry of Fisheries (1971) *Annual Report*. Technological Laboratory, Danish Ministry of Fisheries, Copenhagen, 52 pp.

Index

abalone 66
  blacklip 66
  dried 318
  giant 66
  perlemoen 66
acetic acid 323, 326
Acinetobacter 222, 223
actin 91, 93
actinins 91
Aeromonas spp. 244, 253
air-blast freezing 293, 299
β-alanine 98
albacore 85
alevin 38
algal toxins 249
Alteromonas 222, 223
amberjack 162, 168
amines 97
amino acids 25, 97, 98
α-amino-*n*-butyric acid 98
ammonia 97, 224, 227, 351
anadromous fish 31
anamnetic shellfish poison (ASP) 250
anchoveta 2, 4, 7
anchovy 37, 81, 85, 162, 168, 192, 197
  in oil 191, 192, 193, 194, 206, 208
angler, American 58
anglerfish 58
Anisakis 250, 326
anserine 98
antioxidants 333, 362, 364
aquaculture 18
arachidonic acid 132
arkshells 6
arsenic 268
arsenobetaine 268
ascorbic acid 175, 177, 182
astaxanthin 26, 160
ATP 98, 220, 236
autolysis 218, 351
ayu 80, 140, 163

bacalao 317, 321
Bacillus cereus 244
balenine 98
Baltic sea 269, 270, 271
barracuda 52
  red 81
bass 5, 138
  rock 168
  striped 46
beam trawls 16
belly burst 221
benzo(a)pyrene 273, 324
betaines 97

bilgy fish 229
biliverdin 88
billfish 5
biocontaminants 262
biogenic amines 224
biomagnification 262
Biot number 292
biotin 168, 170, 171, 173, 179, 180
'bloom' 321
bluefish 46
bones, composition 190
bonito 5, 85
  Pacific 161
bottom trawling 17
brevetoxins 250
brill 55
brisling 168, 179
bromelain 332
brown kelp 268
burbot 137, 138

cadaverine 98
cadmium 263
caesium-137 273
calamary 67
calcium 190, 192, 193, 194, 206, 208, 319, 330
Campylobacter 244
canned seafood products 179, 327
canola oil 119
canthaxanthin 26, 160
capelin 85, 118
  Atlantic 118
carbohydrate hydrolases 84
cardiovascular mortality 146
carnosinase 78
carnosine 98
β-carotene 160
carotenoproteins 88, 338
carp 6, 7, 22, 23, 40, 81, 82, 83, 90, 137, 138, 161, 163, 170, 193, 198
  bighead 40
  grass 20, 40
  silver 7, 23, 40
catadromous fish 31, 41
catalase 87
catches 4, 5, 7
catching conditions 229
catching techniques 8
catfish 20, 21, 23, 85, 138, 193, 198
  African 20, 24
  Atlantic 165
  Channel 20, 40, 121, 124, 170, 177
  lake 40
  sea 162, 168
  spotted 40
  white 40
cathepsins 84, 85
caviar 322, 336
cephalopods 6, 7
chilling 304, 308, 352
chill-stored fish 232
chitin 338
chitosan 338
chlorine 195
cholera 245, 248
cholesterol 119, 124, 128, 138, 142, 146
chromium 197, 198, 199, 204
chub mackerel 7
chymotrypsin 85
cichlids 44
ciguatera toxins 249
civenche 248
clam 6, 20, 141, 163, 171, 194, 199, 265
Clarias 41
Clostridium botulinum, botulism 244, 246, 247, 256, 257, 325, 328
Clostridium perfringens 244
Clostridium spp. 222
coagulation, by heat 83
coalfish 43
cobalamin 168, 170, 171, 174, 177, 179, 180, 181, 182
cobalt 204
cockles 6, 20, 65, 265
cod 6, 42, 81, 83, 86, 87, 90, 161, 165, 180, 192, 197, 226, 263
  Atlantic 7, 119, 120, 122, 162, 168
  Pacific 42, 162, 168

cod-like fish 5
cod liver oil 157
cold chain 310
cold shock 228
coliforms 253
collagen 94, 95, 96, 328
  amino acid composition 95
  content 96
collagenases 84, 85, 337
condition of fish 229
conger eel 5, 41
connectin 91, 94
convenience foods 145
cooking 181, 207
copper 197, 198, 199, 201
coryneforms 222, 223
crab 6, 74, 85, 87, 143, 144, 163, 171, 194, 199, 206
  blue 74
  brown 74
  rock, European 74
  spider 75
  spinous 75
crab paste 179
crayfish 86, 87, 163, 171, 194, 199, 264
  red swamp 177
creatine kinase 91
creatine phosphate 97, 99
creatinine 97, 99
croaker, white 85
crustaceans 5, 6, 20, 144
cryoprotectants 329, 330
cultured fish 18, 80, 140, 175, 177, 274
cusk 168
cuticle 87
cuttlefish 6, 87
cyclodienes 269, 270
cyprinids 20, 21, 90, 166
Cytophaga 222, 223

dark muscle 188, 196
DDT 269, 270
dehydration 296, 297, 298, 303, 307, 310
7-dehydrocholesterol 164
dehydroretinol 160
demersal fish 5, 7, 216
demersal trawling 17
denaturation 294
deodorization 355, 366
deterioration 216, 294
  measurement 232
1,2-diacylglycerol 124
diadromous fish 5, 20
diarrhoetic shellfish poison (DSP) 249
digestive tract 86, 87
dimethylamine (DMA) 231, 237
dinoflagellates 249
'dioxins' 269, 271
diseases of fish 26
docosahexaenoic acid (DHA) 119, 127, 134, 137, 142, 145, 323, 373
dogfish 85, 161, 165, 168, 264
  spiny 85, 162, 168
dolphin 162, 168, 192
dredges 16
dried fish 316
drift nets 12
drip 308
drug residues, in cultivated fish 274
drum 162, 168, 192, 197

earthiness 232
echo sounding 13, 18
eel 20, 21, 23, 24, 41, 90, 161, 163, 170, 178, 180, 193, 198, 206, 266, 270, 272, 275
  American 140
eicosapentaenoic acid (EPA) 119, 127, 134, 137, 142, 145, 323
elastin 97
enthalpy 288, 290
environmental contaminants 262
enzymatic ripening 322
enzymes, from fish 83, 336
  from plant origin 332
erythrocytes 87
escallop 63
Escherichia coli 253
ethoxyquin 364
eutrophication 8

farmed fish *see* cultured fish
fat content 118, 119, 120, 122, 124, 127, 128, 130, 133, 136, 138, 140, 141, 143, 144
fatty acid composition of lipids 122, 127, 128, 133, 136, 138, 140, 141, 143, 144
fatty acids, free 119, 124
  in blood platelets 135
  monounsaturated 132
  non-methylene-interrupted dienoic (NMID) 139
  polyunsaturated (n-3) 25, 125, 371, 373, 374
  polyunsaturated (n-6) 25, 132
  saturated 129
feeding, in fish culture 25
fermented fish 257
filefish 85
fish balls 317, 329, 331
fish freshness meter 235
fish liver oil 157, 161, 165
fish meal
  in animal diets 370
  proximate analysis and specification 370, 371, 372
  storage 364
  world production 368
fish oil 372
fish protein concentrate 348
fish protein hydrolysate 335
fish solubles 353, 361
fish tester 235
flake ice 308
flatfish 5, 266, 268
flavins 88
flavour 219
  uncharacteristic 231
flavour score 221
Flavobacterium 222, 223
flora, spoilage 243
flounder 86, 87, 162, 168, 192, 197, 265, 266
  winter 119
fluorine 197, 198, 199, 204
folate 168, 170, 171, 173, 177, 182
food poisoning 244, 255
formaldehyde 231, 237, 323, 324, 330, 352, 358
formic acid 323, 333
freeze denaturation 330
freezer burn 231, 296
freezers, air-blast 299
  belt 300
  cryogenic 301
  fluidized bed 301
  plate 301
freezing, at sea 2
  partial 307
  damage 295
  methods and equipment 298
  rate 292, 293, 295, 297, 308
  zone 288
furazolidone 275
fyke nets 10

gaffelbiter 322
gaping 296
garfish 86
gastric juice 86
gastric mucosa 85
gastropods 65
gelatin 337
geosmin 232
gill 87
gill nets 12
'glass crystals' *see* struvite
glazing 297
gloss 324
glutamate dehydrogenase 87
D-glyceraldehyde-3-phosphate dehydrogenase 87
glycogen phosphorylase 86
glycolysis 218, 307
goldfish 85, 87
green discoloration 89
grouper 168
  white 46
grub fish 81
guanidino compounds 97, 99
gurnard
  grey 54
  red 54

haddock 43, 161, 162, 165, 168, 180, 192, 197, 226
haem protein 88
haemocyanins 89, 201
haemoglobin 88
hake 44, 168, 266
  Argentine 44
  Cape 44
  North Pacific 44
  Senegal 44
  silver 44
  South Pacific 44
halibut 56, 128, 161, 180, 192, 197
  Atlantic 161, 162, 168
  Greenland 125, 127
  Pacific 56
halophiles 319, 321
hand lining 11
hazard analysis critical control point (HACCP) system 287
head 87
heat pasteurization (*sous vide*) 256
heat-gelled products 330, 331
hepatitis 244, 245
hepatitis A virus 254
hepatopancreas 85, 87
herring 36, 85, 87, 119, 128, 170, 192, 197, 206, 326
  Atlantic 7, 162, 168, 178
  Baltic 178
  maatjes 322
  North Sea 7, 8
  Pacific 162, 168
  soft roe 180
herring-like fish 6
herring oil 119
herring worm 250, 326
hexachlorobenzene (HCB) 269, 270
hexachlorohexanes (HCH, BHC) 269, 270
high quality life (HQL) 287
histamine 98, 224, 234, 251
histidine 98
horse mackerel 47, 81, 162, 168
humpback 38
hydrogen sulfide 352
hydrogen peroxide 327
hydrolases 83
hydroxyproline 95
hypocholesterolemic effects 100
hypolipidemic effects 100
hypoxanthine 99, 221, 236

icing of fish 304, 308
indigoids 88
individually quick frozen (IQF) products 301
industrial fish 348
ink gland 87
inosine monophosphate 220, 224
INQ, total 100
intestine 85
invertebrates 143
iodine 197, 198, 199, 203, 205, 208
Irish Sea 264
iron 195, 197, 198, 199, 206, 207, 208
irradiation of fish 256

jack mackerel,
  Chilean 7
  Japanese 47

K value 236
kamloops 39
kelt 38
king crab 73
  Alaska 73
  blue 73
  golden 73
  Hanasaki 73
  Kamchatka 73
  Pacific 73
  red 73
klipfish 321
krill 68
kryddersild 322

lactate dehydrogenases 86, 87
lampara nets 14
lamprey 162, 168
larval and insect infestation 316, 318
latent heat 288
lead 263

lemon sole 168, 264
Lepomis sp. 90
lignins, chlorinated 269
ling 161, 162, 165, 168, 192, 197
lingcod 161, 162, 168
lipases 84, 86
lipid breakdown 351
lipid peroxidation 165
lipids, body distribution 118
  classes 119, 122
  content 128, 130
  seasonal variation 120
lipoxygenases 86, 87
Listeria monocytogenes 245, 251, 326
liver 87, 160, 263, 265
  canned, from Gadidae 328
lizard fish 81
lobster 86, 87, 144, 163, 171, 194, 199
  American 71
  common or European 72
  homarid 71
  Norway 72
LT (low temperature) fish meal 369, 370, 371, 372
lumpsucker 35
lungfish 90
  African 85
lutefisk 317
lutein 160
lysine 318, 324
lysozyme 337

mackerel 48, 81, 161, 178, 180, 192, 197, 206
  Atlantic 49, 118, 119, 162, 168
  horse *see* scad
  jack *see* scad
  North Sea 7
  Pacific 82
macro element composition 192, 193, 194
magnesium 191, 192, 193, 194, 206, 208, 319, 330
Maillard reaction 318, 324
malachite green 275
manganese 197, 198, 199, 201
marinated fish 326
marlin
  black 52
  blue, Atlantic 52
  blue, Indo-Pacific 52
  striped 52
  white, Atlantic 52
mebendazole 275
melanins 88
mercury 264
mercury/selenium interaction 266
metabolism 275
methyl isoborneol 232
methyl glyoxal 323, 324
methylhistidine 98
methylmercury 265, 266
microbial ripening 322
microorganisms
  pathogenic 245
  spoilage 222
Micrococcus 222, 223
milkfish 21, 24, 140, 163, 170
minces, from fish 83, 329
modified atmosphere packaging 305
molluscs 5, 6, 20, 87, 141, 200, 201, 203, 264, 265, 268
molybdenum 197, 198, 199, 203
monk, Cape 58
Moraxella 222, 223
mullet 53, 170, 193, 198
  grey 23
  flathead 53
muscle 79, 85, 86, 87
  dark 117, 119
  light 118, 119
  softening of 84
  protein 80, 91
  pH 218, 228
mussel 6, 20, 87, 143, 163, 171, 178, 194, 199, 206, 263, 264, 265, 326
  blue 59
  brown 59
  Chilean 59
  green 59
  Korean 59
  Mediterranean 60

River Plata 59
myogen *see* proteins, sarcoplasmic
myoglobin 88
myosin 91, 92

n-3/n-6 ratio 373, 374
nematodes 250
nets 12
neurotoxic shellfish poison (NSP) 250
niacin 167, 168, 170, 171, 177, 179, 180, 181, 182
nicotinamide 167
nicotinic acid 167
nitrite 322, 352, 361
N-nitroso compounds 322, 325, 352, 361
non-*ortho*-substituted PCBs 271
non-protein nitrogen (NPN) 77, 78, 97
non-sensory tests for freshness 235
North Sea 264
Norway haddock 53
nucleic acids 338
nucleotides 97, 98
nutritional value 100

ocean perch 5, 53, 125, 126, 128, 162
octachlorostyrene 269, 270
octopine 97
octopus 6, 87, 142, 143, 163, 171, 194, 199
offals 350, 367
omega-3 fatty acids 125, 145
onion cuttings, in marinades 327
organobromine compounds 272
organohalogen compounds 269
organotin compounds 272
ormer, green 66
Oryzias sp. 90
osmoregulation 98
otter trawls 17
oxidoreductases 86, 87
oxytetracycline 275
oyster 6, 20, 141, 163, 164, 171, 199, 200, 201, 206, 264, 265
cupped, American 62
cupped, mangrove 62
Pacific 62
Portuguese 62
rock 62
dredge, New Zealand 61
Eastern 177
flat, Chilean 61
European 61

packing, under carbon dioxide 305
pancreas 85, 86
panthotenic acid 168, 170, 171, 172, 177, 179, 180
paralytic shellfish poison (PSP) 249
paramyosin 92
paratropomyosin 91
parr 38
parrot fish 168
partially frozen fish 307
parvalbumins 81, 89, 90
PCBs 269, 271
planar 271
pectens 6
pelagic fish 5, 7, 216, 347, 350
pepsin 85, 336
peptides 97, 98
perch 45, 183, 198
ocean *see* redfish
yellow 163, 170
peroxidases 86, 87
peroxide number 323
Pham's equation 292
phenol oxidase 87
phenols 323, 324
phosphatase, alkaline 337
phosphate, phosphorus 190, 192, 193, 194, 206, 208
phosphatidylcholine 121
phosphatidylethanolamine 121
phospholipases 84, 86
phospholipids 119, 122, 124
pigments 88
pike 90, 137, 138, 161, 163, 170, 193, 263, 266
pike-perch 263, 266
pilchard 36, 162, 168
European 7
Japanese 7
South American 7

plaice 56, 162, 192, 264, 284
Plank's equation 291
plankton 4, 189
plankton-feeding fish 189
plutonium-239, -240 274
pollack 43, 86, 90, 161
  Alaska 5, 6, 7
pollock 43
pollution, from fish meal plants 365
polonium-210 274
polybrominated phenyl ethers 272
polychlorinated naphthalenes 269
polycyclic aromatic hydrocarbons (PAHs) 273, 324
polyphenol oxidase 78
pomfret 168, 192, 197
pompano 47, 158
  Florida 128
pond fish culture 19
porbeagle 168
potassium 191, 192, 193, 194, 205, 206, 207
pots 10
pound nets 11
practical storage life (PSL) 288
prawn 6, 72, 163, 171, 179, 194, 199
  caramote 70
  deep sea 171
  deepwater 70
  Dublin Bay 72
  fleshy 70
  Japanese 69
  Kuruma 69
  tiger
    giant 70
    green 69
praziquantel 275
pre-straining 357
processed seafoods 178, 180, 229
processing, effect on vitamin composition 176, 181
proteases 84, 320, 337
proteases, alkaline 84, 85
protein, breakdown 351
  coagulation 319, 323, 356
  connective tissue 77
  content 79, 117
  cross-linking of 88
  crude 77, 79
  denaturation 319, 326
  depletion 80
  efficiency ratio 100
  filament 91
  index of nutritional quality (INQ) 100
  muscle 80, 91
  myofibrillar 77, 90, 94
  regulatory 90
  sarcoplasmic 77, 78, 80, 81, 83
  scaffold 91, 94
  stroma 77, 94
proteoglycans 95, 97
proteoglycan degrading enzymes 84
Pseudomonas 222, 223
pterins 88
purines 88
purse seins 14
putrescine 98
pyloric caecae 85
pyridoxamine 172
pyridoxine B_6 168, 170, 171, 172, 177, 179, 180, 181, 182

quality loss during freezing 293
quaternary ammonium compounds 97
queenie, queen's scallop, 64
quinolones 275

radionuclides 273
radium-226 274
rancidity, oxidative 225, 296, 319, 320, 323, 348, 349
Ratkowsky equation 230
raw fish odour 219
ray 5, 90, 168, 192, 197
  thornback 34
rearing systems 22
recommended dietary allowances 182, 190, 208
redfish 5, 53, 125, 126, 128, 162
retinol 158, 160, 161, 162, 163, 177, 182

riboflavin 167, 168, 170, 171, 177, 179, 180, 181, 182
rigor mortis 219
rockfish 85, 161, 162
roe 263
roker 34

sailfish, Atlantic 52
 Indo-Pacific 52
saithe 43, 121, 168, 226
salmon 20, 23, 85, 87
 Atlantic 20, 38, 118, 119, 121, 128, 139, 170, 176, 177, 178, 180, 193, 198, 206
 cherry 39
 chinook 39, 85, 161, 163, 170, 193, 198
 chum, 39
 coho 39, 163, 170, 177, 193, 198
 dog 39
 keta 39
 king 39
 Pacific 20, 179, 206
 pink 38
 red 39
 silver 39
 sockeye 39, 163, 170, 193, 198
  Pacific 121
 spring 39
Salmonella enteritidis 251
Salmonella spp. 244, 251, 361, 365, 369
salmonellosis 244
salmonids 19, 90
salt 318, 319, 326
salt curing 320, 324
salt-water ice, 309
salted fish 318
salting with brine 319
sarcosine 98
sardine 36, 85, 178, 179, 192, 266
 European 119
 Japanese 118
 oil, Indian 118
 Pacific 179
sardines in oil 191, 206
sauce, from fish 331
scad 47, 81, 162, 168
scallop 6, 20, 141, 171, 178, 194, 199
 great 63
 Japanese 63
 queen's 64
scombroid poisoning 98, 238, 244, 252
score sheets for white fish 226
sea bass 45, 81
sea bream 47, 82, 168
 gilthead 48
 red 139, 140
sea cucumber 171
sea perches 45
sea water pollution 263
seafood poisoning 244
seaweeds 20
seine nets 13
selenium 197, 198, 199, 201, 207, 208, 266
sensible heat 288
sensoric tests 232
set nets 12
shark 5, 128
 sandbar 32
 soupfin 161
 liver 160
sheatfish 137
shelf-life
 of chilled fish 305
 of dried fish 318
 of frozen fish 293
shell 87
shellfish, lipid content 139
 macro element content 194
 trace element content 199
 vitamin content 177
Shewanella 222, 223
shigellosis 244
shoaling fish 5, 7
shrimp 6, 20, 23, 87, 124, 144, 163, 171, 177, 194, 199, 206, 264, 268, 297
 banana 69
 brown 71
  northern 69
 common 71

shrimp *contd*
crangonid 70
crystal 70
deepwater 70
Japanese 69
Kuruma 69
northern 70
pandalid 70
penaeid 68
pink 70
deepwater 70
northern 69
southern 70
tiger 85
white, western 70
siganid fish 168
silage 333
skate 86, 87, 168
skin 87, 337
slush ice *see* salt-water ice
smelt 162, 168
smoke-curing 323
smolt 38
snails 163, 171, 194, 199
snapper 46, 128
red 168
sodium 191, 192, 193, 194, 205, 206, 207
sodium/potassium ratio 195
sole 57, 162
butter 168
Dover 57, 168, 192, 197
English 168
lemon *see* lemon sole
Pacific 85
soups, from fish 317
soya protein concentrates 371
species identification 81
spider crab 75
spoilage, microbial 222
sprat 37, 168, 178
spurdog 33
squid 6, 85, 87, 142, 143, 171, 194, 199
European 67
Staphylococcus aureus 253, 332
Staphylococcus spp. 244
steelhead 39
stern trawls 18
sterol esters 122, 124
stews, from fish 317
stickwater 353, 359, 367, 371
stockfish 317
storage conditions 229
in ice 217
in refrigerated sea water 217
streptococcal infections 244
strontium-90 273
struvite 320
stunning, stupefying 9
sturgeon 35, 161
sucker 170, 193
sulfides 224, 229, 231, 232
superchilling 217
superoxide dismutase 86, 87
surimi 6, 83, 88, 100, 172, 205, 330, 331
swordfish 51, 161, 266
broadbill 168
broadtail 162
synthetic fibres 1, 18

taint 232
taste 219
taurine 98
tetrachlorobenzyl toluenes 269
thaw rigor 296
thawing 307
air-blast 309
by electrical resistance heating 310
by high-frequency heating 310
vacuum 309
thiamin 166, 168, 170, 171, 177, 179, 180, 181, 182, 268, 328
thiamin losses 167, 181
thiaminase 166
thiobarbituric acid (TBA) value 323
tidbits 32
tilapia 21, 23, 45, 90, 95, 141
toadfish 90
tocopherol 162, 163, 164, 178, 182, 201
total volatile bases (TVB) *see* volatile basic nitrogen

'toxaphene', PCTs 269, 270
trace element composition 197, 198, 199
trammel nets 12
transferases 86
transglutaminase 86
traps 9
traveller's diarrhoea 254
trawls 16
triglycerides 119, 122, 124
trimethylamine (TMA) 224, 225, 236, 237, 247, 351
trimethylamine oxide (TMAO) 97, 99, 224, 225, 230, 236, 237, 247, 328, 330
trimethylamine oxide (TMAO) demethylase 78, 88
tropomyosin 91, 93
troponin 91, 93
trout 20, 21, 85, 86, 87, 137, 138
  rainbow 24, 39, 85, 124, 163, 170, 177, 193, 198
trypsin, from cod 337
tryptophan 95
tuna 5, 7, 11, 48, 85, 87, 178, 192, 197, 266
  albacore 161, 163, 168
  bigeye 168
  bluefin 50, 161, 162, 168, 263
  in oil 206
  Pacific 179
  skipjack 6, 7, 49, 161, 162, 163, 168
  yellowfin 50
tunaxanthin 160
turbot 24, 55, 161, 162, 168, 192
turbot, Newfoundland 127
Tyrrhenian Sea 265

urea 97, 99, 227
  enzymic degradation 34
uricase 87

vacuum packaging 323
vendace 138
verotoxin 254
Vibrio cholerae 244, 246
Vibrio parahaemolyticus 244, 248
Vibrio spp. 222
Vibrio vulnificus 248
virginiamycin 275
viruses 254
vitamin A 158, 160, 161, 162, 163, 177, 182
vitamin A_2, dehydroretinol 160
vitamin B_1 166, 168, 170, 171, 177, 179, 180, 181, 182, 268, 328
vitamin B_2 167, 168, 170, 171, 177, 179, 180, 181, 182
vitamin B_6 168, 170, 171, 172, 177, 179, 180, 181, 182
vitamin B_{12} 168, 170, 171, 174, 177, 179, 180, 181, 182
vitamin C 175, 177, 182
vitamin D 158, 162, 163, 178, 182
vitamin E 162, 163, 164, 178, 182, 201
vitamin K 165, 182
vitamin contents 161, 162, 163, 168, 170, 171, 177, 178, 179, 180
vitamin losses 176
volatile basic nitrogen (TVN, TVB, TVB-N) 233, 237, 351, 369

water activity 316, 318, 323
water-binding capacity of proteins 307
wels 41
wet rendering process 353
white fish 216
whitefish 161, 193, 198
  Great Lake 194, 198
whiting 44, 161, 162, 168, 192, 197, 226
wild fish 140, 175
wolf fish 162, 168
wreck fishery 12

xenobiotics 262, 268

yellowtail 20, 21, 24, 81, 85, 162, 168
Yersinia enterocolitica 245

zeaxanthin 160
zinc 196, 197, 198, 199, 206, 207, 208